NICKEL IN THE ENVIRONMENT
Jerome O. Nriagu, Editor

ENERGY UTILIZATION AND ENVIRONMENTAL H
Richard A. Wadden, Editor

FOOD, CLIMATE AND MAN
Margaret R. Biswas and Asit K. Biswas, Editors

CHEMICAL CONCEPTS IN POLLUTANT BEHAV
Ian J. Tinsley

RESOURCE RECOVERY AND RECYCLING
A. F. M. Barton

QUANTITATIVE TOXICOLOGY
V.A. Filov, A.A. Golubev, E.I. Liublina, and N.A. Tolokontsev

ATMOSPHERIC MOTION AND AIR POLLUTION
Richard A. Dobbins

INDUSTRIAL POLLUTION CONTROL—Volume I: Agro-Industries
E. Joe Middlebrooks

BREEDING PLANTS RESISTANT TO INSECTS
Fowden G. Maxwell and Peter Jennings, Editors

NEW TECHNOLOGY OF PEST CONTROL
Carl B. Huffaker, Editor

THE SCIENCE OF 2,4,5-T AND ASSOCIATED PHENOXY HERBICIDES
Rodney W. Bovey and Alvin L. Young

INDUSTRIAL LOCATION AND AIR QUALITY CONTROL: A Planning Approach
Jean-Michel Guldmann and Daniel Shefer

PLANT DISEASE CONTROL: RESISTANCE AND SUSCEPTIBILITY
Richard C. Staples and Gary H. Toenniessen, Editors

AQUATIC POLLUTION
Edward A. Laws

MODELING WASTEWATER RENOVATION: Land Treatment
I. K. Iskandar

Continued on back cover

Air Pollution Control,   Part IV

Werner Strauss

1930–1978

AIR POLLUTION CONTROL

Part IV

EDITED BY

GORDON M. BRAGG

University of Waterloo
Waterloo, Canada

and

WERNER STRAUSS (1930–1978)

University of Melbourne
Victoria, Australia

A Wiley-Interscience Publication

JOHN WILEY & SONS

New York • Chichester • Brisbane • Toronto

Library of Congress Cataloging in Publication Data (Revised):
Main entry under title:

Air pollution control.

  (Environmental science and technology)
  Vol. 3 has special title: Measuring and monitoring
air pollutants.
  Vol. 4 has title: Strauss' air pollution control
and is edited by G. M. Bragg and J. Strauss.
  Includes bibliographical references.
  1. Air—Pollution. I. Strauss, Werner, ed.
II. Bragg, Gordon M., 1939–   III. Strauss, Werner,
1933–   IV. Title: Measuring and monitoring air
pollutants.

TD883.A474   628.5'3   79-28773
ISBN 0-471-07957-X (vol. 4)

Printed in the United States of America

10 9 8 7 6 5 4 3 2 1

To the memory of
Werner Strauss
leader, innovator, and teacher in air pollution control and
founding editor of this series

SERIES PREFACE

Environmental Science and Technology

The Environmental Science and Technology Series of Monographs, Textbooks, and Advances is devoted to the study of the quality of the environment and to the technology of its conservation. Environmental science therefore relates to the chemical, physical, and biological changes in the environment through contamination or modification, to the physical nature and biological behavior of air, water, soil, food, and waste as they are affected by man's agricultural, industrial, and social activities, and to the application of science and technology to the control and improvement of environmental quality.

The deterioration of environmental quality, which began when man first collected into villages and utilized fire, has existed as a serious problem under the ever-increasing impacts of exponentially increasing population and of industrializing society. Environmental contamination of air, water, soil, and food has become a threat to the continued existence of many plant and animal communities of the ecosystem and may ultimately threaten the very survival of the human race.

It seems clear that if we are to preserve for future generations some semblance of the biological order of the world of the past and hope to improve on the deteriorating standards of urban public health, environmental science and technology must quickly come to play a dominant role in designing our social and industrial structure for tomorrow. Scientifically rigorous criteria of environmental quality must be developed. Based in part on these criteria, realistic standards must be established and our technological progress must be tailored to meet them. It is obvious that civilization will continue to require increasing amounts of fuel, transportation, industrial chemicals, fertilizers, pesticides, and countless other products; and that it will continue to produce waste products of all descriptions. What is urgently needed is a total systems approach to modern civilization through which the pooled talents of scientists and engineers, in cooperation with social scientists and the medical profession, can be focused on the development of order and equilibrium in the presently disparate segments of the human environment. Most of the skills and tools that are needed are already in existence. We surely have a right to hope a technology that has created such manifold environmental problems is also capable of solving them. It is our hope that this Series in Environmental Sciences and Technology will not only serve to make this challenge more explicit to the established professionals, but that it also will help to stimulate the student toward the career opportunities in this vital area.

Robert L. Metcalf
Werner Stumm

PREFACE TO PART IV

Early in the preparation of this volume the field of air pollution control lost one of its strongest advocates and a man responsible for some of the most widely read contributions in the field. Werner Strauss, in addition to his own contributions to research, also provided in his writing and editing a series of works that have given substance and proportion to the field.

Dr. Strauss did his early work under Professor M. W. Thring at Sheffield University on the control of air pollution from oxygen-lanced open-hearth furnaces. His Ph.D. was awarded in 1959 and he returned to Australia to work at C.S.I.R.O. at the University of Melbourne. He served as Senior Lecturer, Reader, and head of the Department of Industrial Science. He held visiting professorships at Harvard, University of North Carolina, and University of Karlsruhe and served as consultant to many firms. He published the classic monograph *Industrial Gas Cleaning* and over 70 papers. He contributed chapters to several other monographs and acted as editor of this present series. His research, writing, and participation in technical societies brought the knowledge of air pollution control to a wide technical and nontechnical audience. He will be missed.

In this fourth volume of *Air Pollution Control,* the contributions are intended to extend the range of subjects covered in Volumes I to III. The effects of urban meteorology on air pollution distribution are discussed in Chapter 1 and the results of some experimental work in the field are presented. Current knowledge of the effect of air pollutants on plants is surveyed by Dr. Parbery in Chapter 2 and on metals and coatings by Dr. Barton in Chapter 3. The effect of pollutants can often be accurately predicted with present knowledge, and this information has been summarized in these chapters. The last three chapters discuss pollution control equipment. Mist eliminators and the particular design problems associated with removal of liquid droplets are presented in Chapter 4. Chapter 5 discusses particulate removal by filtration, with an emphasis on equipment selection. Chapter 6 presents a review of air and gas cleaning by incineration of contaminants, a subject which, despite its considerable importance, has not been treated in great detail previously.

I would like to acknowledge the help of several people who contributed greatly to the preparation of this volume. I owe a particular debt of gratitude to Mrs. Jennifer Strauss, coeditor, who at all stages has given me help and encouragement in furthering her late husband's work. The cooperation of the authors has been particularly appreciated, especially since they were working under the burden of a change in editors. My thanks and gratitude also go to Georgia Smith of Wiley-Interscience, who has sustained this project through many months of preparation.

GORDON M. BRAGG

Waterloo, Canada
March 1981

CONTENTS

1

URBAN METEOROLOGY AND RELATED AIR POLLUTION DISTRIBUTION

Fred M. Vukovich

*Research Triangle Institute, Research Triangle Park,
North Carolina*

Four fundamental aspects of urban meteorology are urban temperature distribution, the urban wind field, the inadvertent modification effects of urban areas, and air pollution distribution in urban regions relative to how the urban meteorology may contribute to that distribution. These four aspects are discussed individually and their interrelationship, which is extremely important, is also indicated. The first two aspects (the temperature and wind field) are the fundamental meteorological characteristics influenced by the urban area. The latter two aspects (the inadvertent weather modification effects of urban areas and air pollution in urban areas) are the fundamental parameters influenced by urban meteorology.

1.1. INTRODUCTION

The fact that an urban region affects the local climate around the center of a city is well known. Probably the best-known effect is the urban complex acting as a heat reservoir, producing what is called the "urban heat island." This perturbation differs from the classical heat island, which takes its name from the phenomena associated with small isolated islands surrounded by large bodies of water. Daytime solar heating produces considerably higher ground temperatures on the island than water temperatures in the surrounding water, because the specific heat of water is considerably greater than that for soil. Through turbulent heat fluxes, the air temperature over the ground increases much more than that over the water producing a discontinuous change of air temperature at the land/water boundary and large temperature differences between the air over the land and that over the water. The urban heat island, on the other hand, has a more uniform change in temperature at the boundary between the urban and suburban regions, and the temperature contrast is considerably smaller.

The classical heat island produces a rather intense circulation with inflow to the center of the island at all the land/water boundaries, and there is a strong compensation updraft over the island. The circulation associated with the heat island is generally the same in character, but because of the weaker temperature gradients, the circulation is weaker.

There have been many studies to characterize the urban heat island and other factors associated with the urban climate, as well as to determine mechanisms for the various factors associated with the urban climate. In the last decade, there have been two major programs, both performed in St. Louis, Missouri, to study different aspects of the urban climate: the Regional Air Pollution Study (RAPS), sponsored by the U.S. Environmental Protection Agency, and the Metropolitan Meteorological Experiment (METROMEX), sponsored by the U.S. National Science Foundation. Although this chapter discusses the results of many different studies in urban regions, a major portion of the materials were gleaned from the results of these two programs.

The RAPS program, a comprehensive field study, was initiated in July

1972 and terminated in June 1977. The objective of RAPS was to meet the need to demonstrate and evaluate the effectiveness of air pollution control strategies, predicting air quality within a small region, and to provide a basis for developing improved control strategies. In essence, the data from the RAPS program were to be used to develop, evaluate, and validate air quality simulation models by providing, for the first time, a suitable amount of high-quality meteorological, air quality, and emission data for an urban scale.

The data acquisition system for RAPS consisted of a detailed emissions inventory and a ground-based sampling network within a 50-km radius of downtown St. Louis. Air quality and meteorological parameters were measured routinely at 25 stations (Figure 1.1). Additional measurements were provided by instrumented aircraft, pibal, and radiosonde observations.

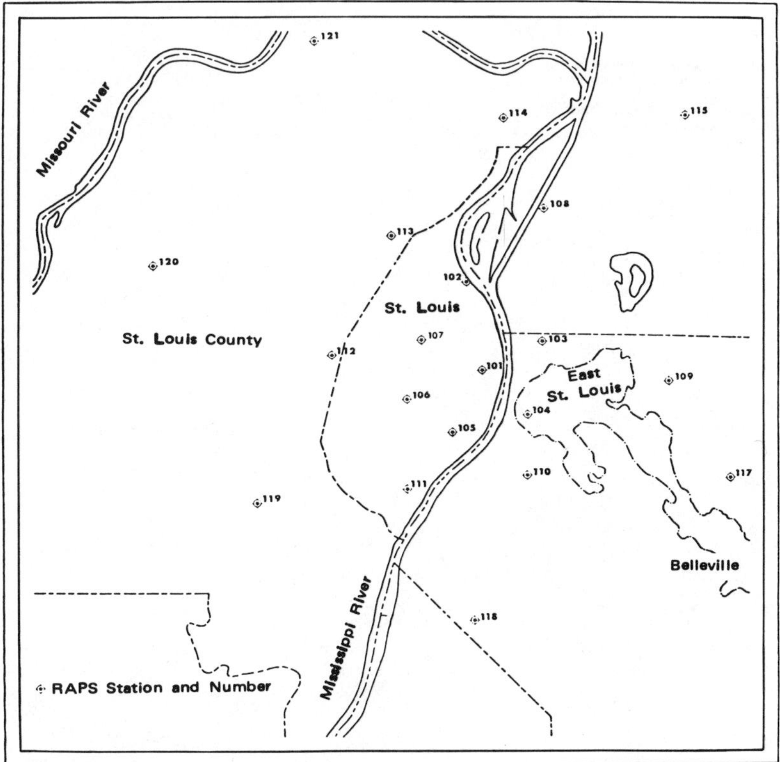

Figure 1.1. RAPS surface network in St. Louis, Missouri. Only 20 of the 25 stations are shown. The remainder were out of the region of immediate interest.

METROMEX, an inadventent weather modification program, was initiated in 1970, and the field experiments, for the most part, ended in 1977. The principal objectives of METROMEX were to determine the extent and location of the effects of an urban area on precipitation and related weather conditions; to discover causes for these effects, both over the city and downwind over suburban and rural areas; and to develop methods for predicting these results and translating them to other urban complexes. The city of St. Louis was chosen for the initial studies because it is situated in the center of a large rural area in which there were no other major sources of contamination within a radius of at least 100 miles.

To accomplish the objectives of this program, a field program was established and data were derived from a surface network (Figure 1.2), remote sensors (lidars and radars), and instrumented aircraft. The surface network included rainfall measurements from 245 recording rain gauges, hail observations from 245 hail pads, temperature readings from 28 sites,

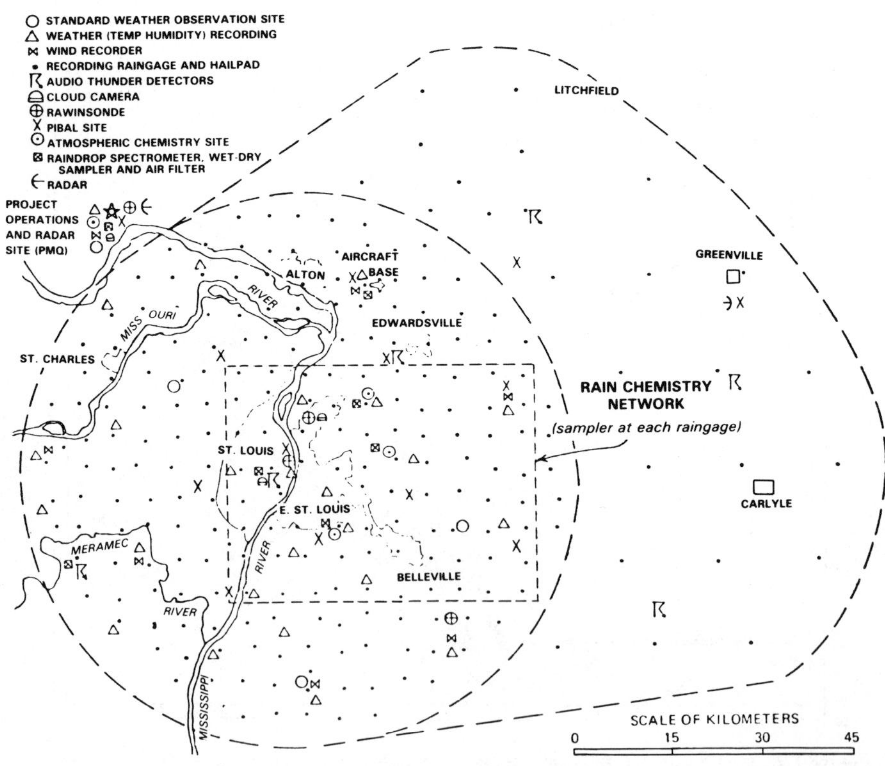

Figure 1.2. METROMEX network.

surface winds from 7 sites, upper-air boundary layer winds from up to 12 sites with pibals and radiosondes, thunder occurrences from 10 sites, and rain and air chemistry from 93 sites.

The METROMEX and RAPS programs contributed significantly to our understanding of urban meteorology and mechanisms that effect the urban climate. The analysis of data from these studies is by no means complete. Some of the early results are presented in this chapter. These results, together with the results of other studies, give a very detailed understanding of the effect of the urban area on local climatic conditions, specific mechanisms related to the urban area that affects weather over and downwind of the city, and the subsequent effect of the urban meteorology on the transport of air pollutants in and around the city.

1.2. TEMPERATURE IN THE URBAN REGION

1.2.1. Introduction

That the urban region affects the thermal structure in and around that region has been known for a long time. Generally, the urban region is warmer than the surrounding environments, and this condition is particularly noticeable at night under clear skys and light winds—the urban heat island.

As early as 1833, Howard[1] documented the heat island of London, England. Later, Schmidt[2] documented the heat island of Vienna in great detail and pioneered use of the automobile to obtain thermal cross sections through the city. Subsequently, there have been numerous studies dealing with statistics on and causes for the urban heat island. A brief description of some of the more significant points concerning the urban heat island is presented next.

1.2.2. Diurnal Temperature Variation

The diurnal variation of the temperature at the surface in an urban region compared with that in the surrounding rural region has been presented by many investigators, notably Mitchell,[3] Hage,[4] Oke and East,[5] and Vukovich.[6] They show that at sunrise the urban region is considerably warmer than the surrounding rural region, and that after sunrise the temperature difference between the urban and rural region begins to decrease, reaching a minimum in the late morning or early afternoon. Around 1500 to 1700 (3 to 5 P.M. local time), the temperature difference in the urban region begins to increase in relation to the surrounding rural region, with the temperature difference reaching a maximum somewhere around 3 to 5 hr after sunset in all season.[4,7]

This is illustrated by the data in Figure 1.3, which show the time variation of the urban and rural temperature difference for the more significant portion of the diurnal cycle for St. Louis and its surrounding environment on June 8, 1976.* The data were obtained by the U.S. Environmental Protection Agency as part of the Regional Air Pollution Study (RAPS). For the center of the city, the data were averaged using four stations in the RAPS network; the rural data were averaged using four stations in the region surrounding the city.

These data agreed with the description of the diurnal cycle in the urban region compared with the surrounding rural region given by many investigators. It should be noted that the minimum temperature difference between the center of the city and the surrounding rural regions occurred at 1000 CST (Central Standard Time), and after sunrise, there was significantly more heating in the rural region than in the center of the city. Cooling began around 1500 CST, with the cooling rate in the rural region being significantly larger than in the urban region.

The cooling rate in the rural region is significantly greater than in the urban region in the late afternoon or early evening. Oke and East[5] and Oke et al.[7] presented data on the cooling rates for Montreal relative to the surrounding urban region. Figure 1.4 shows the average results of Oke et al. for both winter and summer and light wind and scattered cloud conditions.

In winter (Figure 1.4a), the transition from warming to cooling begins about 2 hr before sunset in both the urban and rural regions, the cooling in the rural region being most intense at sunset: approximately $-1.0°C$ hr^{-1} at sunset, diminishing to approximately $-0.7°C$ hr^{-1} for the next 3 hr. During this period the urban cooling rate is almost constant with time, the magnitude being approximately $-0.3°C$ hr^{-1}. Approximately 4 hr past sunset, the rural and urban cooling rates are, for all practical purposes, identical.

In summer (Figure 1.4b), the rural cooling rate is very intense at sunset and in the next 3 hr declines to a value comparable to the urban cooling rate—approximately $-2.3°C$ hr^{-1}. The urban cooling rate ranged between -0.5 and $-1.0°C$ hr^{-1} during this period. The production of the

---

* St. Louis is a city with a population of approximately 1 million people. The city and suburbs are concentrated in an area of about 400 km^2. The city is found in relatively flat terrain, with no major body of water nearby, although the Mississippi River passes immediately east of the city. There is no other major anthropogenic source within 100 miles of the city.

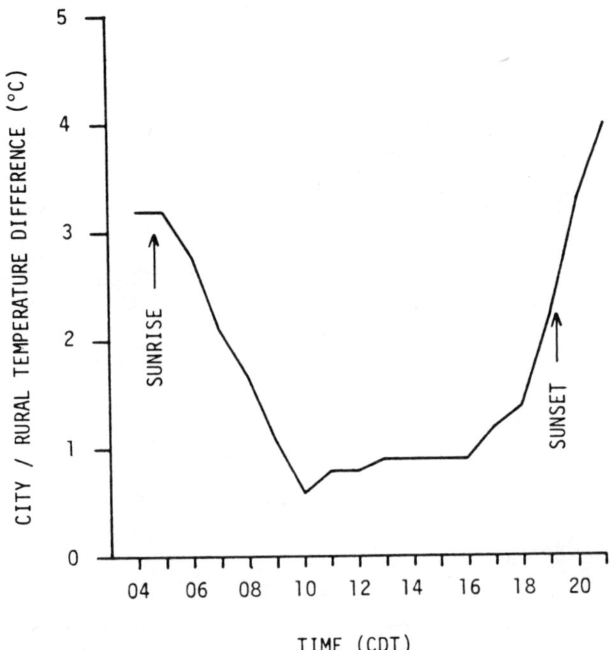

Figure 1.3. Variation of the temperature difference between St. Louis, Missouri, and the surrounding rural region for the period 0400 to 2100 CDT on June 8, 1976. The data represent the temperature difference between the average temperature obtained from four stations in the center of the city and four stations in the surrounding rural region (Vukovich et al.[8]).

nocturnal heat island largely depends on the rural cooling. Reasons for the large differences in cooling rates in the rural and urban regions will be discussed later.

General characteristics of temperature trends in urban regions compared with rural regions have shown that in winter, the cities have a tendency to remain warmer than the surrounding environment in both day and night; however, daytime temperature differences are still smaller than nighttime differences. Daily maximum and minimum temperatures in the city lag temperatures in the suburbs; the daily range in the city is also somewhat less.

In the summer, the urban–rural temperature difference is greatest at night and insignificant during most of the day. In many cases, it has been found that the daytime temperatures in the city are a fraction of a degree cooler than the surrounding environments, which may be due to attenuation of sunlight over the city by airborne particles and other pollutants. As in the winter, the maximum and minimum daily urban temperatures

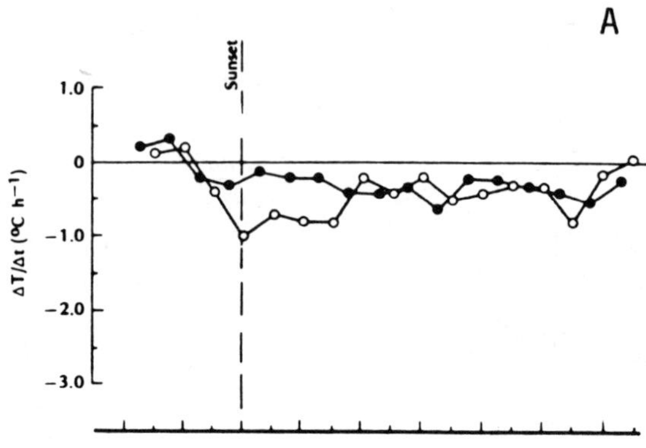

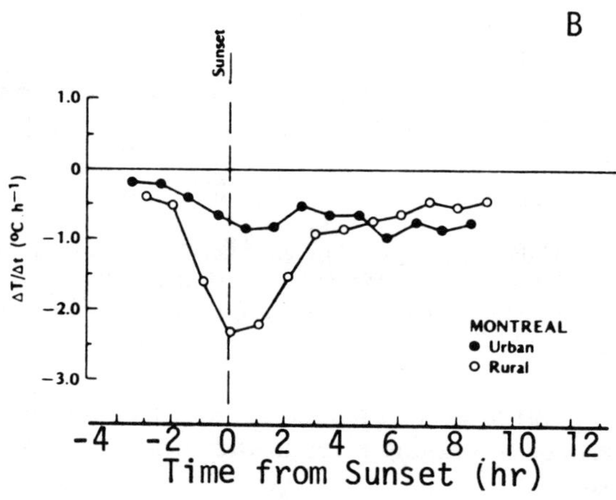

Figure 1.4. Time variations of temperature in the urban (filled circles) and rural (open circles) regions under light winds and scattered clouds at night for a summer period (*a*) and winter period (*b*) in Montreal, Canada (Oke et al.[7]).

lag those for the suburbs. The diurnal range is also smaller in the city during the summer.

The recent U.S. Environmental Protection Agency's RAPS program has provided data whereby the diurnal variation of the horizontal distribution of temperature in an urban region may be studied.[8] The following is a description of that variation in St. Louis for June 8, 1976, a day with partly cloudy skies and light winds. Figure 1.5 shows the area of interest with topographic contours added. The topographic data were derived relative to the Mississippi River height (~100 m MSL).

On this day, the sun rose at 0445 CST, and by 0600 CST the daytime warming was well under way in both the urban and rural regions. Analysis of the surface temperature at 0600 CST (Figure 1.6*a*) indicates the presence of a heat island centered over the builtup regions of the city, with the maximum temperature difference between the center of the city and the rural regions at 3.7°C. There appears to be an extension of the urban heat island to the north, but this is not due to the transport of heat from St. Louis northward, since the synoptic scale flow was from the northwest at this time.

The daytime warming trend continued, with the rate of temperature increase in the rural regions exceeding that for the urban region. At 1100

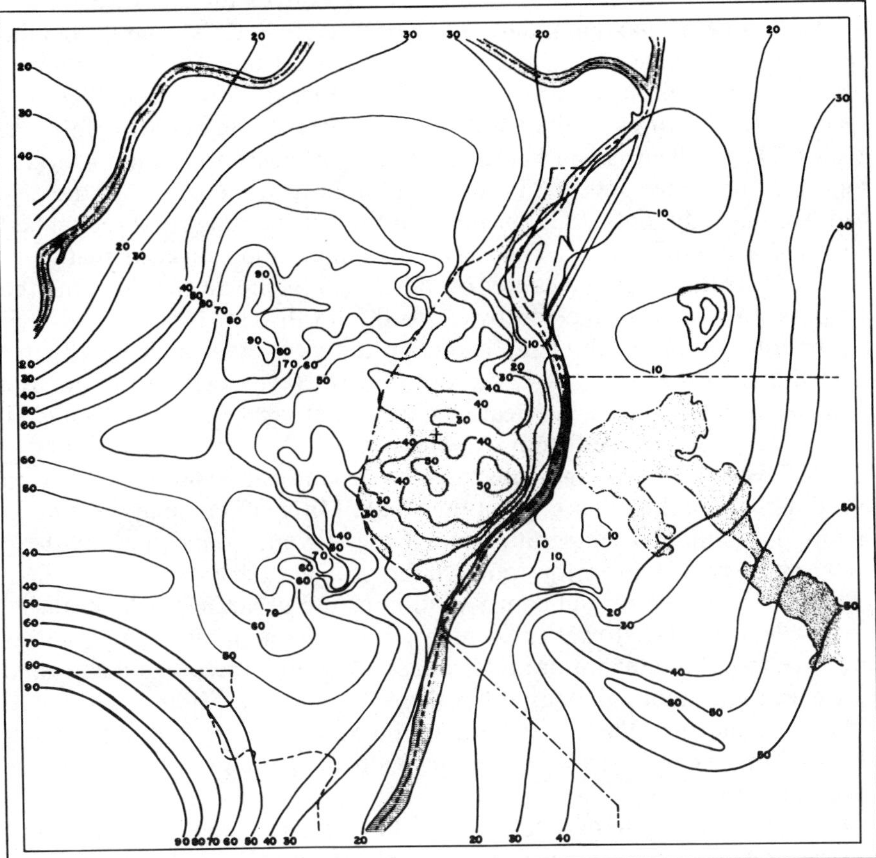

Figure 1.5. Topographic map of the St. Louis region. Contours are in meters relative to the Mississippi River height (100 m).

CST (Figure 1.6*b*), the heat island over the builtup region of the city was still evident, but the temperature difference between the city center and the surrounding rural region had decreased to approximately 1.4°C. An area of minimum temperature exists along the northwest boundary of the urban complex and was on an average more than 1°C cooler than the center of the city to the east and about 0.3 to 0.4°C cooler than the rural areas to the north, south, and west. This region of slightly cooler surface temperatures along the northwestern boundary of the urban complex persisted until late afternoon, when the temperatures began to decrease. The surface isotherm pattern has a configuration that suggests transport of heat from the center of the city to the northeast. That heat transport from the center of the city to the northeast may have occurred is supported by the synoptic flow, which, at this time, was from the southwest.

At the time of maximum daily temperature (1600 CST), the horizontal contrast had increased slightly, to 1.7°C (Figure 1.6*c*). The isotherm pattern still suggests heat transport to the northeast. After 1600 CST, the surface temperatures in the rural regions began to decrease rapidly. At 2100 CST, the heat island had increased in intensity to around 4°C (see Figure 1.6*d*). Since 1600 CST, the temperatures in the center of the city had decreased by only 3°C, but those in the surrounding rural region had decreased as much as 7°C. The isotherm configuration suggests that there is heat transport from the center of the city to the north. At this time the synpotic flow is from the southwest to the northwest. It is possible that the regions to the north of the center of the city retain their heat better than those to the west, east, or south. However, exact causes for the northward extension of the heat island seen at 0600 CST and 2100 CST could not be determined from existing data.

There is evidence to indicate that the urban heat island is weaker on Sunday than on the other days of the weak. Mitchell[3] compared the maximum and minimum temperatures for New Haven, Connecticut, with an outlying airport station for four winter seasons (1939–1943). His data (Table 1.1) show no significant variation of the urban heat island during the daylight hours. However, at night, when the minimum temperatures are reached, the temperature differences show that the urban heat island is weaker on Sunday than on other days of the week. Mitchell attributed this to the fact that the city is dormant on Sunday compared to the other days of the week when it is teeming with people and has significantly more activity.

Lawrence,[9] in a 20-year study (April 1949–March 1969), compared the maximum and minimum temperatures in London to an outlying rural station. His results for the winter period (October–March) (Table 1.2) also show no significant variation of the urban heat island during the daylight

**Table 1.1.   Mean Differences in Maximum
and Minimum Temperatures in Four
Winter Seasons (1939–1943) between New
Haven, Connecticut, and a Neighboring
Airport versus Day of the Week**[3]

Day of the week	Maximum temperature (°C)	Minimum temperature (°C)
Sunday	0.1	1.2
Monday	0.0	2.0
Tuesday	0.2	2.4
Wednesday	0.0	2.1
Thursday	0.0	2.3
Friday	0.0	2.4
Saturday	0.0	2.1

hours relative to days of the week. Examination of the data depicting differences in minimum temperatures showed a weak heat island occurring on Sunday and Monday, with the weakest heat island on Monday. The lack of a persistent weekly pattern suggests that the Sunday effect, as depicted in Mitchell's data, is influenced by factors other then the lack of industrial activity in the city on Sunday. London is much more industrialized than is New Haven, Connecticut and thus has tremendous air pollution problems. Air pollution layers over a city can markedly affect the radiation balance, which, in turn, affects the urban heat island intensity.

Lawrence points out that in London there is a different kind of effect on Sunday, which may be a result of the effect on the radiation balance of a shroud of air pollution over the city. The data in Table 1.3 yield the mean difference in maximum temperatures based on 14 years of data (1957–1970) between London and an outlying rural station for the summer (May–July). This shows a maximum heat island intensity on Sunday and a minimum on Thursday, which Lawrence attributes to the fact that there is a weekly cycle of domestic and industrial habits that causes weekly patterns of air pollution and sunshine. Existing evidence has shown that, on the average, a minimum of smoke and sulfur dioxide[10,11] occurs on Sunday. Apparently, the shroud of air pollution over the city during the week reduces the amount of radiation reaching the surface which can then be used to heat the city, while considerably more radiation reaches the surface on Sunday.

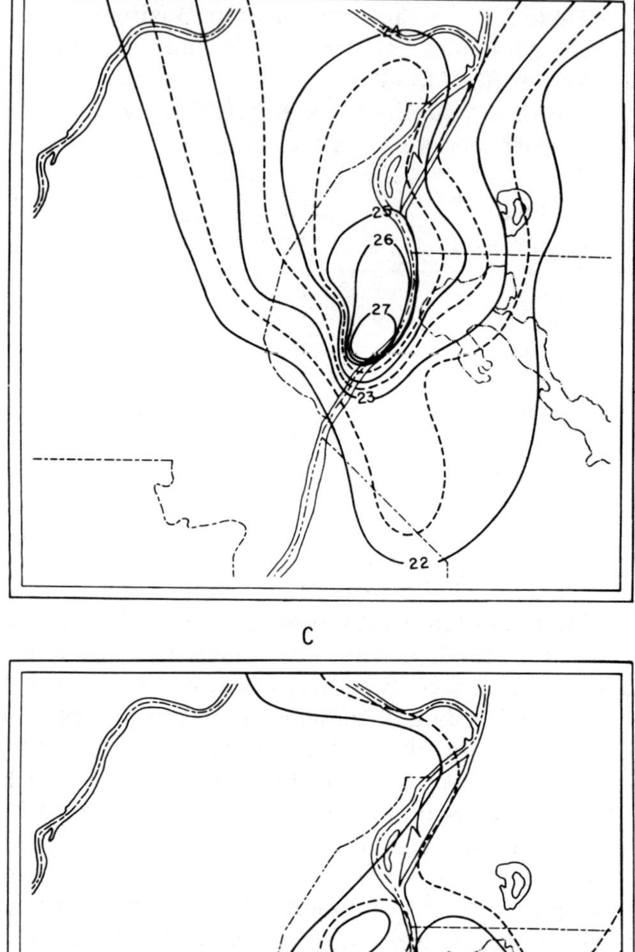

Figure 1.6. Analysis of the surface temperature (°C) distribution CST on June 8, 1976 in St. Louis, Missouri, at (*a*) 0600 CST; (*b*) 1100 CST; (*c*) 1600 CST; (*d*) 2100 CST (Vukovich et al.[8]).

B

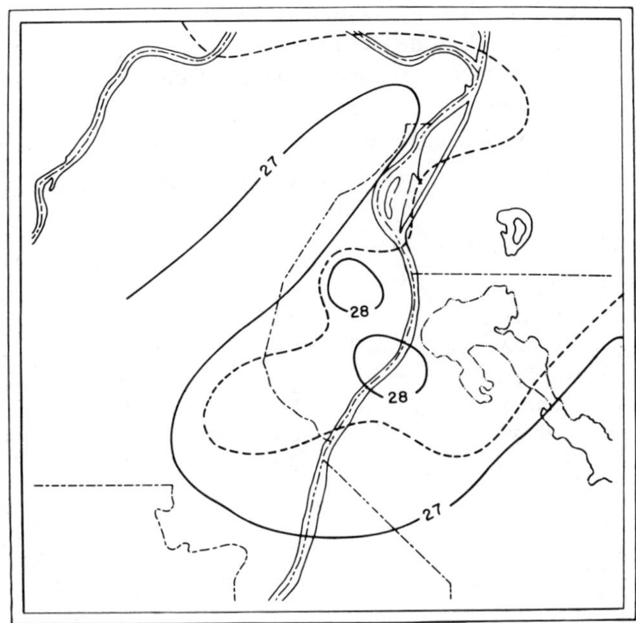

D

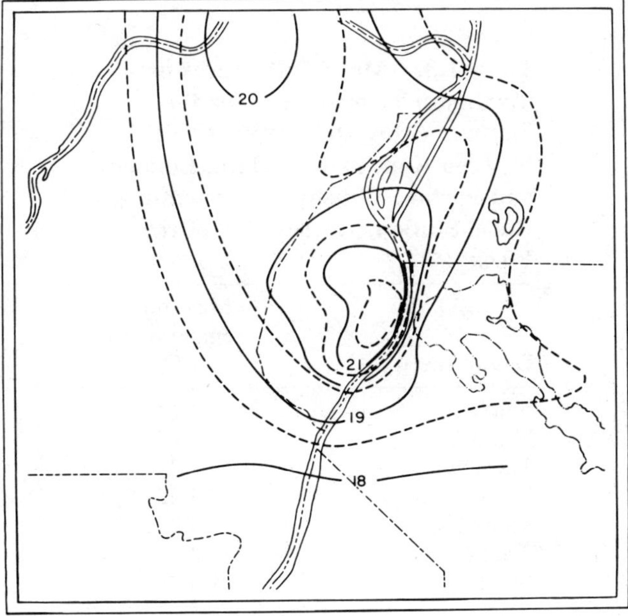

Table 1.2. Mean Differences in Maximum and Minimum Temperatures for the Winter Period (October–March 1949–1969) between Kensington Palace, London, and Porton, Wiltshire (Command Rural Station), versus Day of the Week[9]

Day of the week	Maximum temperature (°C)	Minimum temperature (°C)
Sunday	0.68	1.68
Monday	0.62	1.65
Tuesday	0.68	1.82
Wednesday	0.59	1.76
Thursday	0.71	1.74
Friday	0.65	1.75
Saturday	0.68	1.82

Table 1.3. Mean Differences in Maximum Temperature for the Summer (May–July 1957–1970) between Kensington Palace, London, and Porton, Wiltshire (Surrounding Rural Station), versus Day of the Week[9]

Day of the week	Maximum temperature (°C)
Sunday	1.55
Monday	1.25
Tuesday	1.17
Wednesday	1.16
Thursday	1.13
Friday	1.18
Saturday	1.30

1.2.3. Vertical Structure of the Urban Heat Island

Duckworth and Sandberg[12] gathered vertical temperature data over the city of San Francisco and surrounding rural regions. They showed isothermal or lapse conditions near the ground over the city at night, whereas simultaneous observations over the surrounding rural regions showed a surface inversion. The air in the region between 30 and 100 m above the surface over the city was generally cooler than that over the surrounding rural regions. Duckworth and Sandberg attributed the elevated cooling over the city to four possible factors: (1) vertical mixing of air over the urban area, (2) the effects of a large-scale urban convection cell, (3) radiative cooling in pollution layers, and (4) variations in the wind field between 30 and 100 m. Since this work was completed, considerably more evidence showing the contrast in lapse rates above the city and surrounding rural areas has been obtained by DeMarrais,[13] Clark,[14] Davidson,[15] Oke and East,[5] Angell et al.,[16] and others.

Atwater[17] performed a theoretical study of the influence of an air pollution shroud over the urban area on the vertical temperature distribution in that region. The primary pollutants used in his model were aerosols and nitrogen dioxide. He showed that scattering and absorption of both solar and infrared radiation also occurs within the layer. The resultant effect of the radiative processes on the vertical temperature distribution during the day depended on the relative magnitude of heating caused by absorption of solar radiation and cooling due to divergence of infrared radiation. At night radiative divergence occurred in the polluted layer, producing marked cooling, and resulting in the formation of an elevated inversion that extended from within the pollution layer to a level above. Atwater's theoretical calculation supports the notion of Duckworth and Sandberg that cooling in the pollution layer may be responsible for the elevated inversion.

The model proposed by Summers[18] has also been used to explain the existence of the elevated inversion. An adiabatic mixing layer is formed through the input of heat into the urban atmosphere combined with enhanced turbulent mixing resulting from the increased roughness in the city. The layer increases in depth downwind of the rural–urban boundary. Inversion conditions over the area became elevated due to the input of heat and the mixing. The model was applied to observations from New York and Montreal and gave good agreement.[5,19] However, theoretical calculations performed by Vukovich et al.,[20] using a three-dimensional primitive equation model, did not indicate the formation of an elevated inversion due to the influx of heat at the surface and enhanced mixing due to increased roughness. More than 100 simulation calculations of the urban heat island in St. Louis were performed under various initial conditions, and in no cases were elevated inversions formed. Radiational processes were omitted in the model.

Tyson et al.[21] showed how different types of vertical temperature structures can develop from basically the same kinds of initial states. The data presented were for Johannesburg, South Africa. On the evenings of September 1 and 21, 1971, the vertical temperature structure and surrounding suburban regions were essentially the same (Figure 1.7a and b). The boundary layer stability was slightly more intense in the surrounding suburban and rural regions on the night of September 1 than on September 21. The upper levels were slightly less stable on the night of September 21 than on the night of September 1. In both cases the only evidence of urban modification of the temperature profiles was the definite retardation in the development of the surface radiation inversion.

On the morning of September 2, a cool layer and elevated inversion over the urban and suburban areas had developed (Figure 1.7c). Northeasterly flow existed at the surface with wind speeds on the order of 6 m s^{-1}. There was also a pronounced wind shear, and the urban heat island was not readily evident. However, the cross section on the morning of September 22 indicated a pronounced heat plume extending from the city upward and downwind (Figure 1.7d). Northerly flow was evident at the surface, with wind speeds on the order of 6 m s^{-1}, but in this case there was no wind shear. Bent-over plumes such as these had also been observed by Clark[14] in Cincinnati. Such differences in vertical temperature structures may be attributed to variations in the intensity of the urban heat island circulation, and in turn, to differences in initial heat input the city or in the initial stability in the surrounding rural or suburban boundary layer. Less stable boundary layers will produce more intense heat island circulations.[23] This will be discussed in more detail in Section 1.4.

1.2.4. Factors Influencing the Urban Heat Island

This section will highlight the major factors that influence the urban heat island. To be considered first is the influence that urban air pollution has on the radiation budget of the urban region. Table 1.4 shows the proportional decrease of the total radiation recevied at the city surface relative to that in the surrounding region as found by various investigators.[10,24,25] Although the absolute amount varies from one case to the next, all indicate that the urban region receives less solar radiation at the

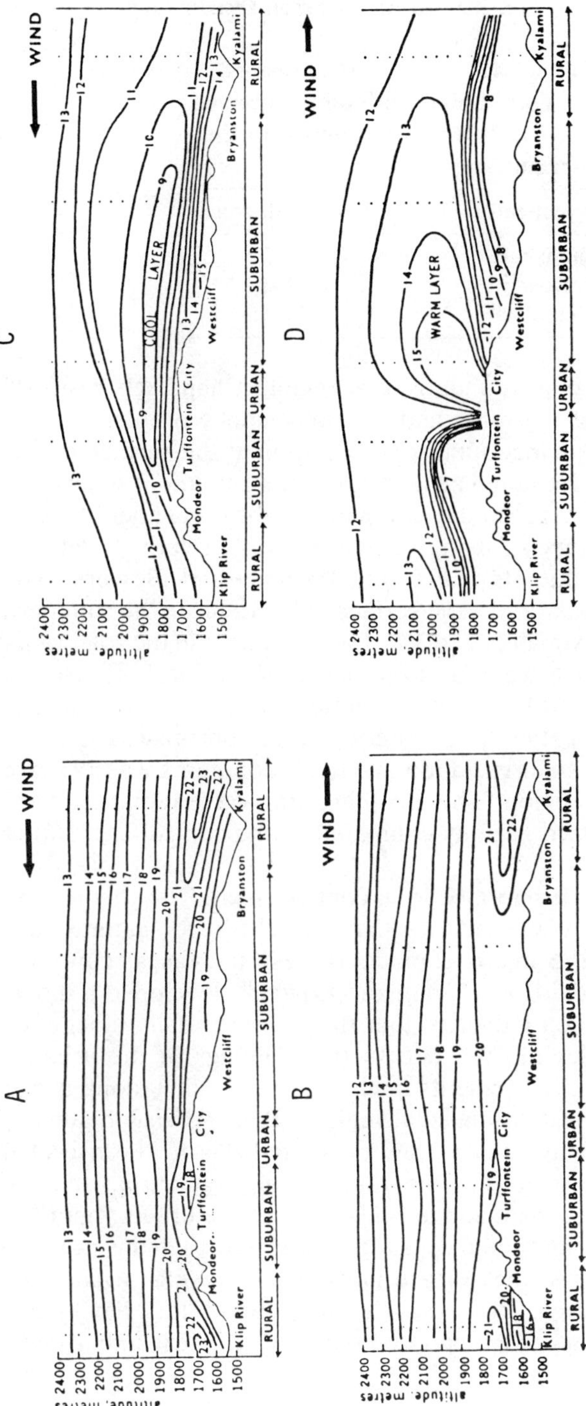

Figure 1.7. Vertical temperature (°C) structure in north–south cross section over Johannesburg, South Africa, around 1800 local time under clear skies on (a) September 1, 1971 (wind direction was northwesterly with a speed of 6 m s^{-1} at an altitude of 1650 m and 11 m s^{-1} at 2400 m); (b) September 21, 1971 (wind direction was southerly with a speed of 6 m s^{-1} at an altitude of 1600 m and 6 m s^{-1} at an altitude of 2400 m); and around 0530 local time under clear skies on (c) September 2, 1971 (wind direction was northeasterly with a speed of 6 m s^{-1} at an altitude of 1650 m and 11 m s^{-1} at 2400 m); (d) September 22, 1971 (wind direction was southerly with a speed of 6 m s^{-1} at 1600 m and 6 m s^{-1} at 2400 m) (Tyson et al.[21]).

Table 1.4.   Percent Decrease of the Total Solar Radiation Received at a City's Surface Relative to the Surrounding

Investigator	Percent decrease
Landsberg[24]	15–20
DeBoer[25]	13–17
Chandler[10]	25–55

surface than does the surrounding rural region. Chandler's results[10] are for London, which at that time had a serious smoke problem.

Mateer[26] studied the incoming solar radiation at the surface in the city of Toronto, Canada. He noted an increase in the amount of solar radiation received at the surface on Sunday compared to the remainder of the week, suggesting the influence of man's activity and the relatively larger concentrations of air pollutants that result. Peterson and Flowers[27] studied the incoming solar radiation at the surface in St. Louis in the summertime, and found that approximately 2 to 3% of the reduction in the solar radiation received at the surface was due to airborne particulates. Rouse et al.[28] compared the solar and long-wave radiation received at the surface and in an industrial part of Hamilton, Ontario, with a control area. They found that the reduction of solar radiation due to air pollutants was 12% greater in the industrialized region. The incoming long-wave radiation was 31% greater in industrialized regions compared to the control site during the daylight hours.

The subject of the urban albedo is controversial. The theoretical calculations of Kondratyev[29] indicate that the albedo for natural surfaces would decrease as the microscopic roughness increases, contradicting some observational evidence. Craig and Lowry[30] developed a model for the urban albedo which indicated that the averaged daily albedo should decrease on the order of 20% due to the buildings in the urban area. However, airborne measurements made by Bray et al.[31] along a transect from Bethel to St. Paul, Minnesota, indicate that the visible albedo was 50% greater over urban areas. They attributed this to the use of light-colored construction material, the removal of vegetation, and the increased amounts of suspended particulates over the urban regions. Data presented by McCormick and Ludwig[32] indicated that the presence of aerosols could increase the surface albedos, producing surface cooling or retardation in heating. However, Charleson and Pilat[33] have indicated that smoke containing carbon and iron oxides could heat rather than cool the surface.

It appears that the character of the urban albedo depends on whether there is a shroud of air pollution over the city. It is also under these conditions that intense heat islands are generally found. The increased albedo usually found under these conditions over the urban region, complemented by the reduced solar radiation at the surface, suggests that these factors may be influential in producing the lag in heating after sunrise found in the urban region compared to the surrounding rural region.

Myrup,[34] Tag,[35] McElroy,[36] Pandolfo et al.,[37] and Nappo[38] have studied the effect of the surface roughness, surface humidity, and soil thermal diffusivity on the urban heat island, and all reached essentially the same conclusion. Nappo's study, the most recent, used a time-dependent one-dimensional model. He allowed the surface roughness to vary from 1 to 1000 cm, the surface relative humidity from 0.25 to 1, and the soil thermal diffusivity from 0.0015 to 0.020 $\text{cm}^2 \text{ s}^{-1}$.

When the soil relative humidity and soil thermal diffusivity were held constant, increasing the surface roughness caused the amplitude of the diurnal surface wave to decrease, the time of surface maximum to increase, and the time of temperature minimum to remain unchanged. When the surface roughness and the soil thermal diffusivity were held constant, increasing the soil relative humidity also caused the amplitude of the diurnal surface temperature wave to decrease, but the time of minimum temperature occurred earlier in the morning and the time of maximum temperature remained unchanged. When the soil relative humidity and surface roughness were held constant, increasing the soil thermal diffusivity had little effect on the amplitude of the diurnal temperature wave. The amplitude of the wave for the smallest value of the thermal diffusivity was only slightly greater than that for the largest value (and had no effect on the time of maximum temperature). However, increasing the soil thermal diffusivity shifted the time of minimum temperature to an hour earlier in the morning.

Nappo constructed an approximation to the classical relationship between the diurnal temperature wave in the urban, suburban, and rural regions, having developed a parameter relationship such that the urban region was rough and the rural region was smooth, the soil moisture in the urban region was the smallest and that in the rural region was the greatest, and the soil thermal diffusivity in the urban region was the largest and that in the rural region was the smallest. The parameter values for the suburban region lie midway between those for the urban and rural regions (see Figure 1.8). Nappo's calculation suggested that soil moisture was the most important parameter.

The soil moisture in cities is considerably less than that in rural regions because the city is generally made up of rocklike substances that do not

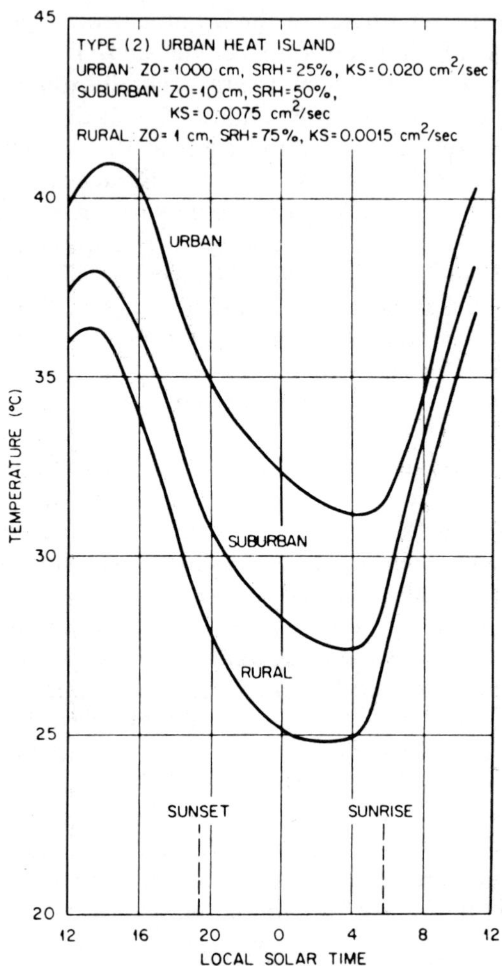

Figure 1.8. Theoretically computed diurnal variation of surface temperature in the urban, suburban, and rural region where the surface roughness (z_0), surface relative humidity (SRH), and the surface heat conductivity (K_s) were different in each region (Nappo[38]).

hold moisture. Precipitation in the form of rain is quickly removed from the surface by drainpipes, gutters, and sewers. Snow is cleared from surfaces by plows and shovels and significant amounts are carried away. However, in the country, most of the precipitation remains in the soil. The water is available for evaporation, which produces cooling.

Further evidence of the importance of the lack of soil moisture in urban regions has been demonstrated in energy balance studies. The net radiation received at the earth's surface affects the sensible heat energy, the latent heat energy, and the ground stored energy. The manner in which the net radiation affects the energy budget components depends on surface characteristics and atmospheric conditions. Maisel[39] found that the latent heat energy from evaporation in urban areas was negligible. To conserve the energy balance, sensible and ground-stored heat energy must have increased. He found that the sensible heat flux in the urban area was approximately 2.7 times greater than that in the rural area. In a qualitative argument for increased heat storage, he suggested that the surface reradiates this thermal energy at night and is used to raise the air temperature, producing the urban heat island.

Oke et al.[7] monitored components of the energy balance in the summer at an urban and rural site in the Vancouver area. The urban station was located on the flat roof of a four-story building in the downtown area. The rural station was at an abandoned airfield at Ladner, approximately 30 km south of Vancouver. They found that during the daytime, the sensible heat flux was higher at the urban location and the heat storage in the building roof was much larger than that of grass at the rural site. At sunset, the sensible heat flux in the urban area did not decline as rapidly as that in the rural area. Actually, the urban area maintained a flow of heat away from the surface throughout the night. This fact must be considered as one of the major sources for the production of the urban heat island. During the daytime, the increased sensible heat flux must be attributed to virtually all of the net radiation at the surface being used in sensible heating and ground heat storage because of the lack of soil moisture for evaporation. Oke et al.[7] noted that in one of their cases involving heavy rainfall, there was a significant decrease in the sensible heat flux in the city, suggesting that much of the available energy was being utilized for evaporation.

Brick, pavement, and the rocklike materials of the city's buildings and streets conduct heat about three times as fast as does wet or sandy soil. When the diurnal temperature is at a maximum, the soil temperature is probably higher than that of a rock wall. However, in the wall, the temperature approximately 3 in. below the surface will probably be higher. Consequently, by sundown the rocklike material will undoubtedly have stored more energy in form of heat than an equal volume of soil, emphasizing the importance of the ground-stored heat, particularly at night.

Wind and stability also influence the urban heat island. Clarke and Peterson[40] used Summers' model,[18] which gives the heat island intensity in terms of the temperature difference between the center of the urban region and the surrounding rural air, to determine the effect of wind and stability on the heat island intensity. The model has the forms

$$\Delta T = \left(\frac{2xH\gamma}{u\rho c_p}\right)^{1/2} \tag{1}$$

where ΔT = heat island magnitude
   x = distance from the upwind edge to the center of the city
   H = excess heat per unit area available to heat the air
   γ = lapse rate of potential temperature
   u = wind speed through the urban boundary layer
   c_p = specific heat at constant pressure
   ρ = density of air

In their calculations, the nonmeteorological parameters were held constant. The diameter of the city was taken as 30 km (x = 15 km) and H was taken to be 10^{-3} cal cm^{-2} s^{-1}. Stability and wind speed were based on climatological estimates of morning mixing heights and transport winds at U.S. Upper Air Stations given by Holzworth.[41] Calculations were of the upper decile, the upper quartile, and the medium ΔT across the continental United States for the summer and winter seasons.

Figure 1.9a shows the results for the summer season. The analysis is for the upper decile value. The calculations indicate the maximum meteorological potential for heat island formation occurs over western Oregon. In that region a relative heat island intensity of 11°C or more can occur on 10% of the mornings. Secondary maxima exist in the north central states and the Appalachian mountains, extending from southern Maine to southern Mississippi. The calculations suggest that the potential for nocturnal heat islands is generally small in the coastal regions and over the south central United States.

Their results also indicated that the intensity of the urban heat islands is generally lower in the winter season (Figure 1.9b), which is due to the increased storm activity. The difference between upper quartile and decile values is often greater during the winter. This indicates that although less intense heat islands are expected during the winter, there are a small number of days with high meteorological potential for a very intense heat island. The meteorological potential for the heat island is greater in the winter along the Gulf coastal area and over much of California. This appears to be a result of lower water vapor content and cloud cover in coastal areas during the winter season.

Oke[42] demonstrated that still another factor—urban population—influences the urban heat island. He showed a positive correlation between

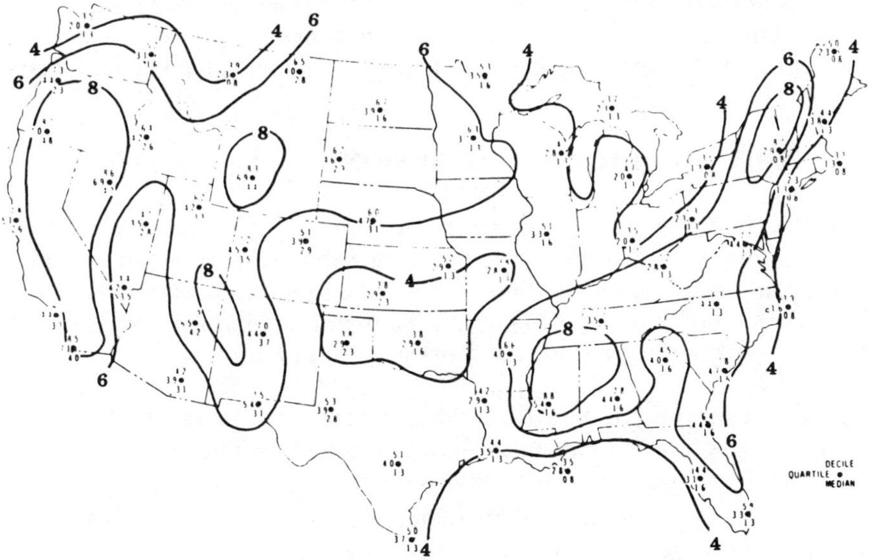

Figure 1.9. Calculated medium, upper quartile, and upper decile heat island magnitudes (°C) for the (*a*) summer and (*b*) winter. Isopleths are for the upper decile values (Clarke and Peterson[40]).

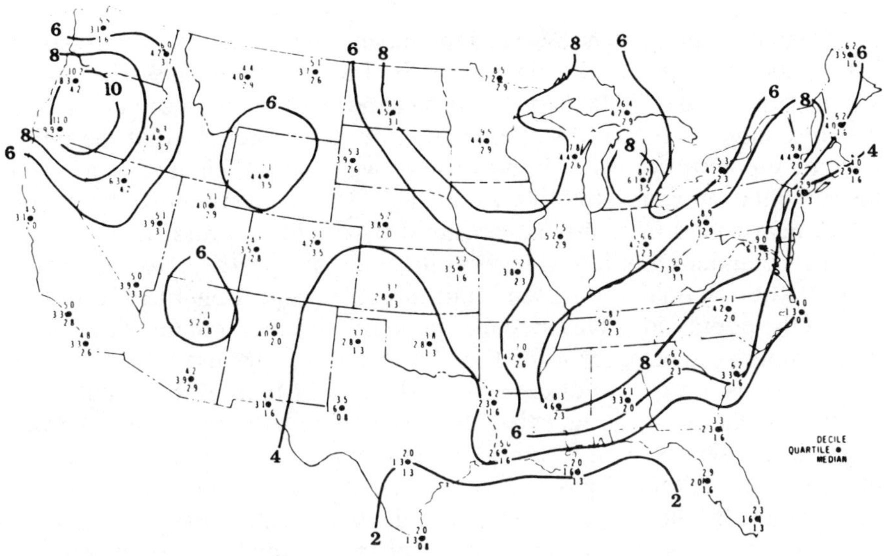

the population of a given town and the magnitude of the urban heat island it produces for 10 settlements on the St. Lawrence lowland with populations from 1000 to 2,000,000. The urban and rural temperature difference was measured in these 10 settlements under the condition that the influence of topography, water bodies (there were none), climate, time, and instrumentation were effectively constant. The analysis showed that the urban heat island intensity as measured by the urban–rural temperature difference under cloudless skies was inversely proportional to the square root of the regional wind speed and directly proportional to the fourth root of the population. This model is qualitatively similar to that developed by Summers (eq. 1) except for the fact that Summers used the diameter of the city as a measure of the size of the city instead of the population.

The last factor is the fact that the city generates its own heat, particularly in the winter, when domestic and commercial heating systems are in operation. Mitchell[3] has suggested that it is the home heating plants that sustain the heat island during the daylight hours in the wintertime, while during the summer, the absence of this factor leads to an equality between the urban and rural surface temperatures. However, even in the summer, the city has many more heat sources than are normally found in the countryside, the more notable being factories, motor vehicles, and air conditioners (which not only remove hot air in order to produce the cooling effect, but in the process produce heat that must then be removed).

1.3. URBAN MOISTURE DISTRIBUTION

Like the differences in urban and rural temperature, there is a contrast in the water vapor content between the urban and rural areas called the urban "dry island," which is not a continuous phenomena but is dependent on time of day and season. Many factors influence the production of the urban dry island, the most important being:

1. The impervious nature of building and pavement materials, which have very limited capacity for water retention relative to the soils surfaces found in the rural environment.
2. Differences in the thermal characteristics of the urban region relative to the rural region, which can effect the moisture balance (e.g., dew formation).
3. The presence of much more vegetation in the rural regions relative to the urban regions, which can be a source of moisture through plant transpiration.

4. The consumption of fossil fuel in the urban areas, which is a source
    of atmospheric moisture.
5. Differences in surface roughness between the urban and rural regions,
    which affect the vertical transport of moisture.

Some of these factors cause the city to be a sink of moisture relative
to the country, whereas others cause it to be a source. Many of these
factors are identical to those that produced the urban heat island.

There have been a number of studies treating the urban and rural moisture difference. One of the earliest was by Kratzer et al.,[43] who studied
the urban–rural moisture differences in three German cities: Berlin, Bres-
lau, and Munich. Kopec[44] studied the dewpoint variations in and around
the city of Chapel Hill, North Caroline. Maisel[39] and Landsberg and
Maisel[45] studied micrometeorological effects (which included urban–rural
moisture differences) on the construction of Columbia, Maryland. Hage[4]
performed similar studies in cities in Canada. The basic results of these
studies are outlined by Ackerman.[46]

Ackerman used 20 years of hourly measurements from Argonne National Laboratory, approximately 25 miles southwest of Chicago in a relatively rural region, and from the synoptic weather station at Chicago's
Midway airport to study the effect of a large city on atmospheric moisture.
The most useful data were from an 8-year period (April 1963–March 1970)
when identical instruments were used to measure moisture at both sites.
The dewpoint difference between the two sites was the basic parameter
used in the study. Figure 1.10 shows the contours of the average hourly
dewpoint difference between the two sites for each hour of each month
over the 8-year period.

There are, on the average, positive dewpoint differences (a moisture
access in the city) during the night in all seasons. These nighttime positive
differences change quite rapidly to negative differences (a moisture deficit
in the city) shortly after sunrise, particularly in the spring and early summer. The average difference between sites in the morning throughout most
of the year was negative and was particularly large in the summer. In the
afternoon, the average difference was negative in the early summer and
spring but positive in the winter.

Ackerman suggested that the positive differences found at night were
a result of greater cooling in the rural regions surrounding the city of
Chicago and that consequently there was greater dew deposition. Therefore, higher dewpoints were likely to be found in the city because it is
a greater sink for moisture than are the surrounding rural regions. The
rapid change from positive to negative values shortly after sunrise prob-

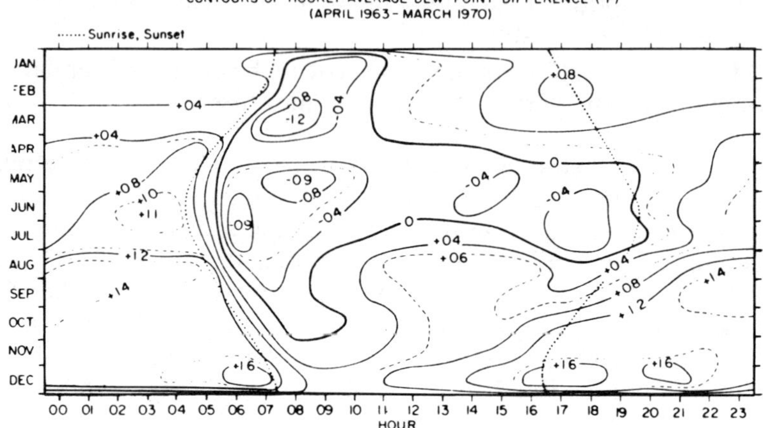

Figure 1.10. Contours of the average hourly Midway Airport, Chicago, versus Argonne National Laboratory dewpoint differences by hour and for each month for the period April 1963–March 1970 (Ackerman[46]).

ably reflected the evaporation of dew and, in summer, the beginning of plant transpiration. These factors increase the moisture concentration in the rural regions relative to the city.

In the spring, the ground moisture content in the rural region is large because of melting snow and high rainfall, whereas in the city, very little moisture is available, owing to rapid runoff. As air temperature increases, evaporation at the surface will be higher in the rural regions. Meanwhile, the contribution to atmospheric moisture in the city from space heating will decrease. In the summer, there is a greater influx of moisture due to plant transpiration because of the greater rural vegetation density as compared to the urban region. The analysis (Figure 1.10) shows a positive difference in the afternoon during the late summer. There are some indications that plant transpiration around Argonne National Laboratories drops dramatically in August.

In the winter, the rural region above 30° latitude is characterized by dormant vegetation and frozen or snow-covered ground, whereas there is an increase in the use of space heating in the city. Thus in the rural region there is very little evaporation at the surface, whereas in the urban region there is an injection of moisture. Generally, the moisture produced by combustion is only a fraction of the amount produced by transpiration. However, in the winter, when transpiration is nonexistent and surface evaporation is negligible, this source of moisture becomes very important and can account for the positive differences.

However, in some cities, observations have been made that are in contrast to those discussed above. Chandler,[10] from a climatological study of urban–rural humidity in and around London, England, in the summertime, found that higher and lower vapor pressures occurred with almost equal frequency in London relative to the rural areas, with only a weak overall tendency toward drier air in the urban area. Such differences must be considered a local urban effect for London because a very dry urban region has been the more general case from observations made in other cities in the summertime.

Jones and Schickendanz[47] showed a time-dependent spatial distribution of dewpoint data gathered in St. Louis, Missouri, as part of the Metromex program (Figure 1.11). Figure 1.11 has been constructed using average midmorning (0700 CDT) data for the month of August 1973.

For partly cloudy and broken conditions, the results are similar. At sunrise, the urban region is characterized by more moisture (higher dewpoint temperatures) relative to the surrounding rural regions. At midafternoon there is a moisture deficit in the urban region relative to the surrounding rural regions. However, under overcast conditions and at sunrise, the urban area has a moisture deficit relative to the immediately surrounding suburban region. However, the moisture content in the urban region is approximately equal to that found in the rural regions to the north, west, and south. In the midafternoon there is a moisture deficit relative to the surrounding suburban and rural regions.

Existing evidence suggests that the urban region, besides being a heat island, is also a dry island. Many of the factors that influence the development of the dry island also influence the development of the heat island; so it is not surprising that Wood[48] found a negative correlation between air temperature and dewpoint temperature in the conditions of heat island. Physically, the correlation indicates that the lower the moisture content, the greater the temperature contrast. This, of course, is in keeping with the results of Nappo[38] (presented earlier), Myrup,[34] and others.

1.4. URBAN CIRCULATION AND WIND FIELD

1.4.1. Introduction

As indicated in Section 1.2, an urban complex acts as a heat reservoir, producing air temperatures greater than the surrounding rural suburban regions. This spatial variation in heating can develop horizontal and vertical pressure variation such that horizontal and vertical acceleration will exist. This should lead to the development of a solenoidal-type circulation

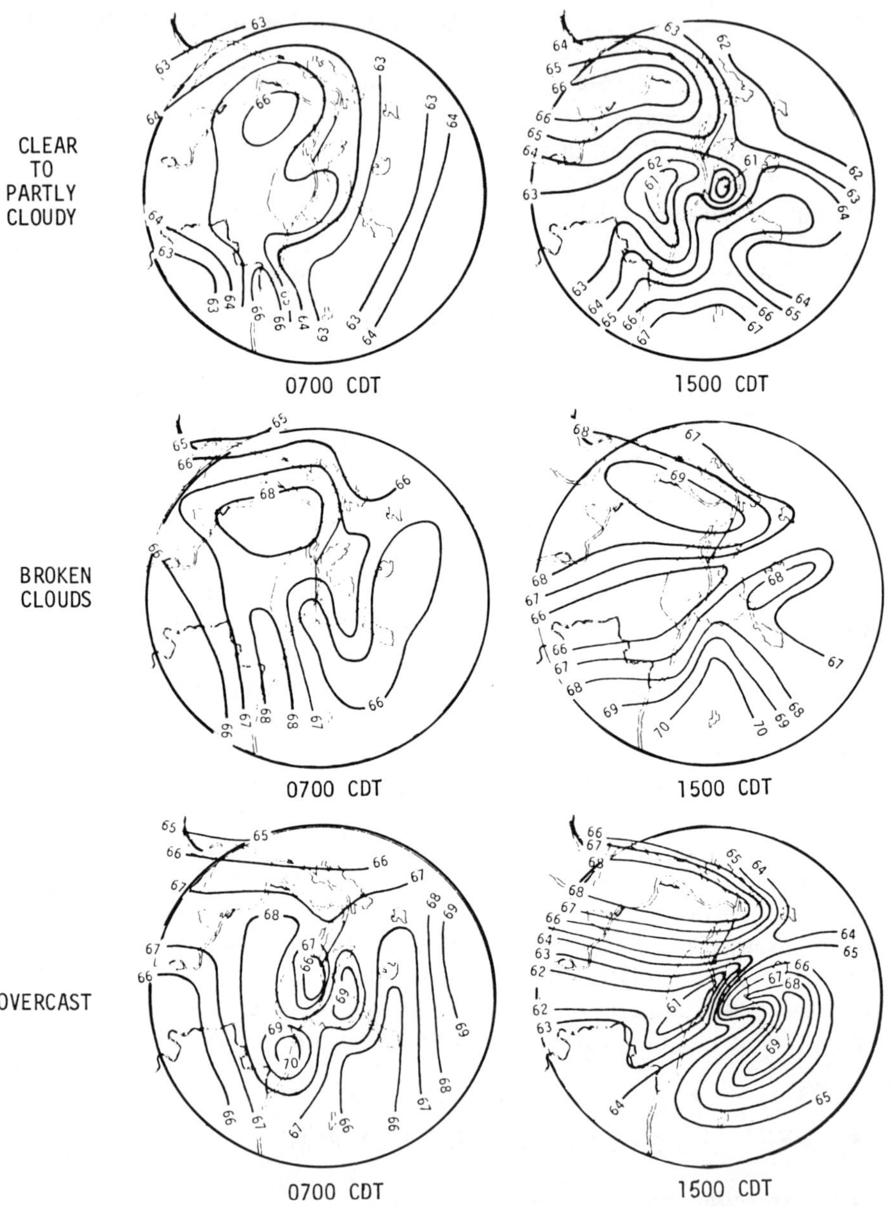

Figure 1.11. Average spatial distribution of dewpoint temperature (°F) at the surface for the month of August 1973 at sunrise (0700 CDT) and midafternoon (1500 CDT) under conditions of partly cloudy skies, broken skies, and overcast conditions (Jones and Schickendanz[47]).

similar to that for the classical heat island (i.e., the circulation associated with a small island in the ocean heated by the daytime insulation). Theoretical studies of the classical heat island circulation indicate that a rather strong wind system should be present (Estoque and Bhumralkar[49]), with an extremely large horizontal temperature gradient existing between the island and the surrounding body of water producing strong winds. In comparison, in the urban heat island, the temperature field varies smoothly, with a minimum temperature in the peripheral regions of the city and the maximum temperature near or at the center. The maximum temperature gradient is generally at the boundary of the city and suburban regions, and along the boundary the temperature gradient is continuous. The pressure perturbations associated with the urban heat island do not produce as intense a circulation system as does the classical heat island. In the following sections, results of theoretical computations and of analyses of observed data on the heat island circulation are presented.

1.4.2. Theoretical Studies

One of the initial theoretical studies on heat island circulation characteristics of an urban region was by Vukovich,[50] who used simple two-dimensional linear models to study the circulation with and without mean wind. The atmosphere was divided vertically into two layers: the upper layer had the standard atmospheric lapse rate at all times while the lower layer was characterized in one case by an isothermal lapse rate and in the other by a near-neutral stratification. The top of the atmosphere was the 10-km level and the top of the lower layer was the 1-km level. The heating rate was such that a 6°C temperature difference existed between the center of the city and a point 20 km away in 3 hr. The heating function varied in space as a cosine function, with the maximum in the center of the town.

A two-cell circulation system characterized both the near-neutral and stable boundary layer cases under the conditions of no mean wind. Low-level horizontal convergence, upward vertical motions, and upper-level divergence characterized the motion field over the city. The surrounding areas were characterized by low-level divergence, downward vertical motions, and upper-level convergence.

The horizontal wind profile at $x = 10$ km, which is the point of maximum temperature gradient and maximum horizontal wind speed, for the stable case (Figure 1.12a) showed greater horizontal speeds at the surface and in the return flow aloft than the near-neutral case (Figure 1.12b). In the

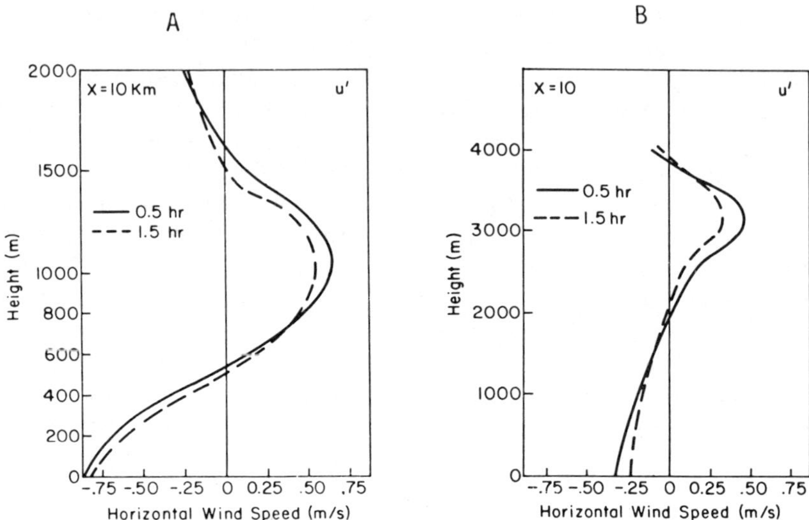

Figure 1.12. Theoretically computed vertical profile of the horizontal wind speed associated with urban heat island circulation for the condition of (*a*) a stable boundary layer and (*b*) a neutral boundary layer taken at the point where the horizontal wind speeds are a maximum ($x = 10$ km) in the model (Vukovich[50]).

stable case the maximum low-level wind was approximately 0.8 m s^{-1}, whereas in the near-neutral case is was 0.32 m s^{-1}. The vertical extent of the circulation is larger for the near-neutral case than for the stable case.

The vertical shrinking produced by the stability in the case of an isothermal lapse rate required larger horizontal velocities to maintain the assumed nondivergent state in the model. Vertical velocities, however, must be larger and exist through a deeper layer in the near-neutral case to satisfy the nondivergent condition (compare Figure 1.13*a* and *b*). In the stable case, the maximum vertical velocity was 5 cm s^{-1}, whereas in the near-neutral case it was 5.7 cm s^{-1}. In the near-neutral case the top of the circulation was at 3.2 km, whereas in the stable case it was approximately 1 km.

The principal effect of adding the mean wind, a constant throughout the extent of the atmosphere, was to displace the circulation downwind of the center of the city ($x = 0$), which was its position when there was no mean wind (see Figure 1.14). Displacements were wholly a function of the mean windspeed. Thus a rather weak circulation exists with a typical urban heat island.

Vukovich[23] studied the effect of heating, stability, and mean wind on

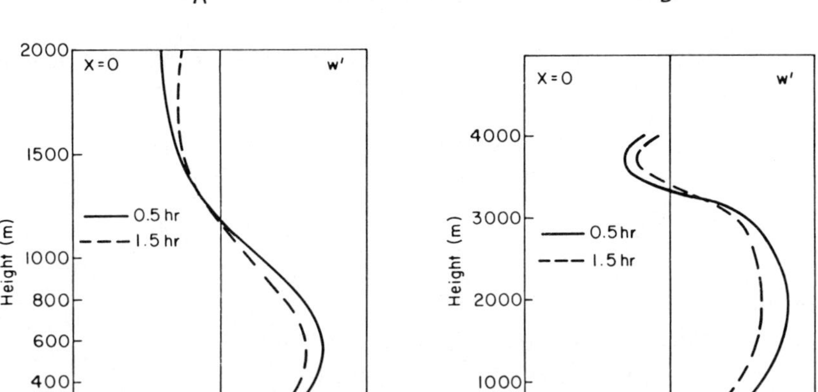

Figure 1.13. Theoretically computed vertical profile of the vertical velocity for heat island circulation under the conditions of (*a*) a stable boundary layer and (*b*) a neutral boundary layer at the point of maximum vertical velocity ($x = 0$) in the model (Vukovich[50]).

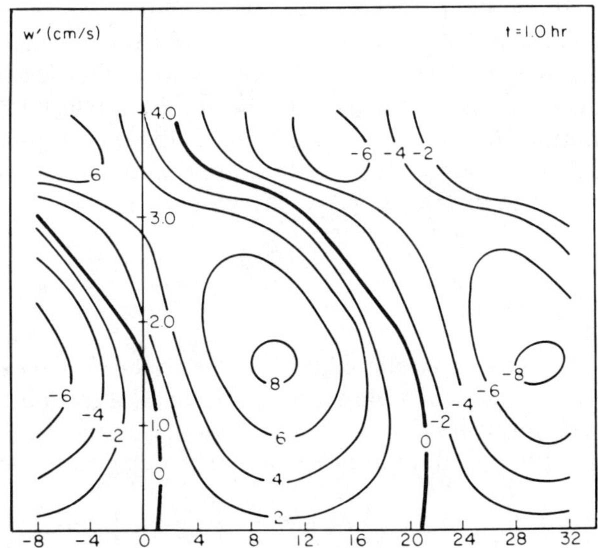

Figure 1.14. Theoretically computed horizontal and vertical distribution of the vertical velocity for heat island circulation under the conditions of a neutral boundary layer and for the mean wind speed through the atmosphere of approximately 10 m s^{-1} (Vukovich[50]).

the intensity of the urban heat island circulation, also using a simple linear model. The equations of motion in two dimensions (x, z) are

$$\frac{\partial u}{\partial t} + Ku = -\frac{1}{\rho}\frac{\partial p}{\partial x} \tag{2}$$

and

$$\frac{\partial w}{\partial t} + Kw = -\frac{1}{\rho}\frac{\partial p}{\partial z} - g \tag{3}$$

where u = horizontal component of the wind velocity
w = vertical component of the wind velocity
ρ = density
p = pressure
g = acceleration due to gravity
K = friction coefficient

Other terms used are: p_o, the surface pressure; p_1, the pressure at the top of the urban boundary layer; p_2, the pressure at the top of the rural boundary layer; $\bar{T}$, the mean temperature of the boundary layer; H, the depth of the boundary layer; U_g, the free stream geostrophic wind; and ω^2, the Brunt-Vaisala factor. The subscript c refers to properties of the urban boundary layer and the subscript E to the rural boundary layer.

For the friction coefficient, it was assumed that the sign and magnitude of the second-order change of velocity is approximately equivalent to the magnitude and sign of the velocity. Therefore,

$$K = \frac{K_m}{H^2} \tag{4}$$

where K_m is the characteristic eddy viscosity. Here the mean wind was assumed to be zero, which eliminated horizontal avection in the linear case, and the Coriolis force has been neglected. Summing eq. 2 and 3 and integrating over a closed interval yields the circulation theorem:

$$\frac{dC}{dt} = \oint\left(\frac{du}{dt}\,dx + \frac{dw}{dt}\,dz\right) = -\oint\frac{dp}{p} - \oint g\,dz - KC \tag{5}$$

where

$$dp = \frac{\partial p}{\partial x}\,dx + \frac{\partial p}{\partial z}\,dz \tag{6}$$

and

$$C = \oint (u\ dx + w\ dz) \tag{7}$$

Since g, the acceleration of gravity, is a single-valued function of z, the closed integral of the function involving the acceleration of gravity has a value of zero. Therefore,

$$\frac{dC}{dt} + KC = - \oint RT \frac{dp}{p} \tag{8}$$

A substitution has been made for the density through the equation of state in the first term on the right-hand side of the equal sign in eq. 5.

Generally, the horizontal pressure difference at the surface between the urban and environmental regions is negligible compared to the pressure difference at the surface and the top of the urban and/or rural boundary layers. If the temperatures at the top of the rural and urban boundary layers are assumed equal, an integration path can be formed similar to that shown in Figure 1.15. Assuming an isothermal rural boundary layer (i.e., T_E is approximately equal to T_c), then when a substitution is made through the hypsometric formulas, integration yields

$$\frac{dC}{dt} + KC = g \frac{\Delta \bar{T}}{\bar{T}c} H_c \tag{9}$$

where

$$\Delta \bar{T} = \bar{T}_c - \bar{T}_E \tag{10}$$

The steady-state solution for eq. 9 is

$$C = g \frac{\Delta \bar{T}}{\bar{T}_c} \frac{H_c}{K} \tag{11}$$

Eddy viscosity in an adiabatic boundary layer should be at least 100 times greater than for the very stable nocturnal environmental boundary layer. However, the depth of the urban boundary layer should be at least 10 times greater than the stable environmental boundary layer at night. According to eq. 4, therefore, the friction coefficient in the urban boundary layer should be approximately equal to the friction coefficient in the rural boundary layer. If the friction coefficient, K, is defined as the average friction coefficient between the rural and urban regions, the equality of the coefficients in the urban and rural regions allows for substitution with the friction coefficient in the urban region.

In an adiabatic boundary layer, the formula for the eddy viscosity is

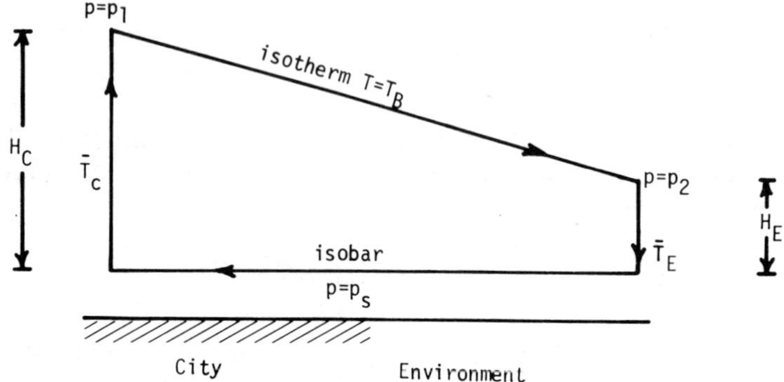

Figure 1.15. Path of integration for heat island circulation (Vukovich[50]).

given by

$$K_m = k_0 z^2 \frac{du}{dz} \tag{12}$$

where k_0 is the von Kármán constant. If H_c, the characteristic height, and U_g/H_c, the characteristic shear, are substituted into eq. 4,

$$K = \frac{k_0^2 U_g}{H_c} \tag{13}$$

Putting this value of K into eq. 11 yields

$$C = k_0^2 g \frac{\Delta \bar{T}}{\bar{T}_c} \frac{H_c^2}{U_g} \tag{14}$$

Equation 14 shows that the intensity of the urban heat island circulation is directly proportional to the temperature difference between the urban and rural region and to the square of the depth of the urban boundary layer, and inversely proportional to the geostrophic wind speed at the top of the urban boundary layer and to the surface temperature in the urban region. That is, the most intense urban heat island circulation will be found when the geostrophic wind speed is small, when there is an intense horizontal temperature gradient between the urban and rural regions, and when the depth of the urban boundary layer is large.

The most intense horizontal contrast between the rural and urban temperatures occurs at night, but the urban boundary layer at night is not generally as deep as that found during the day. Consider the case where the geostrophic wind at the top of the urban boundary layer is the same both during the day and at night, where the temperature contrast between the urban and rural region is approximately 7°C during the night and 1.5°C during the day, and where the depth of the urban boundary layer is 300 m at night and 1000 m during the day. Since temperature is measured in degrees absolute, variations of the surface temperature in the urban region by as much as 5° or 10°C between day and night should not significantly affect the intensity of the urban heat island circulation. However, application of the foregoing parameter values will show that the intensity of the urban heat island circulation is 2.5 times greater during the day than during the night, even though the horizontal contrast at night is about 4.5 times greater than that during the day, the main factor influencing this computation being the depth of the urban heat island.

In a more sophisticated theoretical approach, Delage and Taylor[51] studied the influence of the eddy transfer coefficients, the heat island intensity, the initial temperature stratification, and heat island size on the steady-state solution. A nonlinear two-dimensional model derived from Estoque's (52) sea-breeze model was employed.

In their study of the effect of the eddy transfer coefficients, the transfer coefficients for heat and mass were set equal and constant and were assigned values ranging from 5 to 50 m^2 s^{-1}. The most significant effects of increasing the eddy transfer coefficient are shown in Figures 1.16 and 1.17. In Figure 1.16, the height of the zero wind level and of the maximum winds in the return flow is shown as a function of the magnitude of the eddy transfer coefficient. It is seen that as the coefficient increases, the

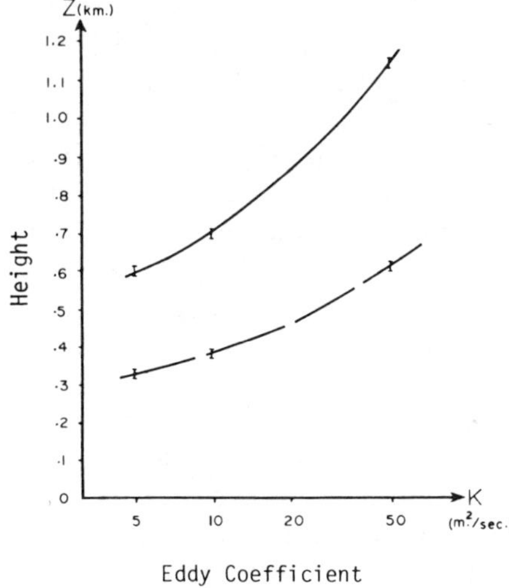

Figure 1.16. Height of the zero wind level (dashed curve) in the horizontal flow and of the maximum winds in the return flow (solid curve) as a function of the eddy transfer coefficient. All other parameters were held constant (Delage and Taylor[51]).

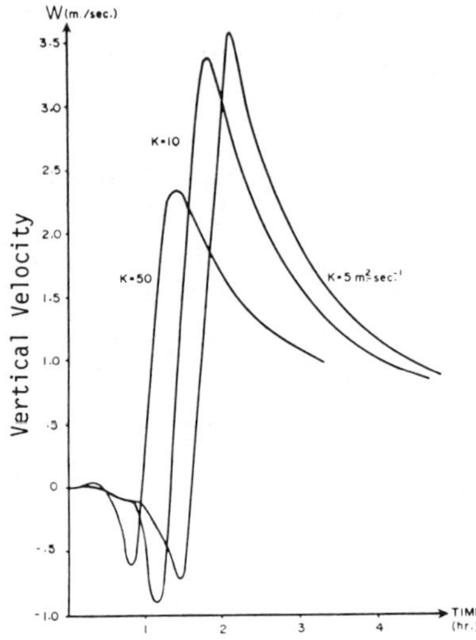

Figure 1.17. Maximum vertical velocity at the center of the heat island ($x_A = 0$) as a function of time and of the eddy transfer coefficients (Delage and Taylor[51]).

depth of the circulation increases both in terms of the height of the zero wind level and of the maximum winds in the return flow. Figure 1.17 shows the maximum vertical velocity at the heat island center as a function of time and of the transfer coefficient (set at 5, 10, and 50 m² s⁻¹). All three curves have the same shape, but the maximum positive vertical velocity is associated with the smallest coefficients (5 m² s⁻¹) used, and the smallest value of the positive vertical velocity is associated with the largest value of the eddy coefficient (50 m² s⁻¹).

The effect of heat island intensity on heat island circulation was studied in two ways. First, the heat island intensity was varied as a function of the temperature difference at the surface between the urban and the rural region (taken as 2.5°C and 5°C), and was the driving force for the circulation, all other parameters being held constant (Figure 1.18). Here, the wind speeds are essentially doubled and the depth to the circulation system increased by about 25% as the heat island intensity increased from 2.5 to 5°C.

Second, they changed the size of the heat island, all other parameters being held constant, including the heat island intensity. The heat island radius was assigned values of 2.75, 4.5, and 8.25 km, and the calculations indicate that the intensity of the circulation near the boundaries of the island was the same in each case and at steady state. The depth of the circulation increased, but only slightly, with an increase in heat island size (Figure 1.19), as did the vertical velocities near the heat island center.

They also showed that as the stability decreased, the depth of the cir-

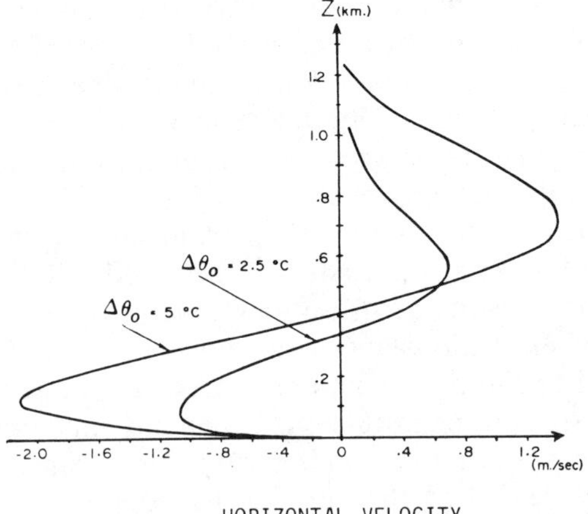

Figure 1.18. Steady-state vertical profiles of the horizontal wind for heat island intensities of 5°C and 2.5°C. All other parameters were held constant (Delage and Taylor[51]).

culation increased, results identical to those of Vukovich,[50] But in contrast to Vukovich, they found that the horizontal velocities increased as the stability decreased.

Vukovich[23] studied the effect of wind shear on the urban heat island circulation, again using a simple linear model where the mean wind speed was zero at the surface and increased linearly with height. The vertical shear of the mean wind ranged from 5 to 20 m s^{-1} km^{-1}, and the temperature lapse rate was held constant and identical to the standard atmospheric lapse rate. The environmental Richardson number therefore became a function of wind shear alone. When holding all other parameters

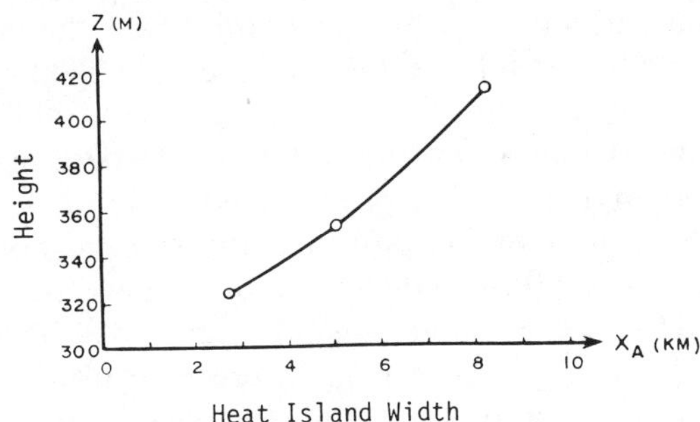

Figure 1.19. Variation of the depth of the heat island circulation versus the size (x_A) of the heat island (Delage and Taylor[51]).

constant, it was shown that a critical environmental Richardson number appeared for which there was a maximum finite circulation, but this was contingent on the boundary conditions used to solve the differential equations. Thus great care should be taken in interpreting any kind of numerical results related to the urban heat island, as they are dependent on the boundary conditions employed.

Other theoretical studies of the urban heat island circulation using two-dimensional models (e.g., Bornstein,[53] Lee and Olfe,[54]) have attained basically the same results as those mentioned above. Vukovich et al.[20] analyzed the urban heat island circulation using a three-dimensional model. They used primitive equations to forecast the wind, temperature, and pressure distribution throughout a volume encompassing an urban region for a given set of initial and boundary conditions. Standard boundary layer theory described the vertical distributions of momentum and heat. The major forcing functions in the model were terrain, surface roughness, and surface heating. Initial conditions were assumed geostrophic and time-independent. The horizontal boundaries, except for the upwind boundary, were open, so that heat, mass, and momentum were allowed to be transported through the boundary. The upstream boundary was closed. The upper boundary was a rigid lid, and the lower boundary conditions were specified as a set of input parameters which included local terrain-height parameters, regional surface roughness parameters, and spatial distribution of ground temperature. The ground temperature distribution was allowed to be time-independent.

The model was applied to St. Louis. The number of integrations and the computer time was reduced by employing a variable grid. This consisted of a 21×21 central grid, grid spacing 1 km, beyond which the grid spacing expanded to 2, 4, 8, and 16 km. At the outer fringes of the grid, three grid points in all directions had 16-km grid spacings. The grid covered an overall area of 144×144 km. There were eight vertical levels in the model: the surface ($z = 0$), the 100-m level, the 300-m level, the 600-m level, the 1000-m level, the 1500-m level, the 2500-m level, and the 4000-m level.

The study showed that when the initial wind speed near the surface was small (less than 1 m s^{-1}) a pronounced zone of convergence was established in the center of the city near the surface possessing strong cyclonic curvature in the flow to the center of convergence (Figure 1.20a). The strongest winds were found upwind relative to the initial wind direction (north) of the city's center and were a result of relatively large pressure gradient accelerations. Counterflow was found southeast of the city's center and was a result of pressure gradient accelerations opposite to the direction of the initial flow. The flow aloft (300-m level) was characterized by divergence (Figure 1-20*b*), and the analysis of the vertical velocities at the 300-m level (Figure 1.20*c*) showed a concentrated region of upward vertical motions over the city's central region, with a maximum upward vertical velocity of approximately 19 cm s^{-1}.

When the initial windspeed near the surface was somewhat larger (1.5 m s^{-1}), flow near the surface was characterized by a zone of convergence and with both cyclonic and anticyclonic curvature (Figure 1.21*a*). The distribution of maximum and minimum wind speed were approximately the same as before. At the 300-m level (Figure 1.21*b*), divergence prevails. The analysis of the vertical velocity distribution at the 300-m level again showed a concentration of upward vertical motions in the central regions of the city, but in this case the maximum upward speed was 16 cm s^{-1} (Figure 1.21*c*).

An interesting result occurred when the initial wind speed near the surface was relatively large (6 m s^{-1}). The horizontal wind speed near the surface indicated a zone of convergence, but the zone was not as pronounced as before (Figure 1.22*a*) and the divergence aloft was not as pronounced as before (Figure 1.22*b*). The interesting result occurred in the vertical velocity distribution. The analysis of the vertical velocities at the 300-m level was characterized by an elongated region of upward vertical motion extending from the center of the city downstream about 20 km and containing three centers of maximum updraft (Figure 1.22*c*). The maximum vertical velocity was found in the center farthest downstream of the city and was 11 cm s^{-1}. The updraft tongue was flanked on either side by centers of downward vertical motion.

The calculations indicated that changes in wind direction did not affect the perturbations produced by the St. Louis heat island until the initial wind speeds became rather large (the order of 6 m s^{-1}). Effects of changes of wind direction were demonstrated using the analysis of the vertical velocities at the 300-m level. They showed that the most intense upward vertical velocities among all the cases studied were found when the initial wind direction was from the northeast (Figure 1.23*a*). The tongue of upward vertical motion extended to a point 30 km downstream of the city's central region. In this case only two updraft centers were found and the maximum vertical velocity occurred in the updraft center farthest from the center of the city (14 cm s^{-1}).

The smallest upward speed was found when the initial wind direction was from the west with the same initial wind speed (Figure 1.23*b*). Again a tongue of updraft velocity extended about 30 km downstream from the city's center, but the maximum vertical speed was no greater than 3 cm s^{-1}. This effect was attributed to the rapid decrease of surface elevation east of St. Louis resulting from the river bottom, which would produce

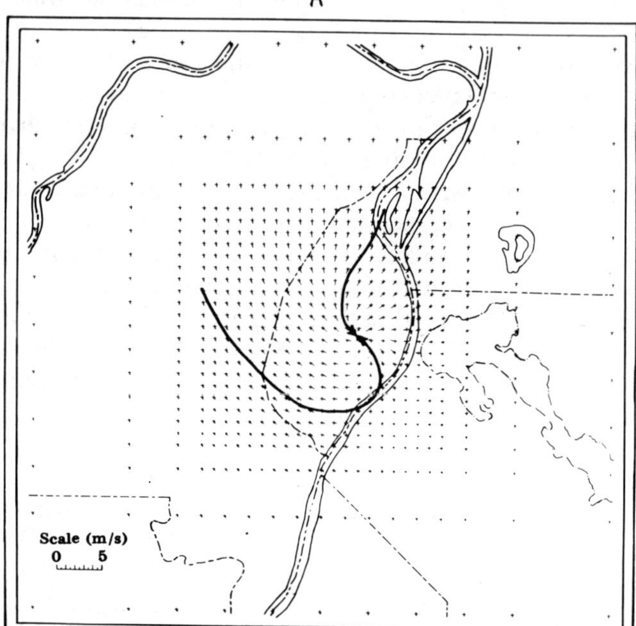

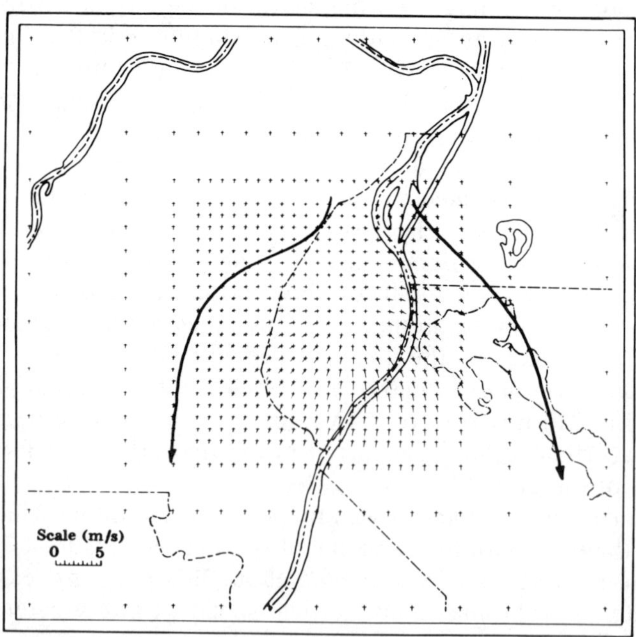

40

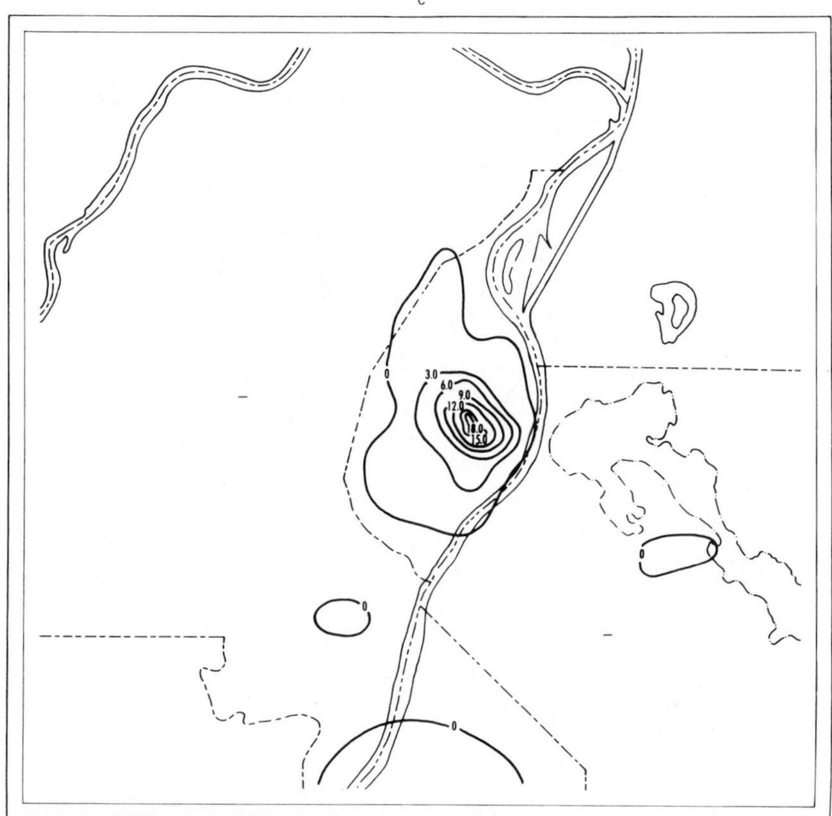

Figure 1.20. Theoretically computed horizontal distribution of the quasi-steady-state horizontal wind vector at the (*a*) 100-m and (*b*) 300-m levels above the river height in St. Louis, Missouri, and (*c*) the horizontal distribution of the vertical velocity (cm s^{-1}) at the 300-m level. The initial wind direction was from the north at a speed of approximately 0.25 m s^{-1} at the 100-m level (Vukovich et al.[20]).

vertical expansion, horizontal divergence, and downward vertical motion in the boundary layer, acting in contrast to the vertical contraction, horizontal convergence, and upward vertical motions produced by the urban heat island.

The most complex vertical velocity pattern was found when the initial wind direction was east and the initial windspeeds again were large (Figure 1.23*c*). The tongue of upward vertical velocities was not well defined in this case. There were two updraft centers, which appeared to be separate entities. The downdraft center was located to the west-northwest of the city over a low topographical region. It suggested that either topograph-

A

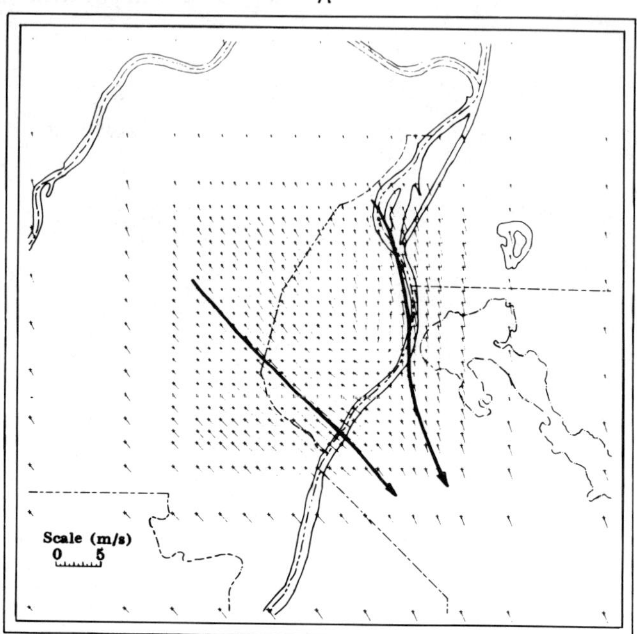

B

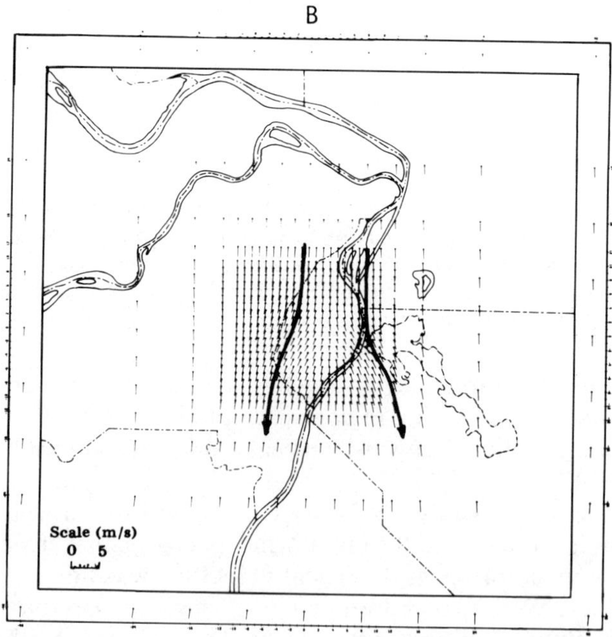

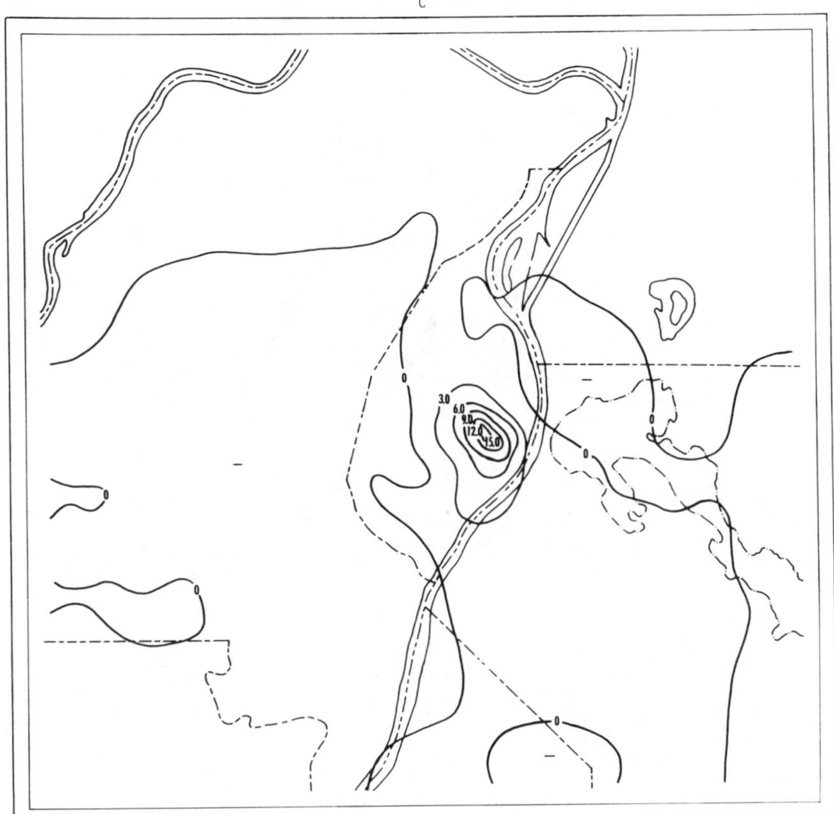

Figure 1.21. Theoretically computed horizontal distribution of the quasi-steady-state horizontal wind vector at the (*a*) 100-m and (*b*) 300-m levels in St. Louis, Missouri, and (*c*) horizontal distribution of the vertical velocity (cm s^{-1}) at the 300-m level. The initial wind direction was from the north at a speed of approximately 1.5 m s^{-1} at the 100-m level (Vukovich et al.[20]).

ically induced downdraft or an intensification of a compensating downdraft associated with the heat island perturbation by topography produced the downdraft center. Thus the downstream updraft center and the zone of upward vertical motion to the west of the city were products of flow over bluffs found in that region.

In the case studies discussed above, the boundary layer was initially stable in both the rural and the urban region. Recently, this model has been used to study the flow in an initially neutral daytime boundary layer in both the rural and urban regions, whose depth is approximately 1 km

A

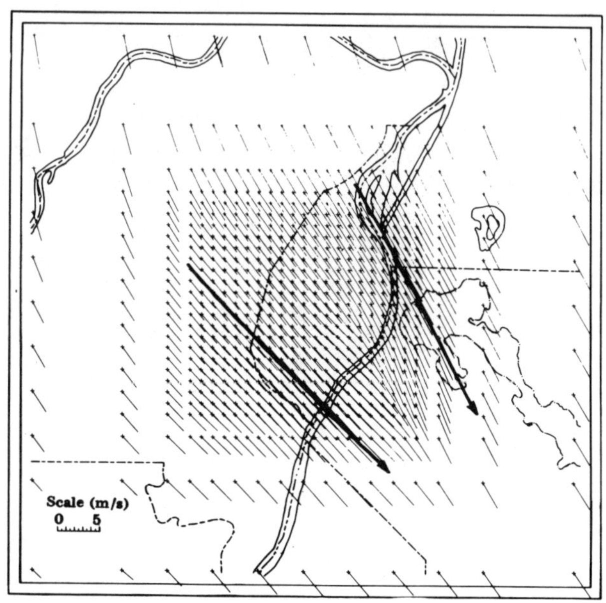

B

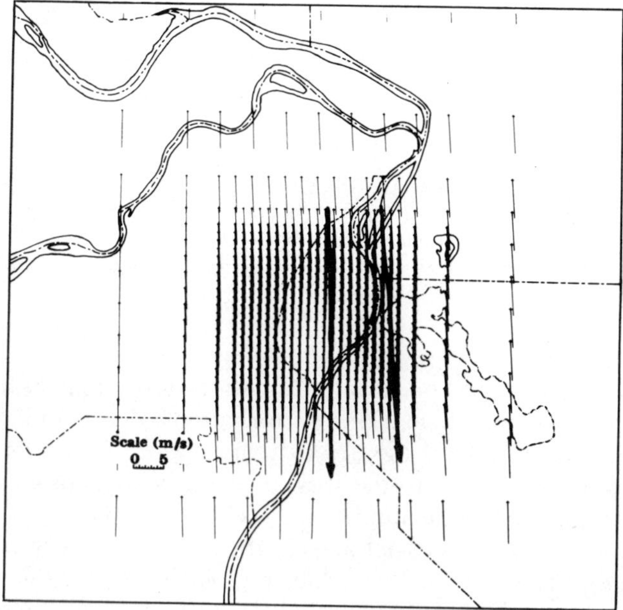

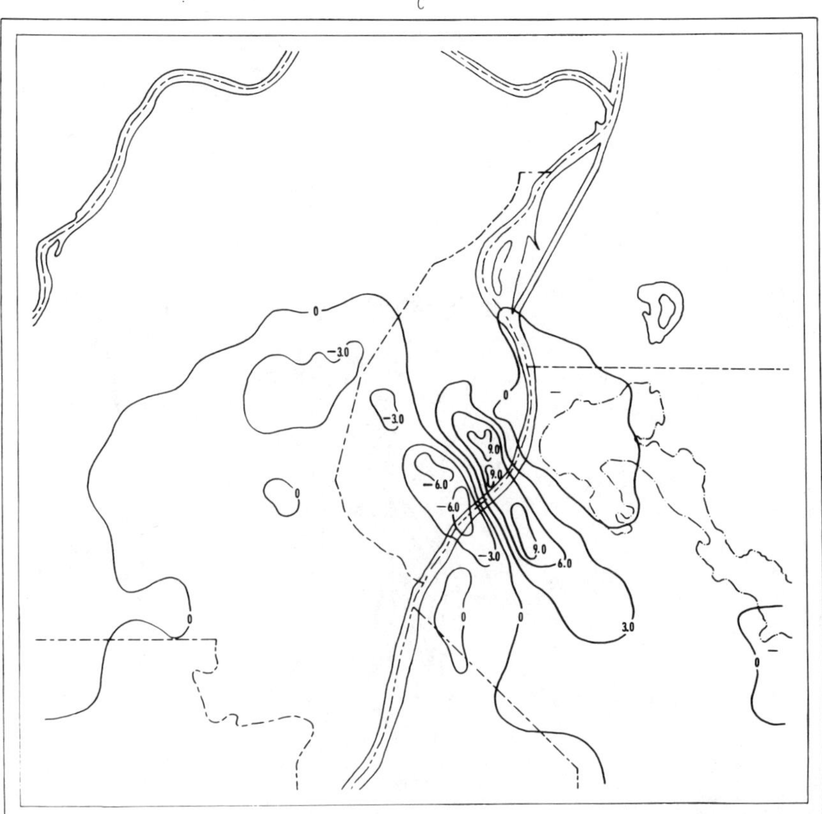

Figure 1.22. Theoretically computed horizontal distribution of the quasi-steady-state horizontal wind vector at the (*a*) 100-m and (*b*) 300-m levels and (*c*) horizontal distribution of the vertical (cm s^{-1}) velocity at the 300-m level in St. Louis, Missouri. The initial wind direction was from the north at 6 m s^{-1} at the 100-m level (Vukovich et al.[20]).

(Vukovich et al.[8]). The initial conditions, except for the boundary layer stability and initial wind direction, were similar to those used in cases discussed previously. The initial wind direction in this case was from the west.

The calculations show that in the case of an initially neutral boundary layer, the heating is distributed to a relatively deep layer (greater than 1 km) and the magnitude of the temperature change through this layer is on the order of 1°C or more, in contract to the case with the same initial conditions except that the boundary layer was initially stable. In that case the heating was distributed through a layer less than 300 m thick and the

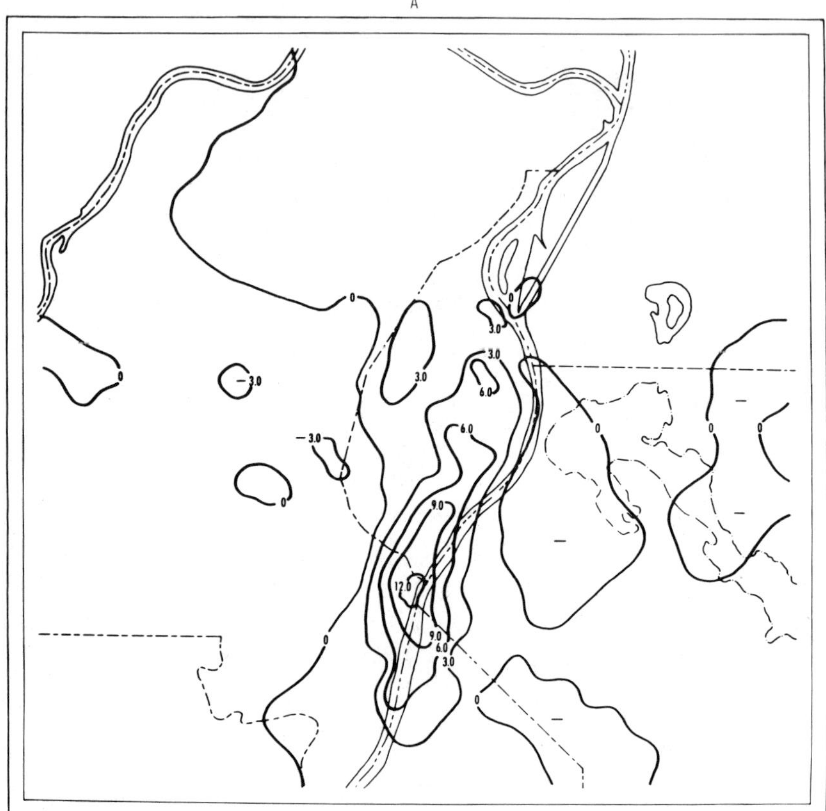

Figure 1.23. Theoretically computed horizontal distribution of the vertical velocity at the 300-m level in St. Louis, Missouri. The initial wind direction was from the northeast (*a*), west (*b*), and east (*c*) at 6 m s^{-1} at the 100-m level (Vukovich et al.[20]).

magnitude of the temperature change was less than 1°C. Dirks'[55] observations indicate that heating through a deep layer can occur in the St. Louis daytime boundary layer.

Analysis of the near-surface wind distribution indicates an intense zone of convergence over and downwind of the city, with a region of inflow centered over the downtown regions of St. Louis extending to the east of the city (Figure 1.24*a*). The figure shows both strong cyclonic and anticyclonic curvature in the flow. The upper-level flow (1-km level) is characterized by an intense zone of divergence over the city (Figure 1.24*b*). Here the top of the urban heat island circulation (essentially where the vertical velocity is zero) is approximately 2000 m, and the maximum

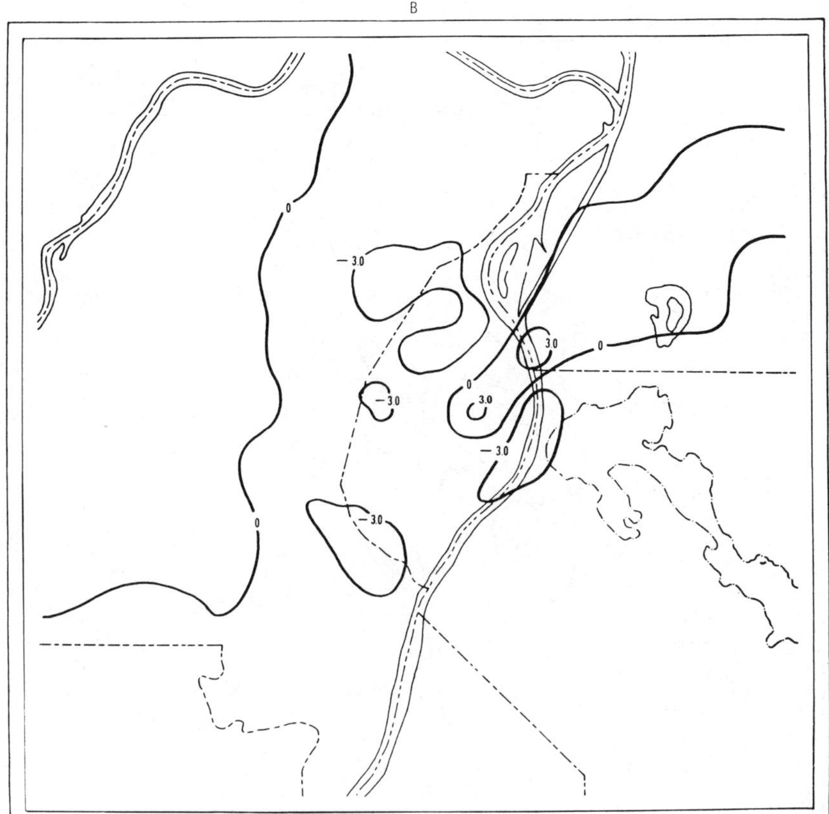

vertical velocities were found at the 600-m level. With an initially stable boundary layer, the top of the urban heat island circulation extended to the 1000-m level, with a maximum vertical velocities found at the 300-m level. At 600 m (Figure 1.24c), the analysis of the vertical velocity shows a rather narrow but intense zone of upward vertical motion centered over the city, with a maximum vertical velocity of 69 cm s^{-1}

The foregoing calculations suggest that if a temperature contrast is maintained between the rural and urban regions during the day, an extremely intense urban heat island circulation will develop. This results from heating being distributed through a deeper layer, which in turn has a profound effect on the pressure distribution. Intense horizontal and vertical accelerations are produced from the resultant pressure perturbation. These results fortify earlier analyses concerning the depth of the

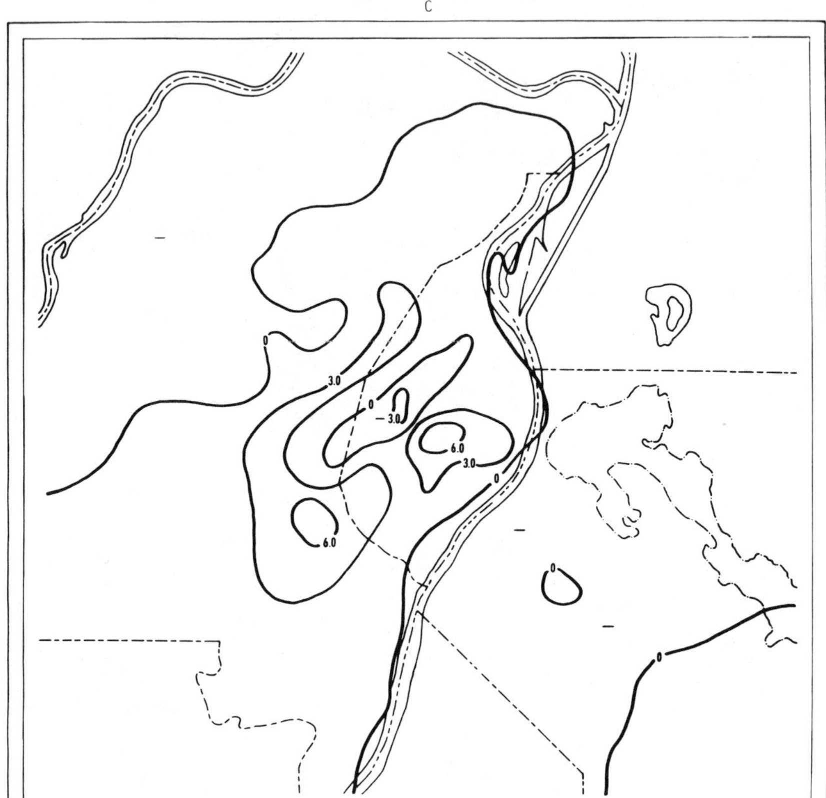

urban boundary layers (essentially the depth through which heat is distributed) and the heat island circulation intensity.

1.4.3.   Observational Data

Numerous observations indicate that an urban-induced circulation exists. The U.S. EPA RAPS program has provided data for the diurnal variation of the wind distribution in a relatively large city in the presence of an urban heat island. A typical case study of the diurnal variation of the wind distribution in St. Louis is presented for June 8, 1976 (Vukovich et al.[8]), over the same time period as the diurnal analysis of the spatial temperature distribution in St. Louis (Figure 1.6).

An interpolation model analyzed the surface wind field in the region. Wind vectors, which are derived for a 4-km grid, were proportional in length to the windspeed and indicate the direction toward which the wind is blowing. The interpolation model employed hourly average wind date from the 25 RAPS stations to produce the analyses. The wind vectors

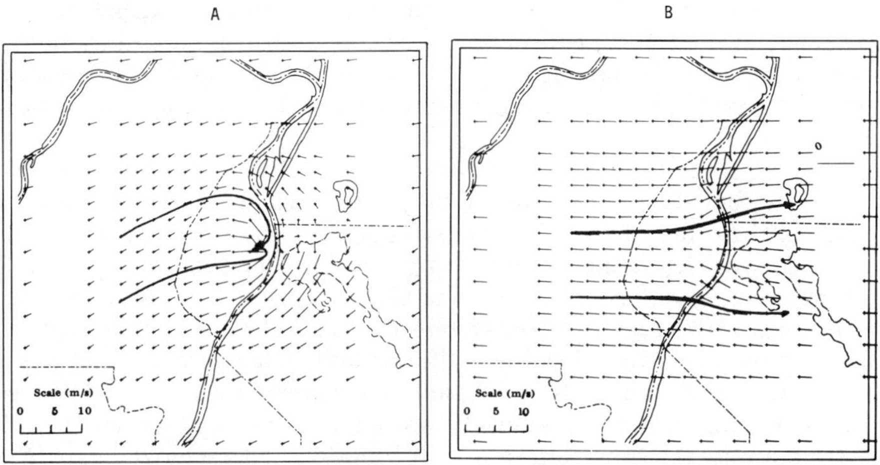

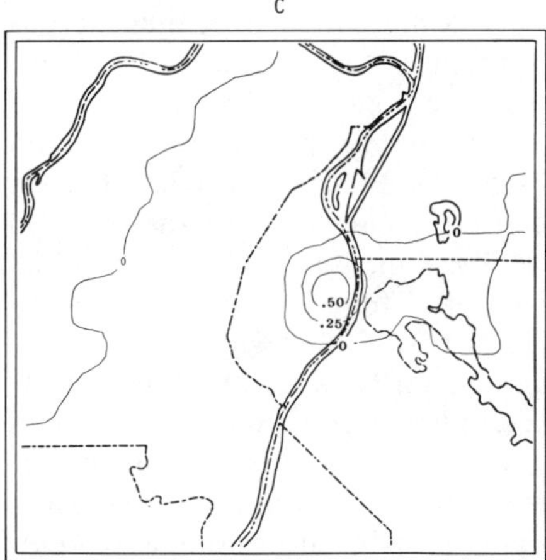

Figure 1.24. Theoretically computed horizontal distribution of the horizontal wind vector at the 100-m (a) and the 2-km (b) level and the horizontal distribution of the vertical velocity (m/s) at the 600-m level (c) in St. Louis, Missouri, after 3 hr of integration. The initial wind direction was from the west at 2 m s^{-1} at the 100-m level and the boundary layer was neutral (Vukovich et al.[8]).

were observed at the 10-m level, and the hourly averages represented sixty 1-min observations from the previous hour.

At 0600 CST on 8 June (about an hour after sunrise) a fairly intense heat island existed in St. Louis (Figure 1.6). The prevailing flow was from the west-northwest and was typically light (of the order of 1 m/s or less) for an early morning period (Figure 1.25a). The strongest winds were west of the city and the weakest winds were in the low area, east of the Mississippi River. This appears to be a center of convergence, suggesting the influence of the urban heat island or possibly of urban-induced frictional effects. This flow pattern agrees with some of the theoretically calculated flow patterns associated with the urban heat island presented in Section 1.4.1.

By 1100 CST, the temperature in the center of the city had increased markedly, but the intensity of the urban heat island had decreased to a value of the order of 1.5°C. Surface windspeeds had generally increased over the area, reflecting the effect of vertical mixing (Figure 1.25b), and strong convergence is noted over the central area of the city, extending to the east with a region of counterflow over and downwind of the central area of the city. By 1500 CST, the maximum temperature had been reached, and subsequently the region was characterized by cooling. At 1600 CST, the heat island intensity was similar to that observed at 1100 CST, but now the flow pattern over the city was characterized by a line of convergence that extended downwind (Figure 1.25c). This suggests, although the evidence is not conclusive, that there is a weakening of the urban heat island circulation associated with the cooling process, which is tending to stabilize the boundary layer.

At 2100 CST, a very intense heat island had developed over the city (Figure 1.6). The windspeeds over the region had diminished to 1 to 2 m s^{-1}, indicating the effect of the stabilization of the boundary layer, which reduces and eventually elminates mixing with the upper parts of the atmosphere (Figure 1.25d). There is only a weak suggestion of confluence over and downwind of the urban region.

There is good agreement between the theoretical calculations and the observed data. They indicate that the most intense heat island circulation will develop when there is deep mixing, which generally occurs during the day, even though the intensity of the heat island may be greater at night when the mixing is reduced or completely eliminated. Evidence in Section 1.2 indicates that the existence of heat islands during the day are infrequent compared to the night. Therefore, the intense circulation suggested by these analyses will be rare.

Evidence of the influence of the urban heat island on the synoptic flow was presented by Findlay and Hirt[56] for Toronto, Canada, which borders

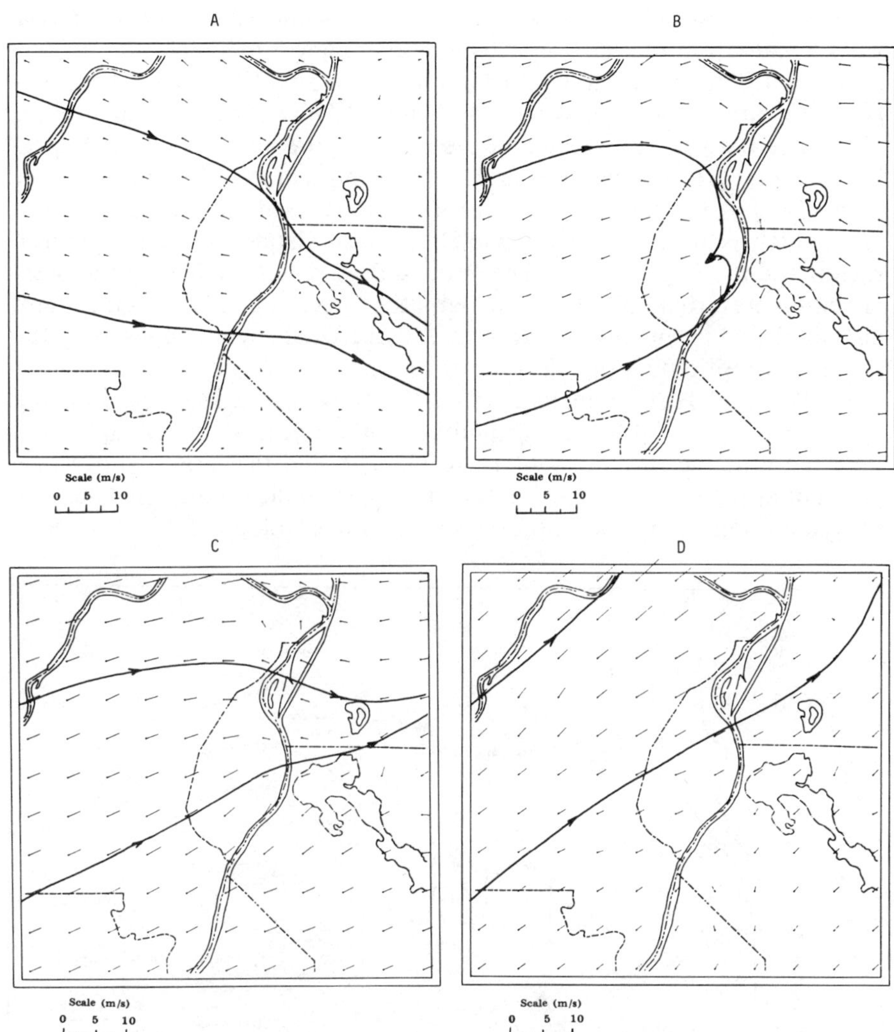

Figure 1.25. Interpolated wind field at the surface based on observations made in St. Louis, Missouri, at 0600 CST (*a*), 1100 CST (*b*), 1600 CST (*c*), and 2100 CST (*d*) on June 8, 1976 (Vukovich et al.[8]).

Lake Ontario. The period was the daylight hours on February 23, 1966, and so essentially represents a winter case study.

On this day, the central portions of the North American continent were characterized by a large-scale high-pressure system centered over the Great Lakes. Toronto, which was dominated by the central regions of the high-pressure system, was characterized by sunny skies and relatively weak winds from the north-northwest, averaging 3 m s^{-1}.

At 1300 EST, an urban heat island having a 2 to 3°C rural–urban temperature difference was observed (Figure 1.26). Here the temperature difference between the center of the city and the air over Lake Ontario is somewhat greater than the rural–urban temperature difference (greater than 3.5°C). An intense temperature gradient exists at the lake/land boundary, which is much more intense than that existing between the rural region and the city's center. The surface flow pattern indicated the presence of a perturbation in the flow, with the onset of winds from the south and southeast in the city (Figure 1.27), these winds being opposed by the general synoptic flow (north to northwest).

Findlay and Hirt[56] concluded that the flow was associated with the urban heat island. Although certainly the urban heat island will affect the flow pattern, the proximity of Lake Ontario and the large temperature gradient at the lake boundary in the city would indicate that the lake also played an important role in the mesoscale meteorology of Toronto. Sep-

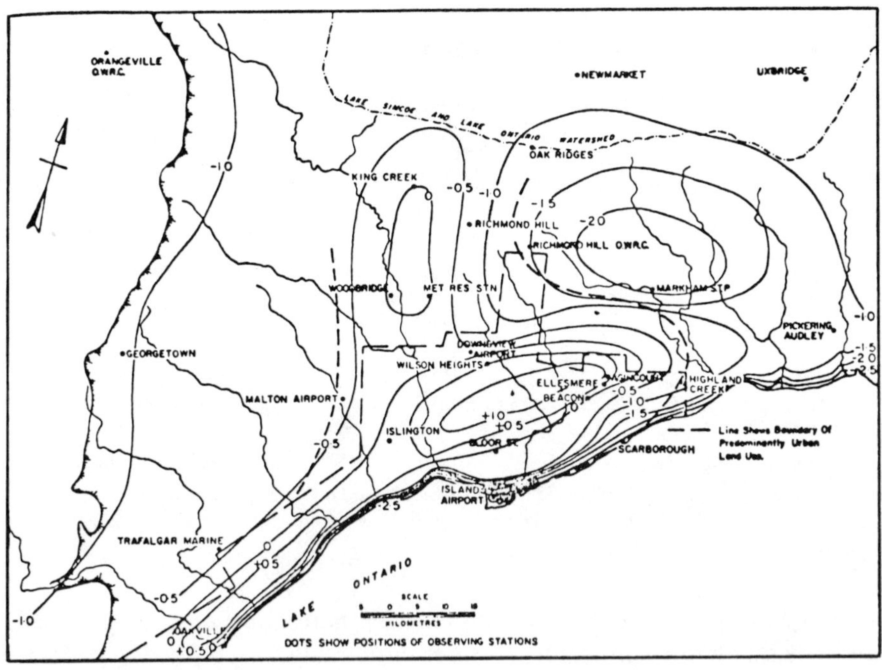

Figure 1.26. Departures of maximum temperature (°C) from a temperature in downtown Toronto, Canada, on February 23, 1966. The heavy dashed line approximates the boundary of the heat island (Findlay and Hirt[56]).

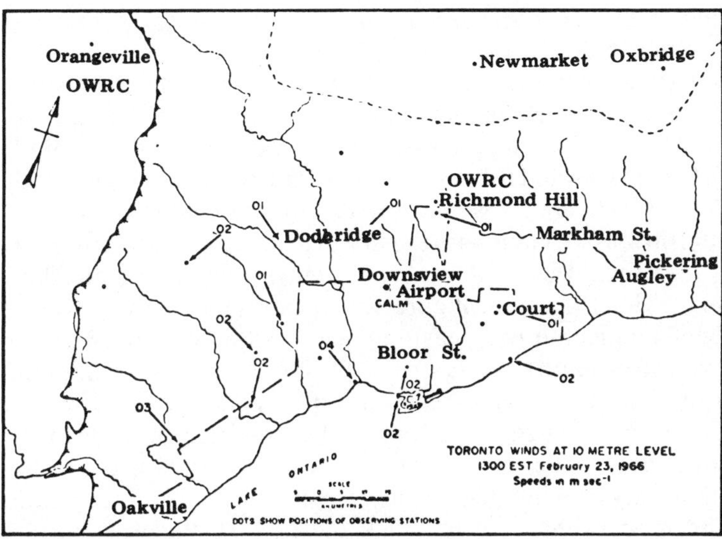

Figure 1.27. Horizontal distribution of winds (m s^{-1}) in Toronto, Canada, at the 10-m level for February 23, 1966, at 1300 EST (Findlay and Hirt[56]).

aration of the components of the circulation due to the lake breeze and urban heat island was not possible, although the net effect was to produce an inflow into the city with speeds as high as 1.7 m s^{-1}.

Pooler[57] studied the surface flow conditions using meteorological data obtained from a network established to study local air pollution in Louisville, Kentucky, an urban area within a region of relatively minor topographic features. It is in an irregular valley, with the Ohio River flowing through it in a northeast-southwest direction. The land rises abruptly to a maximum elevation of approximately 130 m above the valley floor to the west. There is a more gradual increase in surface elevation to the east, reaching values of 75 to 100 m above the valley floor in about 16 km. To the north, the land slopes are insignificant. About 40 km to the south there is again an abrupt elevation change of approximately 100 to 130 m above the valley floor. Also to the south there is a mass of high ground extending northward toward Louisville, which splits the end of the valley into two parts: a narrow channel to the west and an irregular elongated bowl to the east.

Pooler separated his wind data into two categories based on stability in the boundary layer, which was determined by temperature differences from hydrothermographs installed at different ground elevations. For each stability group, hourly wind data from five surface stations were classified according to the wind speed and direction and observed at a 170-m television tower in downtown Louisville.

Flow with an unstable boundary layer showed no variations that might be attributed to the urban heat island. In the stable boundary layer and with the TV tower's winds classified as calm, the data indicated effects of the urban heat island as well as downslope drainage (Figure 1.28). Data from all sites indicated a weak flow toward the center of the city.

Counterflow associated with the urban heat island circulation was found when a mean wind existed, and the intensity of the counterflow was a function of wind direction. Figure 1.29a shows the analysis when the TV tower winds indicated northeast and east-northeast flow. Weak counterflows (from the south) associated with the urban heat island circulation were infrequent. The counterflow was observed at sites downwind (south) of the urban area, suggesting that in the presence of a mean flow, the center of convergence was displaced downwind of the city. Figure 1.29b shows the analysis when the TV tower winds were from the south-southeast. The topography acts to block the flow to the west, producing a counterflow such that along the west edge of the valley, the counterflow seems to be the rule.

Pooler's analysis indicated that the heat island in Louisville produced an inflow of air to the center of the city near the surface. The convergence in the urban area was the dominant feature of the surface flow pattern when the mean wind was weak, or when the topography blocked the motion induced by the synoptic-scale pressure gradient.

The flow field over a metropolitan area under the condition of strong localized temperature gradients has been studied in St. Louis by Ackerman[58] as a part of the METROMEX program. The data were obtained from a series of nighttime observations of pilot balloons tracked by double

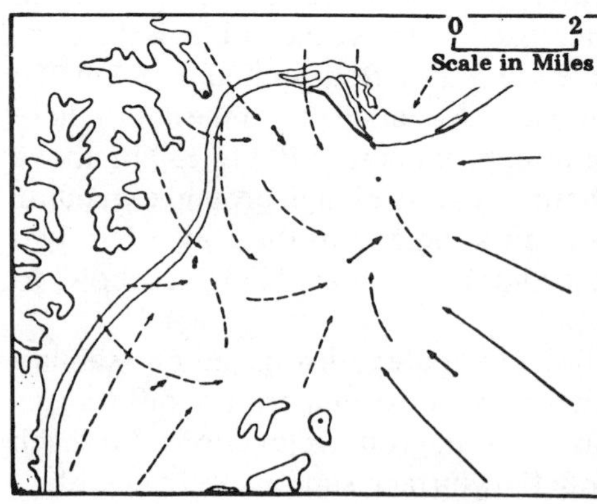

Figure 1.28. Vectors indicating average flow under the conditions of calm winds and a stable boundary layer (Pooler[57]).

A

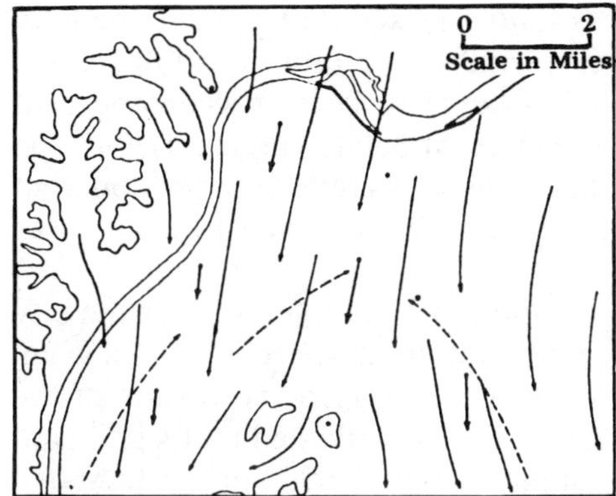

B

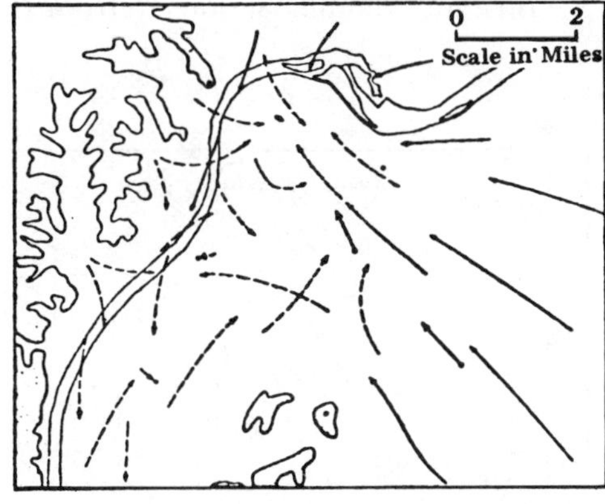

Figure 1.29. Vectors indicating average flow under the conditions that the initial wind was from the northeast and east-northeast (*a*) and south-southeast (*b*) and the boundary layer was stable (Pooler[57]).

theodolite. Data were obtained between 2100 hours and 2400 hours on August 17, 1971.

High pressure dominated the midwestern part of the United States, and St. Louis was in the southwestern quadrant of that system. Geostrophic winds were of the order of 2.0 m s^{-1} and the actual surface wind was less than 1.5 m s^{-1}. Under clear skies, the surface temperature in the downtown regions of the city was about 4.5°C higher than in the surrounding rural regions.

The upper wind data indicated that the city affected the wind field to considerable heights but that the deformation was most pronounced below

300 m. At 100 m (Figure 1.30*a*), the prevailing wind direction was from the northeast. The wind speeds on the upwind side of the city were about 3 m s^{-1} and decreased to values less than 1 m s^{-1} downwind. The flow pattern indicates cyclonic curvature upwind of the city and anticyclonic curvature over and slightly downwind of the city. There was an apparent zone of convergence downwind of the city. At 350 m (Figure 1.30*b*), the cyclonic and anticyclonic curvature in the flow was reduced. The apparent zone of convergence found at the 100-m level downwind of the city was replaced by a zone of divergence.

Ackerman used the wind data to compute net divergence over the city and rural regions. Because the values of net divergence represented small differences between two large terms, Ackerman used the results of the computations only as a guide (Table 1.5). These indicated that significant convergence occurred over the city at and below the 350-m level. Above 350-m, the city was characterized by divergence. Divergence generally characterized the rural region.

The air flow at night (March 1969) over Columbus, Ohio, was studied by Angell et al,[16] using radar to track constant-volume balloons (tetroons)

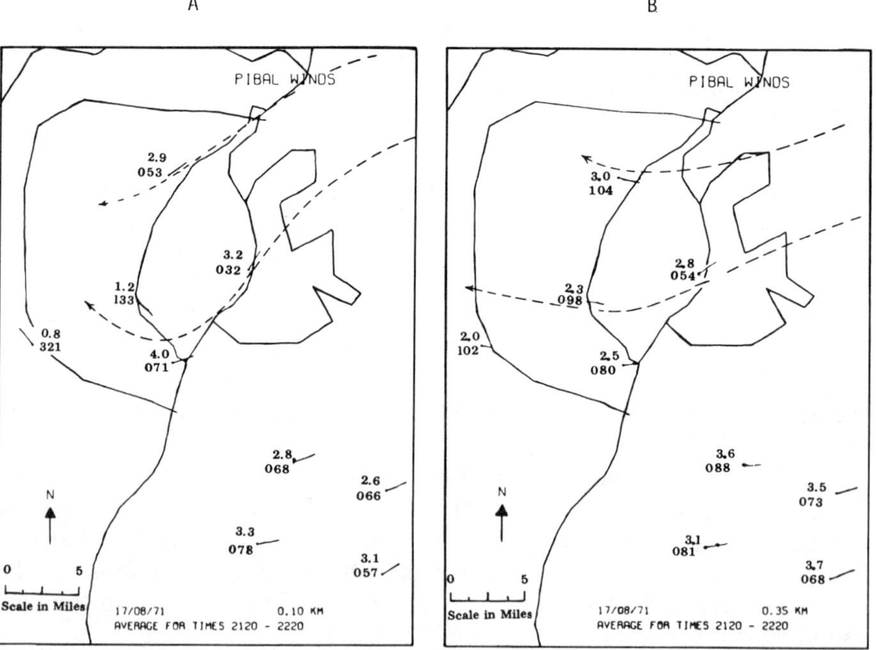

Figure 1.30. Horizontal distribution of the horizontal winds at the 100-m (*a*) and the 350-m (*b*) levels on August 17, 1971 at approximately 2200 CST in St. Louis, Missouri (Ackerman[58]).

Table 1.5.   Calculated Divergence 10^{-5} s^{-1} over St. Louis for August 17, 1971 (2120 to 2220 CDT) (Ackerman[58])

Height (m)	City	Rural
100	-12.7	7.1
200	-5.7	10.4
350	-5.8	4.4
500	4.2	0.9
1000	3.5	-3.1
1500	12.8	7.1

at heights of 200 m or less. They found anticyclonic turning of the wind at 200 m, which was greatest on nights when the urban heat island was strongest. It was also found that the anticyclonic turning was greater under inversion conditions than under lapse conditions.

Three hypotheses were examined to find potential causes for the anticyclonic turning: (1) ascending motions in the wind field varying with height, (2) a change in fictional drag due to enhanced vertical mixing over the city, or (3) formation of a mesoscale high due to the urban heat island effect.

They found that the correlations between tetroon-derived vertical velocity and the anticyclonic turning were not significant and concluded that the upward motions in the wind field varying with height were not likely to be the primary cause for the anticyclonic turning. To examine the effect of frictional drag, they estimated the frictional force by computing the residual of individual accelerations, specifying the pressure gradient and the Coriolis force, and found that the frictional force was of the same magnitude as the other forces. They concluded that some of the anticyclonic turning observed over the city, at least under lapse condition, was due to an increase in the frictional force brought about by increased vertical mixing over the city. This effect may also explain most of the anticyclonic turning observed under inversion conditions.

They noted that when the urban heat island is present, the high temperatures over the city compared to the surrounding rural region generally produce a mesoscale low-pressure system at the surface and a mesoscale high-pressure system at greater heights, with ascending motion over the city providing the connecting link between the inflow at the surface and the outflow aloft. The pressure at the 200-m level, which was assumed to be in the upper high-pressure region was calculated, and it was assumed that the pressure access declined to zero 10 km from the downtown region.

They used these data to examine the turning of trajectories resulting from two extreme flows: geostrophic flow and Eulerian flow (i.e., pressure gradient force balanced by the acceleration). Assuming an ambient geostrophic wind of 10 m s^{-1}, the assumption of geostrophic flow yielded an anticyclonic turning of approximately 15°, whereas the assumption of Eulerian flow yielded a turning of only 2°. So the actual anticyclonic turning due to the heat island effect lies somewhere between these two extremes. The heat island appears to play a fairly important role in causing the observed anticyclonic turning over the city.

Angell et al.[16] also found that the windspeeds decreased across the city relative to the upwind wind speed. A region of maximum deceleration, which was more pronounced than under inversion condtions, was found upwind of the downtown region, but there was a downward tilt of this region under a lapse condition. Decreases in wind speed amounted to as much as 20% of the upwind wind speed in some cases and were considered to be due to increased vertical mixing, which caused the transport of momentum downward.

Their analysis of the average tetroon-derived vertical velocity across the city under lapse and inversion conditions (Figure 1.31) indicated that upward vertical motion occurs above the city with maximum vertical speeds ranging from 4 to 8 cm s^{-1}. Theoretical studies have indicated that with small vertical velocities, the actual vertical velocity is twice the tetroon vertical velocity, and the region of maximum upward vertical motion is displaced downwind of the region of maximum upward tetroon motion by 2 to 4 km.

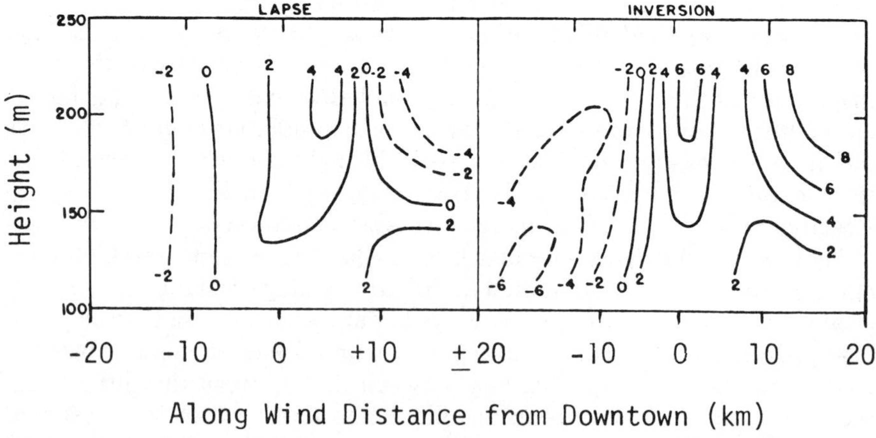

Figure 1.31. Vertical velocity (cm s^{-1}) distribution determined from flights passing across Columbus, Ohio, under lapse and inversion conditions (Angell et al.[16]).

Angell et al,[59] also studied the flow over Oklahoma City. Tetroon flights across the city determined the influence of an isolated urban area on both the horizontal and vertical mesoscale air flow at 400 m under the conditions of strong daytime flow (of the order of 13 m s^{-1}). They analyzed the Lagrangian motion obtained from their 56 tetroon flights in tandem with fixed-point measurements of horizontal and vertical velocity made at a 460-m instrumented television tower located in the northern part of the city in the predominant southerly flow between September 25 and October 12, 1971.

In the morning the downtown city region induces a trajectory turning of approximately 10 ° toward lower pressure. In the afternoon, the mean trajectory across the downtown region turned only about 4° toward lower pressure. This was attributed to the angle of indraft being two to three times greater in the morning, while in the afternoon the mean height of the tetroon above the city had increased. Also, in the morning, the depth of the mixed layer was only a few hundred meters and there was an intense effect of frictional drag, caused by the city and confined to a shallow layer. However, in the afternoon, the mixing depth often exceeded 1000 m with the increased frictional drag effect of the city spread through a much deeper layer.

The tendency for air to flow around the city was more prevalent in the afternoon than in the morning, and it was suggested that this is due to a barrier effect, uncompensated by vertical motions. In terms of the vertical velocity over the city, an average plume of ascending motion extended at least 30 km downwind of the city's center (Figure 1.32). This plume had two centers of upward motion; one slightly upwind of the city and the other far downwind. The maximum vertical velocity in these centers was about equal (0.4 m s^{-1}, with compensating downward vertical motion on either side. The most intense downward motion was to the left, downstream from the city.

Data were grouped into morning, afternoon, and evening periods (flights) to determine the diurnal distribution of the vertical velocities. In the morning (Figure 1.33) a plume of ascending motion (maximum updraft 0.5 m s^{-1}) was located upwind of the city's center. Downwind, the maximum updraft velocity was approximately 0.3 m s^{-1}, with compensating downward vertical motions to the right and left. In the afternoon, the upward vertical motion plume covered a broader area and the vertical velocities were generally larger. Again, the maximum updraft velocity (0.7 m s^{-1}) was upwind of the city's center, but a secondary maximum (0.5 m s^{-1}) was downwind of the city with compensating downward vertical motions to the left and right of the plume, with that on the left (in the downwind direction) being most intense. In the evening, the vertical

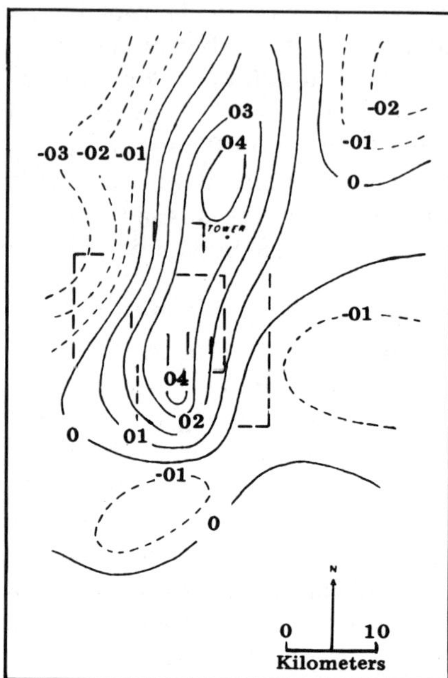

Figure 1.32. Smooth daytime distribution of air parcel vertical velocity (m s^{-1}) at heights above 400 m above Oklahoma City, Oklahoma, based on 32 twin flights on days with strong southerly flow with speeds of the order of 13 m/s (Angell et al.[59]).

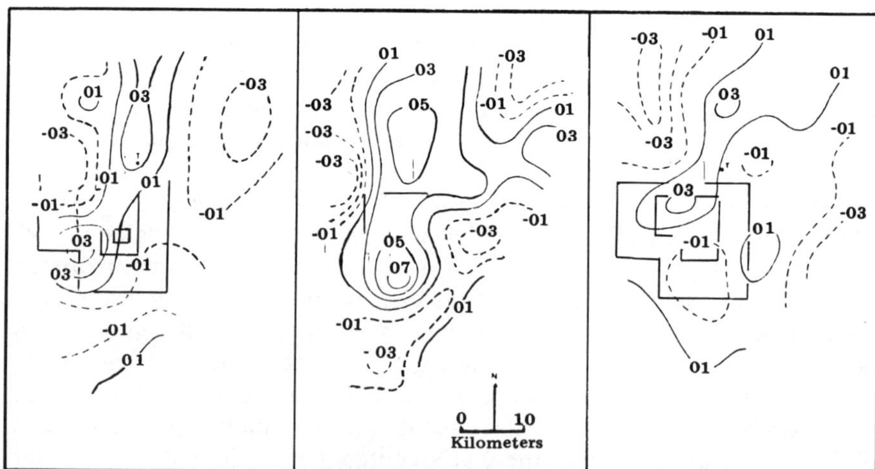

Figure 1.33. Smooth distribution of derived air parcel vertical velocity (m s^{-1}) at about the 400-m level above Oklahoma City, Oklahoma, during the morning (left), afternoon (center), and evening (right) of days with strong southerly flow. Positive values and solid lines indicate upward vertical motion.

60

motion pattern had changed from the morning and afternoon. The plume was oriented northeast-southwest, in contrast to north-south in the morning and afternoon. Two centers of upward vertical motion (0.3 m s^{-1}) were found in this plume, both being located upwind. The authors suggested that the out-of-phase relationship between the evening vertical motions may, in part, be due to the dip of the isopicnic (constant density) surfaces associated with the urban heat island affect.

The observed data agree qualitatively with the earlier theoretical results, showing that the urban region affects the flow in the mesoscale either by increasing the frictional drag or by induced pressure gradient accelerations. Under certain circumstances, the perturbations produced can be quite intense.

1.5.   URBAN CLIMATE AND WEATHER MODIFICATION

1.5.1.   Introduction

Initial evidence that the anthropogenically produced urban climate may, in turn, modify mesoscale weather phenomena in and around the given urban region was first presented in the 1960s, when it was suggested that urbanization and industrialization may influence local precipitation patterns. The most important investigation was the "La Porte Weather Anomoly" (Changnon[60]), which indicated the effects of urbanization on downwind weather phenomena. A local precipitation maxima relative to the surrounding region was found in and around LaPorte, Indiana, immediately downwind of Chicago. This study was a precursor to the METROMEX study described in Section 1.1.

1.5.2.   The Urban Weather Anomaly

Figure 1.34 shows the total rainfall for the St. Louis region obtained from data collected through the METROMEX rain gauge network for the summers of 1971 and 1972 (Huff and Schlessman[61]). This is far more detailed than usual, but the general features are similar to those obtained in other large cities, such as Chicago, Washington, D.C., Houston, New Orleans, and Cleveland (Huff and Changnon[62]).

The June–August network average rainfall is about 66% of the long-term normal for the summer. The analysis shows a major high east of Edwardsville,* Illinois, which is downwind of St. Louis, and the

---

* Edwardsville is a small town in Illinois with a population of about 11,000. The town is about 30 km northeast of St. Louis and about 15 km east-southwest of the Alton–Wood River area.

Figure 1.34. Total rainfall over St. Louis, Missouri, and surrounding region from data obtained during the summers of 1971 and 1972 (Huff and Schlessman[61]).

Alton–Wood River area, which is heavily industrialized. The Alton region is also downwind of St. Louis when storms move from the southwest to the northeast. The prevailing wind direction in this region in the summertime is from the south to southwest; however, southeasterly winds also occur frequently. The region is, however, downwind of the Alton–Wood River area for storms moving from the west-northwest, suggesting that this industrialized region may also affect storms. Precipitation maximum at Edwardsville exceeded the network average by 60% during the two summer periods and was the only long-term station exceeding its normal rainfall.

A secondary high rainfall was found in the region where the Missouri River bends northeast of St. Louis, and this region is a potential breeding location of convective storms.[61] These are the hot, moist bottomlands of the Missouri River and are close to the source of the Mississippi lowlands, a region with rainfall in excess of 30% of the network mean over the 2-year period.

Another region of moderately high rainfall is located about 20 km southeast and is downwind of St. Louis for storms that move from the northwest. It received over 20% more rainfall than the network average for the 2 years.

A major low center, west of the city, had rainfall 70% less than the network average. It is located in the lee of the Ozark Hills, to the southwest. The low may be produced by the warming and drying of the low-level air flow across these hills. The METROMEX temperature and humidity data have indicated that warmer and dryer air during the midafternoon is common when convection is usually the greatest (Jones and Schickendanz[47]).

Huff and Schlessman[61] divided the METROMEX rainfall data for the period 1971–1972 into rain days with *potential effect* and *no effect* on the basis of low-level wind analysis and the location of maximum storm rainfall. They found that a much higher percentage of the total rainfall occurred in the major high centers found east of the Mississippi on days when this region was in the "potential effect" classification. In the Edwardsville–Collinsville region (downwind of St. Louis), about 65 to 75% of the total rainfall for two seasons occurred in the area designated "potential effect storms." The "no effect" pattern over the period was similar to the dry season pattern in its distributional characteristics. This suggested that in the absence of urban effects or conditions of urban suppression of rainfall during dry summers, climatic patterns developed similar to dry seasons.

The relationship between prevailing surface winds prior to the onset of rain and the distributional characteristics of resulting rainfall showed that the rainfall was greatest with winds from the northwesterly quadrant in the summer of 1971. This region of maximum rainfall was near Edwardsville, downwind of the Alton–Wood River industrial complex. Over 50% of the rainfall near Edwardsville was associated with winds from the west-northwest, west, and northwest directions. In 1972, most of the rainfall was associated with winds out of the southwesterly quadrant, and over 70% of the summer total rainfall in the major maximums occurred when winds were out of that quadrant. This frequently places the maximum rainfall downwind of St. Louis in that year.

Schickendanz[63] delineated surface rain cells over the same two summers (1971–1972) using the METROMEX rain gauge network data and determined the character of any urban-induced changes in precipitation. Comparison of 605 potential effects cells and 870 noneffect cells indicated that the average rainfall volume was 167% greater for cells that occur in the urban industrialized region of St. Louis than for cells in the control sample. The increase in volume was present in both moving and stationary cells. Even more dramatically, the average rainfall volume was 262%

greater for cells in the industrialized Alton–Wood River area than for the control sample. Again the increase in volume was found in both moving and stationary cells.

By comparison, the rainfall volume was 61% greater for cells in the hill region southwest of St. Louis than for the control sample, the increase in volume in this region being approximately one-third of the increase in the urban and industrial cells. The increase in volume was very small for cells occurring in the bottomland regions to the north, averaging only 15%.

Huff and Schlessman[61] showed a close correspondence between the total rainfall over the two-summer period and squall line and cold frontal rain patterns. Squall lines and cold frontal storms accounted for over 80% of the total network rainfall over the period. Schickendanz[63] showed that the largest percentage increase occurred with squal line cells over St. Louis, which were also the heaviest rain producers. In the case of the Alton–Wood River region, the largest percentage increase occurred with cold frontal cells, which were also the heavist rain producers in this region.

Schickendanz found that the greatest cell rainfall occurred during the maximum heating periods of 1200 to 1800 CDT, and the lightest rainfall, during the cooling period (0000 to 0600). In the Alton–Wood River region, the largest percentage increase in cell rainfall was found to occur in the two periods 0600 to 1200 and 1200 to 1800 CDT. As in the case of St. Louis, the least cell rainfall was found during the period 0000 to 0600 CDT.

The METROMEX data indicated both a major increase in precipitation in and east of an urban–industrial area associated with squall lines, squall zone activity, and an increase (25%) in thunderstorm activity and in hailstorn and hail intensity (about 80%). Table 1.6 gives statistics on hailstreaks and hailfalls in the METROMEX study area during the summers of 1971 and 1972 for cases when the urban effect was noted and when no urban effect was indicated. To be classified as a hailstorm affected by the urban area, the hail had to be produced from a rain cell that developed or passed over St. Louis or the Alton–Wood River industrialized area. All of the urban effect hailstreaks occurred downwind (northeast, east, or southeast) of these two areas. The statistics show that the characteristics of hailfalls and hailstreaks over the period increased in affected storms.

Analysis of data from an RHI radar also provided information about the urban weather anomaly, specifically the location of and character of first echoes, which identifies ways in which urban areas modify clouds and precipitation (Braham[65]). The data base consisted of 1302 first echoes from 87 hours during 19 days in July and August 1972, favorable for pre-

Table 1.6.   Comparison of Urban Effect and No Urban Effect Hailstreaks and Hailfalls in the METROMEX Study Area during the Summers of 1971 and 1972 (Semonin and Changnon[64])

	Urban effect	No urban effect
Number of hailstreaks	73	52
Average hailstreak duration (min)	15.1	12.7
Average hailstreak area (km^2)	17.1	14.8
Median hail energy (kJ m^{-2})	2.5×10^{-5}	0.4×10^{-5}
Average point number of hailstones	62.0	37.5
Average point rainfall with hail (cm)	1.78	1.60
Average point hailfall duration (min)	3.6	3.6

cipitation initiation. The analysis (Figure 1.35) shows a distinct plume of first-echo formation about 32 km wide extending 72 km to the northeast from the center of the city of St. Louis. In the center of the plume, the first-echo frequency was about three times that found in the rural area south and east of Greenville, Illinois, where the radar was located (the center of the figure). A secondary maximum is noted to the south and slightly east of St. Louis across the river. Braham suggested that the primary plume indicates the influence of the city of St. Louis, and the secondary plume, the influence of smelting activities located south of St. Louis.

Braham then stratified his data to contrast first echoes on weekdays with first echoes on weekends. Precisely 806 first echoes from 14 weekdays and 496 first echoes from weekends were available for the analysis. The results showed that the plume was almost unchanged on the weekday map but was essentially absent on the weekend map. The increased number of first echoes during the weekday may, therefore, be a result of urban activity, which occurs predominantly on weekdays.

Braham also analyzed the convective first-echo data that formed over and immediately northeast of the city of St. Louis relative to those that formed elsewhere at equal radar range to reduce any effect of attenuation. With the 63 urban first echoes and 268 rural first echoes available, the cumulative frequency of first echoes in the urban and rural areas were compared with the cloud-top temperature (Figure 1.36), which showed that the urban first echoes occurred at lower levels in the atmosphere (warmer temperatures) than did the rural first echoes.

Figure 1.35. Location of first-echo formations. Units are the number of first echoes that formed in a 10-mile range by 20 degrees azimuth box in the summer of 1972 (Braham[65]).

Further evidence of the urban anomaly was presented by Changnon,[66] who studied radar echo top histories for 702 echoes from 19 rain periods in the summer of 1973 for the St. Louis area during near normal conditions. The results obtained from these limited data were in agreement with rainfall results based on the longer periods.

He divided his data into urban and nonurban echo classes. Of the 702 echoes available for the summer of 1973, 137 echoes (20%) fell into the urban class, which was characterized by echoes that developed or passed over the urban area. Of the 137 echoes in the urban class, about half (61 echoes) merged with other echoes. The duration of the urban merged echoes was about 110 min. Changnon showed, as a part of the same study, that the average lifetime of merged echoes in a category that combined both urban class mergers and nonurban class mergers was about 94 min, whereas all other echoes had durations on the average of 43 min. The 76 nonmerged echoes in the urban class had durations of 44 min. The average lifetime of the nonurban echo was 73 min (Table 1.7). The value given for the average lifetime of echoes in the nonurban class included merged as well as unmerged echoes. The merged echoes were over the urban area approximately 50% longer. Only 10% of the merged echoes dissipated over the urban area. More than 30% of the nonmerged echoes dissipated over the urban area.

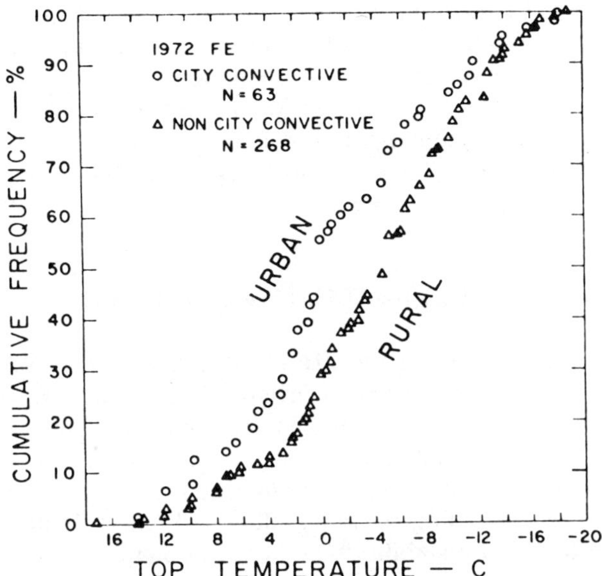

Figure 1.36. Accumulated frequency distribution for temperatures of tops of urban and rural first echoes for the summer of 1972 (Braham[65]).

Changnon also showed that 65 to 75% of the urban echoes were taller at all stages of life than were the nonurban echo. The difference in the average maximum height in the lifetime of the echoes between the two classes was approximately 3900 m (see Table 1.7). The maximum 5-min growth rate in the urban class was twice that in the nonurban class. The

Table 1.7.   Some Statistics on Difference between Urban and Nonurban Radar Echoes (Changnon[60])

	Urban class	Nonurban class
Percent of class total having vertical growth	84	53
Percent of class total having growth durations ≥30 min	75	40
Average duration (min)	110	73
Average maximum height in lifetime (m)	14,100	10,200
Average of maximum 5-min growth (m)	4800	2400

heights of the 137 echoes in the urban class had a bi-model distribution which suggested that the effects of the urban area resulted in two classes of convection. The overall result of Changnon's research indicated that the urban region influences convection in rain processes to the point where there is enhancement of vertical growth, dimensions, and longevity of echoes that pass over the city and merge over and downwind of the urban region.

1.5.3.   Factors Influencing the Urban Weather Anomaly

Changnon et al.[67] examined the METROMEX data for 1971–1973 and developed a hypotheses to explain urban-related rain in severe weather anomalies in St. Louis. An average 25% increase in summer rain, more particularly an increased initiation of rain entities over the urban–industrial areas, was found to be due to the culmination of several factors: (1) thermodynamic effects leading to more clouds, higher clouds bases, and slightly more in-cloud instability; (2) thermodynamic and mechanical effects that produce confluent zones where clouds and rain initiate; and (3) the effect of adding giant nuclei from the urban area, leading to vigorous and more frequent coalescence in urban clouds than rural clouds. With the production of additional cells, nature, coupled with the increased probability for merger with more storms per unit area, allows for the production of heavier rainfalls.

A potential source for condensation nuclei seed clouds downwind of an urban area are the air pollutants emitted by the urban area. Gatz[68] investigated the air pollutant potential emitted in St. Louis that would reach cloud levels soon enough to participate in the deposition of anomalous rainfall at the time and distance scale involved. A lithium tracer was used to measure particle removal efficiencies and to follow particles ingested in convective storm updrafts. The lithium concentration in rain upwind and downwind of the city was used to determine where and when urban pollutants appear in the downwind precipitation.

In this case, 746 g of lithium was released from an industrial stack in light rain (July 14, 1971) and the deposition pattern determined (Figure 1.37). Analysis indicates that the total integrated lithium deposition was 188 g, approximating a 25% removal efficiency, which must be attributed to an in-cloud scavenging process. Particle scavenging by rain impaction below the clouds case is negligible, because impaction collection efficiencies for submicrometer particles are small. If the collection efficiency by impaction below the cloud base was large, a large deposit would exist near the release point, but this was not shown. Gatz has, therefore, dem-

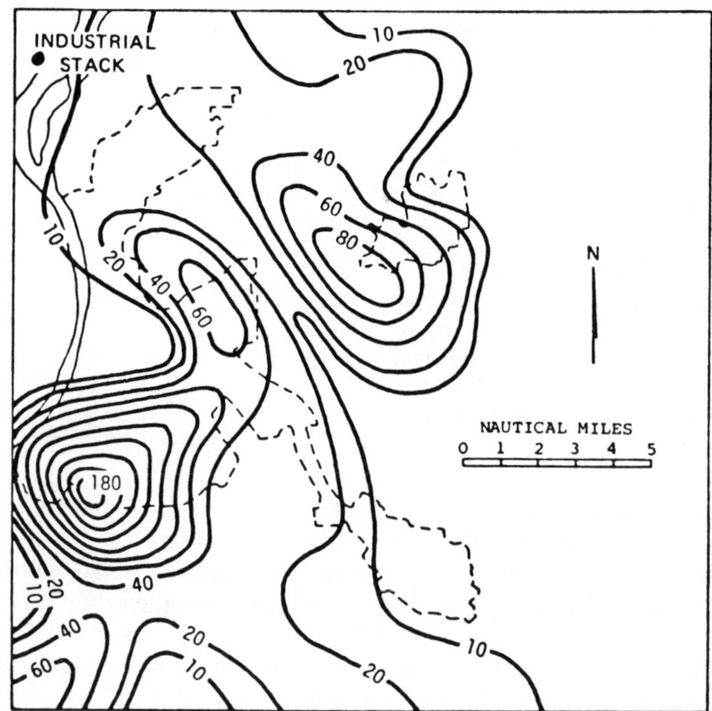

Figure 1.37. Deposition pattern in rain for Li tracer released from an industrial stack on July 14, 1971. Units are pg Li cm^{-2} (Gatz[68]).

onstrated that materials (such as air pollutants) released near the ground can enter convective clouds and be deposited in rain within small time and space scales from the release point.

Critical in the production and enhancement of rain are cloud condensation nuclei (CCN), which are small, soluble hygroscopic aerosol particles that have critical supersaturations attainable in updrafts of natural clouds. Braham[65] studied the CCN concentration upwind, over, and downwind of St. Louis as part of the METROMEX program (Table 1.8). His results showed that the average downwind CCN concentration was larger than the average upwind concentration, indicating that the city acts as a source. This distribution was maintained despite the level of supersaturation. Furthermore, the fractional increase in the CCN concentration downwind of the city was greatest when the level of supersaturation was largest, which indicated that relatively more urban nuclei are emitted in smaller size ranges. Table 1.8 also shows a pronounced increase in winter,

Table 1.8. Mean Cloud Condensation Nuclei Spectra Upwind and Downwind of St. Louis for Summer and Winter (Braham[65])

Sampling period	Sampling level	Number of upwind and downwind comparisons	Average spectra	
			Upwind	Downwind
July–August 1971	610 m MSL	19	2210	3890
January–March 1972	460 m MSL	13	848	1678
July–August 1972	Spiral downward from top of mixed layer to 460 m MSL	25	1166	1625
March–April 1973	460 m MSL	10	977	1451

so that the concentration of the smaller nuclei in the urban plume was even more pronounced in the winter than in the summer.

The urban CCN may be largely from coal-burning power stations or from automobile exhausts. Calculations indicate that the urban production rate for the CCN is of the order of 1 to 5 $\times$ 10^4 cm^{-2} s^{-1}.

Braham also studied the cloud droplet distribution across the city of St. Louis. He surmised that if the urban-induced CCN was affecting clouds, it would affect the size and concentration of cloud droplets in the urban clouds. These would contain high concentrations in proportionally smaller drops when compared to the upwind clouds of similar types. He analyzed the droplet spectra at nine points along a cross-wind section over St. Louis during the early morning hours of March 3, 1972 (Figure 1.38). In the southeast, the droplet spectra was essentially rural (east-northeast wind), but to the northwest and over the central portions of the city of St. Louis the drop concentration increased and the spectra narrowed. Farther to the northwest, the spectrum regains its rural character, which corresponds to the air flow through the industry-free gap between Alton and Granite City. At the northwest end, downwind of the Alton industrial area and two major power stations, high droplet concentration with a narrowing of the size spectrum occurred. Overall, the data show a clear example of urban modification of stratus clouds.

Auer[69] found evidence of ingestion of industrial effluents by cumulus clouds. A cumulus cloud formed over the oil refinery complex at Wood River, Illinois (northeast of St. Louis) on August 10, 1973. This cloud was in a relative steady state and was nonprecipitating. A meteorologically instrumented aircraft penetrated the cloud and its subcloud environment 12 times between 11:30 and 12:15 local time to obtain the Aitken nucleus

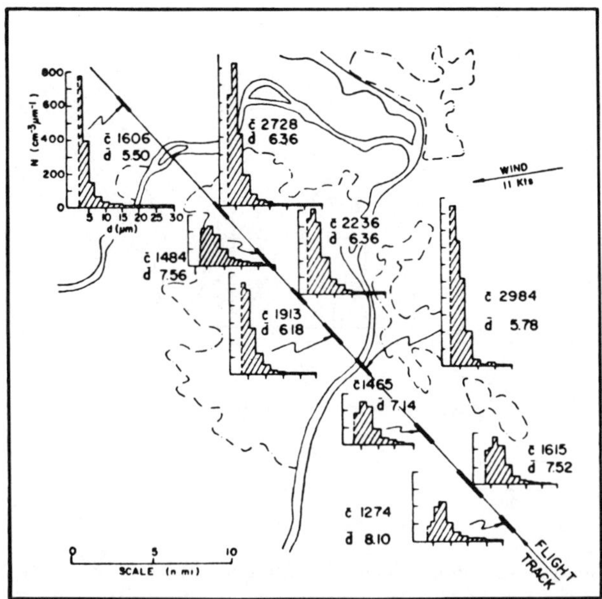

Figure 1.38. Cloud droplets spectra showing the effect of urban CCN upon low stratus (Braham[65]).

concentrations (Figure 1.39). Analysis of these data indicated that large concentrations existed from the cloud base to near the top, the source of the nuclei clearly being the oil refinery complex. The environment surrounding the cloud has significantly smaller concentrations of Aitken nucleus.

In spite of their limitations, Auer's observations clearly show the ingestion of the industrial effluent by cumulus clouds. Auer's studies, and those discussed earlier, indicate the potential for significant cloud modification by urban and industrial effluents, supporting the hypothesis that these enhance rainfall downwind of a city.

The other factors suggested by Changnon et al. for producing anomalous urban rainfall are confirmed by the following facts. It has been shown that owing to the thermal effects of the urban heat island, a horizontal pressure distribution is established that produces the urban heat island circulation. Because the urban heat island is most prevalent in the evening and night, it is expected that heat island circulation should also be most prevalent at that time. However, at night, the dampening influence of the stable boundary layer produces very weak circulation, whereas during the day, if only a weak heat island exists, rather intense

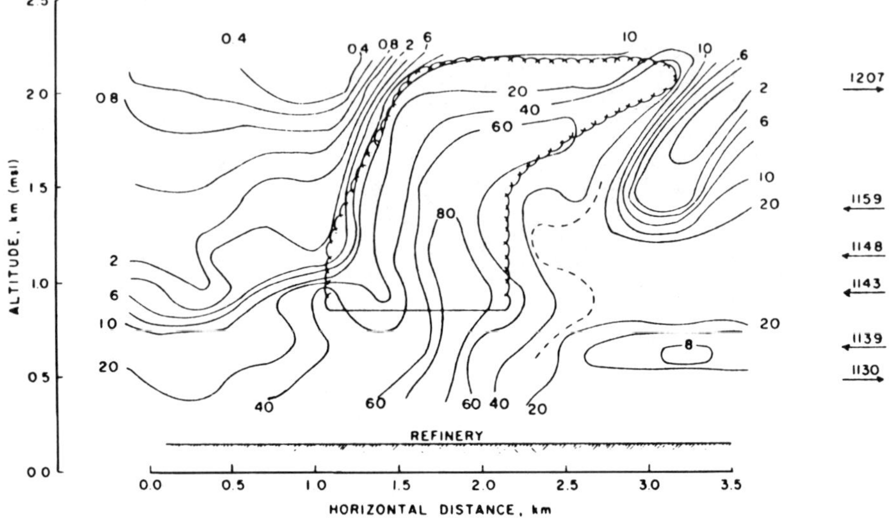

Figure 1.39. Anaiysis of Aitkin nuclei concentration (10 particles cm^{-3}) along the north-south orientation in a cloud that develops over an oil refinery at Wood River, Illinois.

heat island circulation may develop because the heating will be distributed through a deeper layer.

Vukovich[70] has shown that nocturnal heat island circulation is expected to be largest between 1500 and 2100 hours local standard time, diminishing during the remainder of the night. If a daytime heat island circulation exists, it may be significant as early as 1100 LT (local time) and persist into the late afternoon hours. Also, the 1971 and 1972 METROMEX data[61] show that the maximum rainfall downwind of St. Louis occurred somewhere between 1200 and 1800 LT. The coincidence in time between the presence of the urban heat island circulation and the maximum rainfall downwind of St. Louis, and the fact that the urban heat island is sometimes significant enough to lift air parcels to the condensation and free convection levels so as to produce clouds, indicate the significant role that urban heat island circulation may play in the production and enhancement of rain downwind of an urban region.

Further influence of urban heat island circulation was shown in the extensive case studies presented in a report edited by Changnon and Semonin.[71] An example of the study of the downstream enhancement of an isolated storm (July 14, 1973)[72] will be discussed briefly here. The data for the case study came from the METROMEX network.

At 1600 CDT, a relatively weak heat island persisted over the St.

Louis–East St. Louis–Granite City urban complex (Figure 1.40), and a secondary hot spot was located in the bottomlands to the north of St. Louis. The flow pattern showed a zone of convergence over and downwind of St. Louis (Figure 1.41) within which, at this time, a number of first echoes were detected by a 3-cm radar. Another isolated echo, located some 30 km to the northwest of the center of St. Louis, was moving east-southeastward toward St. Louis. About 1635 CDT, this echo merged with one of the echoes present in the convergence zone over and downwind of the city. Afterward the merged echo intensified significantly and produced intense rainfall (as much as 4 in./hr) downwind (east-southeast) of the city.

Although this study does not show direct enhancement of rain processes

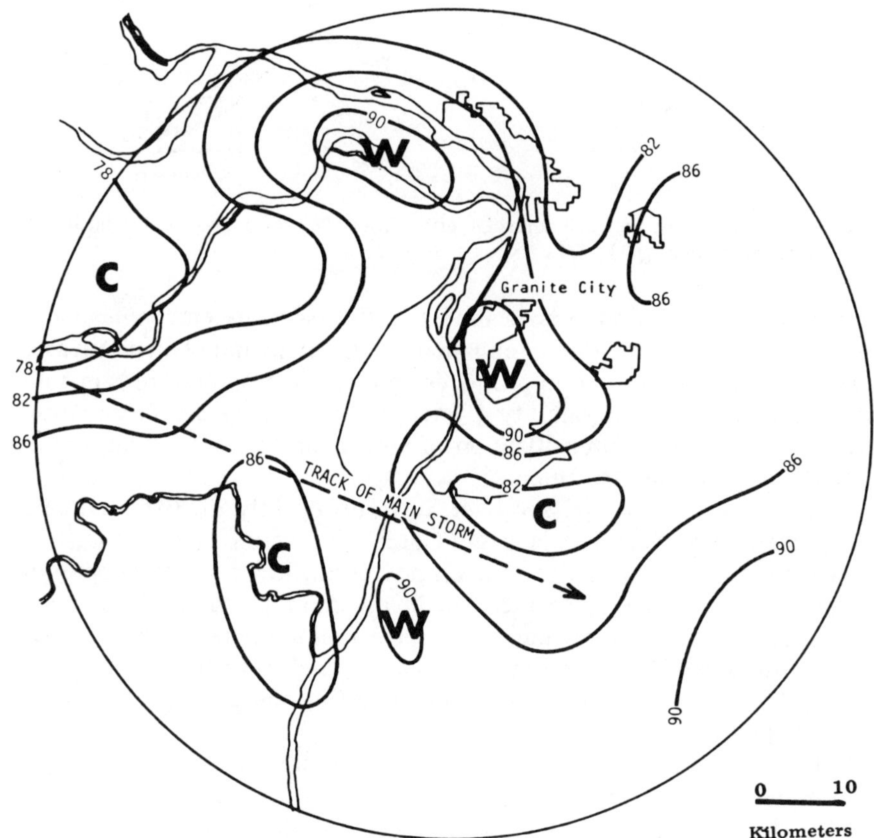

Figure 1.40. Surface temperature (°F) distribution over St. Louis, Missouri, and vicinity on July 14, 1973, at 1600 CST (Semonin[72]).

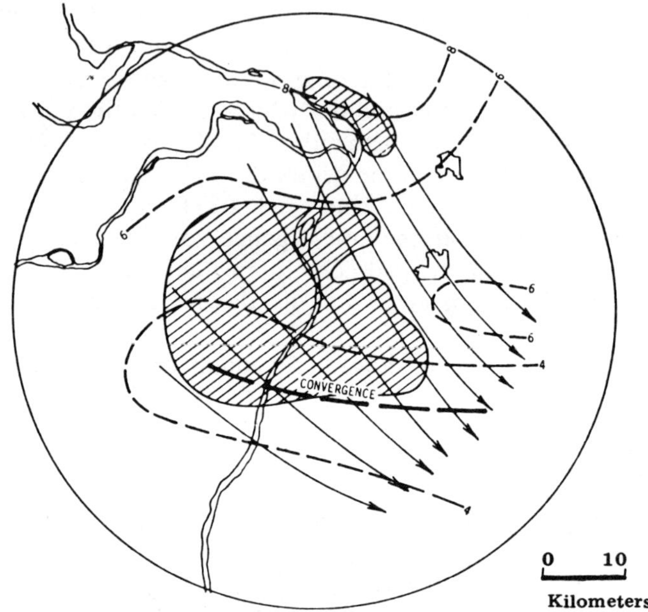

Figure 1.41. Air flow at the 1000-m level over St. Louis, Missouri, and vicinity on July 14, 1973, at approximately 1600 CDT.

by the urban circulation, it does show that the associated mechanism may have the highest probability of enhancing rain downwind of an urban area; that is, the urban circulation produces an increased number of rain cells, which increases the probability of merger, since there are more storms per unit area. Rapid intensification of storms and heavy rainfall are associated with storms that merge.

When a severe storm passed over St. Louis on July 23, 1973, evidence was presented by Schickendanz and Gatz[73] that the intensity of the storm increased as a result of ingestion of air possessing low equivalent potential temperature, causing some increase in the rainfall rate and downdraft intensity. The air with low equivalent potential temperature was a result of the low dewpoint temperatures found in and near the center of St. Louis. The low urban moisture content during the daylight hours was discussed earlier.

Ochs[74] developed a theoretical model to study cumulus convection, comparing the results with METROMEX data for August 7 and 10, 1973. The model suggested that on these two days, the surface temperature distribution was a significant factor in determining the location of cumulus initiation. The air near the thermal hot spots, which could not be directly associated with the urban heat island in all cases, reached convective instability sooner than did the surrounding air. This is in qualitative agreement with the observed results, suggesting that in some cases, only the heat island is required to initiate cumulus convection over the city.

1.6. URBAN CLIMATE AND AIR POLLUTION

In previous sections it was shown that the urban air pollutants affect the urban climate, reducing radiation (Landsberg,[24] DeBoer,[25] Chandler,[10] Charleson and Pilat,[33] and Atwater[17]) and producing anomalous weather affects (Gatz,[68] Braham,[65] and Auer[69]). However, there is very limited evidence that the urban air pollution distribution is affected by the urban climate.

Padmanabhamurty and Hirt[75] performed a study using data from Toronto, Canada, correlating the urban heat island intensity, as measured by the maximum surface urban–rural temperature difference, to the "soiling index" or "COH" values. A COH unit is defined as the quantity of particulate matter that produces an optical density of 0.01 in filter paper with light at the 400-m wavelength when 1000 linear feet of air is drawn through filter paper.

As the heat island intensity increases, the soiling index also increases (Figure 1.42), perhaps as a result of urban heat island circulation. This

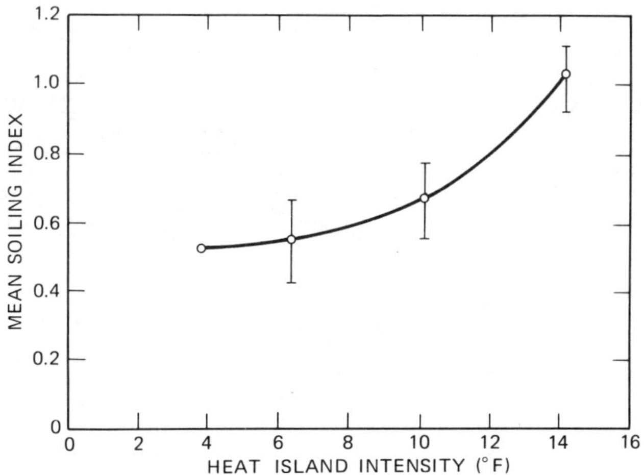

Figure 1.42. Relationship between the heat island intensity and the mean soiling index (Padmanabhamurty and Hirt[75]).

condition could result in a well-marked difference in pollution concentrations between the urban and rural regions.

In a recent unpublished study by Vukovich et al.[8] it was found that the urban flow may affect the air pollution distribution. Figure 1.43, an analysis of the surface SO_2 distribution in St. Louis at 1600 CST on June 8, 1976, is extracted from the U.S. Environmental Protection Agency's Regional Air Pollution Study network. SO_2 is a primary pollutant and is therefore source-oriented. The surface wind analysis at 1600 CST is given in Figure 1.25*b*. When comparing these figures, it is noted that the region of maximum SO_2 falls within the zone of convergence associated with urban heat island circulation, suggesting mass convergence within the region to produce the SO_2 maximum.

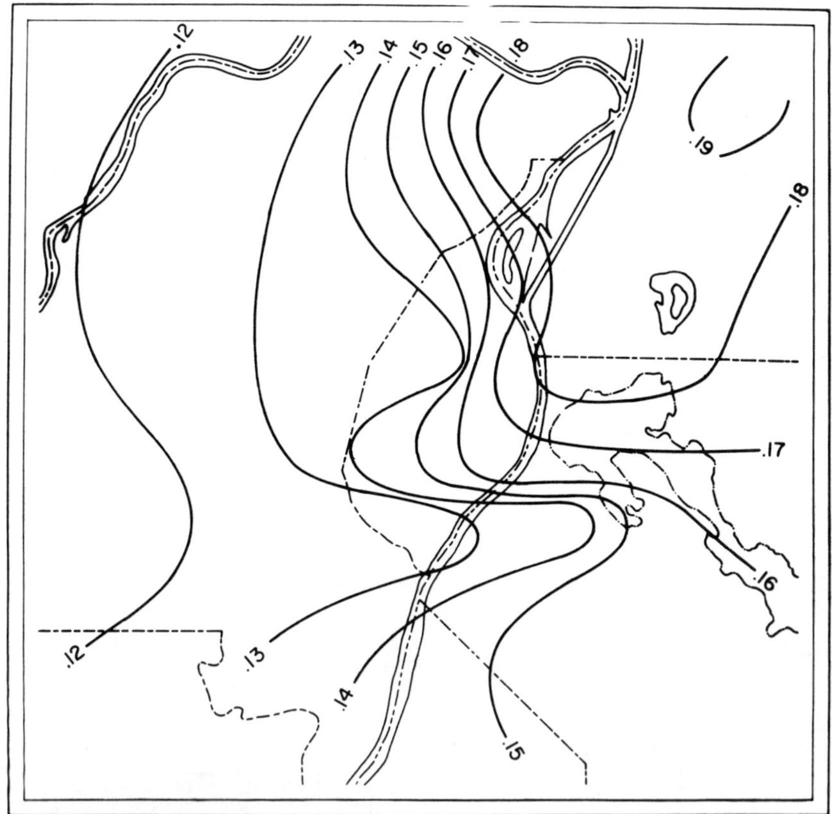

Figure 1.43. Analysis of the SO^2 distribution (ppm) at the surface in St. Louis, Missouri, on June 8, 1976, at 1600 CST (Vukovich et al.[8]).

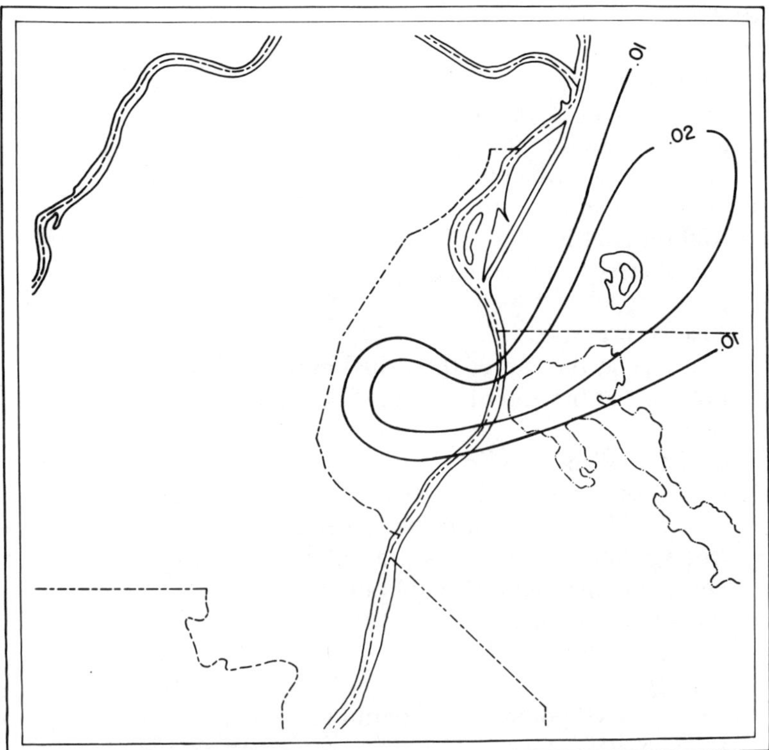

Figure 1.44. Analysis of the surface distribution of ozone (ppm) in St. Louis, Missouri, on June 8, 1976, at 1600 CST (Vukovich et al.[8]).

Figure 1.44 shows the ozone (a photochemical pollutant) distribution in St. Louis at 1600 CST on June 8, 1976. Ozone is the plume downwind of the city is undoubtedly produced by the precursors emitted from the St. Louis region. Here with horizontal advection, ozone should be found relatively far downwind of the urban region. When comparing Figure 1.44 with the wind distribution (Figure 1.25b), it is noted that although the high ozone concentrations are found relatively far downwind from the center of the city and from the zone of convergence, high concentrations of ozone are also found in the zone of convergence, and there is a crowding of ozone concentration (isopleths) at the boundaries of the convergence zone.

Very little other evidence exists so far to demonstrate the influence of the urban climate on the air pollution distribution. Thus the principal effect of the air pollution distribution observed with any continuity is the effect on the urban climate and the urban weather anomaly.

SYMBOLS

C	circulation
c_p	specific heat at constant pressure
g	acceleration due to gravity
H	excess heat available per unit area to heat the air
k_0	von Kármán constant
K	friction coefficient
K_m	characteristic eddy viscosity
p	pressure
p_0	surface pressure
p_1	pressure at the top of the urban boundary layer
p_2	pressure at the top of the rural boundary layer
t	time
ΔT	heat island magnitude
$\bar{T}$	mean temperature of the boundary layer
u	horizontal component of wind velocity
U_g	free stream geostrophic wind
w	vertical component of wind velocity
x	horizontal distance
z	vertical distance
ρ	density
γ	lapse rate of potential temperature
ω^2	Brunt-Vaisala factor

Subscripts

c	properties of the urban boundary layer
E	properties of the rural boundary layer

REFERENCES

1. Howard, L., 3rd ed., Harvey and Darton, London, 1833.
2. Schmidt, W., *Fortsch. Landwirtsch.* (1929).
3. Mitchell J. M., *Air over Cities*, 1971.
4. Hage, K. D., *J. Appl. Meteorol.*, 11, 1 (1972).
5. Oke, T. R., and East, C., *Bound.-Layer Meteorol.*, 1 (1971).
6. Vukovich, F. M., *Mon. Weather Rev.*, 101, 6 (1973).
7. Oke, T. R., Yap, D., and Maxwell, G. B. Proc. Conf. Urban Environ. Sec. Conf. Biometeorol., Philadelphia, 1972.

8. Vukovich, F. M., Dunn, J. W., and King, W. J. Unpublished report for the Environmental Protection Agency, 1977.

9. Lawrence, E. M., *Atmos. Environ.*, 5 (1971).

10. Chandler, T. J., Hutchinson, London, 1965.

11. Lawrence, B. N. *Atmos. Environ.*, 3 (1969).

12. Duckworth, F. S., and Sandberg, J. S., *Bull. Am. Meteorol. Soc.*, 35, 5 (1954).

13. DeMarrais, G. A., *Bull. Am. Meteorol. Soc.*, 42, (1961).

14. Clark, J. F., *Mon. Weather Rev.*, 97 (1969).

15. Davidson, B., *J. Air Pollut. Control Assoc.*, 17 (1967).

16. Angell, J. K., Pack, D. H., Dickson, C. R., and Hoecker, W. H., *J. Appl. Meteorol.*, 10 (1971).

17. Atwater, M. A., *J. Appl. Meteorol.*, 10, 2 (1971).

18. Summers, G. W. Proc 1st Can. Conf. Micrometeorol., Toronto, 1965.

19. Leahey, D. M., Final Rep., Geophys. Sci. Lab., School of Engineering and Science, New York University, 1969.

20. Vukovich, F. M., Dunn, J. W., and Crissman, B. W. *J. Appl. Meteorol.*, 15, 5 (1976).

21. Tyson, P. D., DuToit, J. W. F., and Fuggle, R. F., *Atmos Environ.*, 6 (1972).

22. Clarke, J. F., *Mon. Weather Rev.*, 97, 8 (1969).

23. Vukovich, F. M., *Mon. Weather Rev.*, 103 (1975).

24. Landsberg, H. E., Great Printing Company, DuBois, Pa., 1960.

25. DeBoer, H. J., Wetenschappelyk Rapport, Netherlands, 1966.

26. Mateer, C. L., *Int. J. Air Water Pollut.*, 4 (1961).

27. Peterson, J. T., and Flowers, E. C., *Bull. Am. Meteorol. Soc.*, 54 (1973).

28. Rouse, W. R., Noad, D., and McCutcheon, J., *J. Appl. Meteorol.*, 12 (1973).

29. Kondratyev, K. Y. A., *Radiation in the Atmosphere*, Academic Press, New York, 1969.

30. Craig, C. D., and Lowry, W. P., Proc. Conf. Urban Environ. Sec. Conf. Biometeorol., Philadelphia, 1972.

31. Bray, J. R., Sanger, J. E., and Archer, A. L., *Ecology*, 47 (1966).

32. McCormick, R. A., and Ludwig, J. H., *Science*, 157 (1967).

33. Charleson, R. J., and Pilat, M. J., *J. Appl. Meteorol.*, 8 (1969).

34. Myrup, L. O., *J. Appl. Meteorol.*, 8 (1969).O

35. Tag, G. M., NSF Grant Rep. Pennsylvania State University, 1969.

36. McElroy, J. L., *Bound.-Layer Meteorol.*, 3 (1973).

37. Pandolfo, J. P., Atwater, M. A., and Anderson, G. E., Final Rep., The Center for Environment and Man, Hartford, 1971.

38. Nappo, C. J., Proc. Conf. Urban Environ. Sec. Conf. Biometeorol., Philadelphia, 1972.

39. Maisel, T. N., Master's thesis, University of Maryland, 1971.

40. Clarke, A. F., and Peterson, J. T., Proc. Conf. Urban Environ. Sec. Conf. Biometeorol. Philadelphia, 1972.

41. Holzworth, G. C., EPA Rep. No. AP-101 (1972).

42. Oke, T. R., *Atmos. Environ.*, 7 (1973).

43. Kratzer, P. A., Vieweg, F., and Braunschweig, S., Air Force Camb. Res. Lab. Rep., Bedford, 1956.

44. Kopec, R. J., *J. Appl. Meteorol.*, 12 (1973).

45. Landsberg, H. E., and Maisel, T. N., *Bound.-Layer Meteorol.*, 12 (1972).

46. Ackerman, B., Proc. Conf. Air Pollut. Meteorol., Raleigh, 1971.

47. Jones, D. A., and Schickendanz, P., METROMEX Report, Ill. State Water Sur., Urbana, 1974.

48. Wood, J. L., Rep. No. 28, University of Texas, 1971.

49. Estoque, M. A., and Bhumralkar, C. M., *Mon. Weather Rev.*, 97, 12 (1969).

50. Vukovich, F. M., *Mon. Weather Rev.*, 99, 12 (1971).

51. DeLage, Y., and Taylor, P. A., *Bound.-Layer Meteorol.*, 1 (1970).

52. Estoque, M. A., *Quart. J. R. Meteorol. Soc.*, 87 (1961).

53. Bornstein, R. D., *J. Appl. Meteorol.*, 14 (1975).

54. Lee, R. L., and Olfe, D. B., *Bound.-Layer Meteorol.*, 7 (1974).

55. Dirks, R. A., *J. Geophys. Res.*, 24 (1974).

56. Findlay, B. F., and Hirt, M. S., *Atmos. Environ.*, 3 (1969).

57. Pooler, *J. Appl. Meteorol.*, 2 (1963).

58. Ackerman, B., Proc. Conf. Urban Environ. Sec. Conf. Biometeorol., Philadelphia, 1972.

59. Angell, J. K., Hoecker, W. H., Dickson, C. R., and Pack, D. H., *J. Appl. Meteorol.*, 12 (1973).

60. Changnon, S. A., *Bull. Am. Meteorol. Soc.*, 49 (1968).

61. Huff, F. A., and Schlessman, E. E., Sum. Rep. METROMEX Stud. 1971–1972, Ill. State Water Sur., Urbana, 1973.

62. Huff, F. A., and Changnon, S. A., *Bull. Am. Meteorol. Soc.*, 54 (1973).

63. Schickedanz, P. T., *J. Appl. Meteorol.*, 13 (1974).

64. Semonin, R. G., and Changnon, S. A., *Bull. Am. Meteorol. Soc.*, 55 (1974).

65. Braham, R. R., *Bull. Am. Meteorol. Soc.*, 55 (1974).

66. Changnon, S. A., *J. Appl. Meteorol.*, 15 (1976).

67. Changnon, S. A., Semonin, R. G., and Huff, F. A., *J. Appl. Meteorol.*, 15 (1976).

68. Gatz, D. F., *Bull. Am. Meteorol. Soc.*, 55 (1974).

69. Auer, A. H., *J. Appl. Meteorol.*, 15 (1976).

70. Vukovich, F. M., *Mon. Weather Rev.*, 101 (1973).

71. Changnon, S. A., and Semonin, R. G., (Ed.), Studies of Selected Precipitation Cases from METROMEX, Ill. State Water Surv., Urbana, 1975.

72. Semonin, R. G., Studies of Selected Precipitation Cases from METROMEX, Ill. State Water Surv., Urbana, 1975.

73. Schickedanz, P. T., and Gatz, D. F., Studies of Selected Precipitation Cases from METROMEX, Ill. State Water Surv., Urbana, 1975.

74. Ochs, H. T., *J. Appl. Meteorol.*, 14 (1975).

75. Padmanabhamurty, B., and Hirt, M. S., *Water, Air Soil Pollut.*, 3 (1974).

2

POLLUTANTS AND PLANT HEALTH

D. G. Parbery

School of Agriculture and Forestry, University of Melbourne, Victoria, Australia

Pollutants are discussed as being additional or abnormal components of the environment in which plants have evolved and grow. The inability of plants to accommodate to the conditions of the changed environment leads to pollution damage. In parts of the northern hemisphere this results in many millions of dollars damage per annum. The genetic capacity of some individual plants to tolerate such changes, however, has enabled some cultivars of plants to be developed for commercial and other purposes where increased or reduced sensitivity to presence of a pollutant is required. The effects of the main pollutants of importance to plant health are described. In addition, an attempt is made to demonstrate that the direct effects of pollution on plant health are not the only effects; that in some instances (e.g., the heavy metals) it may be the indirect effects of the pollutants on the well-being of microorganisms, important components of the plant growth support system, that are likely to influence plant health the most. The influence of pollution on whole floras, on plant communities, and on individual plants is considered.

2.1.   INTRODUCTION

The environment in which plants have evolved includes many factors that regulate their life cycles and contains no extremes which local, indigenous species cannot tolerate or survive. For example, fire, which may be a natural component of regions of the globe such as southern Australia, will be tolerated, if not required, by plants native to the region, but not by plants that have evolved in the absence of fire, such as rain forest species.

Similarly, plants that have evolved in the presence of high levels of toxic chemicals in soils, such as common salt or heavy metals, will tolerate those toxic levels of the chemical in that soil or a similar soil under similar conditions, whereas other plants will not tolerate them under any conditions. Consequently, in defining what constitutes a pollution situation regarding plant health, it is necessary to know the species involved, their susceptibility to the toxin in question and the conditions of exposure.

The growth and reproduction of plants are influenced and frequently controlled by two interacting sets of factors. The first set, the physical-chemical factors, include the components of air, temperature, light, water, minerals and soil pH, structure, and aeration. Allied to these are changes in weather patterns and seasonal variation. The second set consists of biological factors. It includes a range of microbes that influence plant growth either beneficially or detrimentally, interactions with other plant species, as well as species of fauna ranging from arthropods to birds and animals. The composition of the air, the nature of the soil, the associated flora and fauna, and the pattern of weather and seasons in which or with which a species has evolved govern the behavior of the individual plant species. Introduction of new factors or the changing of existing factors are potentially capable of affecting the growth of plants by changing the environment in which they have evolved. Not all changes will influence a particular species. The nature, magnitude, and duration of the change have to be considered.

2.1.1. The Pollutants

There are many forms of pollution and there are many pollutants. There are various ways of classifying them, according to whether they are air pollutants, water pollutants, or soil pollutants, depending on where they accumulate and how they cause problems. Some may be classified as "deliberate pollutants," being chemicals such as agrochemicals, released deliberately into the environment to achieve a specific purpose but which sometimes have secondary effects of an undesirable nature. Others may be described as "nondeliberate pollutants," since they are unwanted by-products of industry released into the environment as a means of getting rid of them. The latter group includes most air pollutants. As in most classifications, there is overlap between categories, so that although we will be considering mainly pollutants, such as components of photochemical smog (sulfur dioxide, etc.), which are the commonly regarded air pollutants, we will also consider aspects of heavy metal pollution, which, although they mostly end up as pollutants of water and soil, frequently begin as airborne pollutants (Goodman and Roberts[1]) and sometimes (e.g., in the case of zinc) may be absorbed through aerial plant parts.

2.1.2. Pollutants and Plant Health

A number of chemicals that are sometimes referred to as "pollutants" may prove beneficial to or even essential for, plant health when supplied in moderate amounts. It is necessary to recognize that such chemicals are not pollutants unless they reach toxic proportions. This is true for sulfur dioxide, nitrogen dioxide, ethylene, most agrochemicals, and several heavy metals, such as iron, copper, and zinc. Therefore, it is not presence that is the critical consideration but such concepts as toxic levels, critical times of exposure, factors predisposing plants to damage, and interactions of various kinds which influence plant health indirectly. While the following pages describe the direct effects that pollutants cause in plants and communities of plants, an attempt is made to show that there are many indirect ways that plants can be affected through changes in the environment. Although it is not the intention to complicate the subject, it is intended to demonstrate that there are many complex relationships within a plant community which are important to the health of individual plants and which are frequently overlooked in discussions of pollution effects. A useful conceptual model of the interactions between the many factors involved in moderating the effects of pollutants on plants, proposed by van Haut and Stratman[2] and modified by Heck and Brandt,[3] is reproduced here (Figure 2.1).

2.2. SULFUR DIOXIDE

One undesirable consequence of the industrial revolution was the wholesale destruction of vegetation surrounding large coal-fired and smelting industries. The association of this with smoke damage led to improvements in smoke dispersal which prevented the complete destruction of the surrounding countryside but not all damage, such as that frequently noted in agricultural crops. The recognition by the late nineteenth century that sulfur dioxide was the main toxic principle in smoke was generally acknowledged by the early twentieth century. In spite of this, because of the vast amounts of sulfur-bearing fuels and ores used annually by industry, sulfur dioxide continues to be a serious cause of plant injury in various parts of the world.

Some countries, such as Australia, are generally fortunate in producing

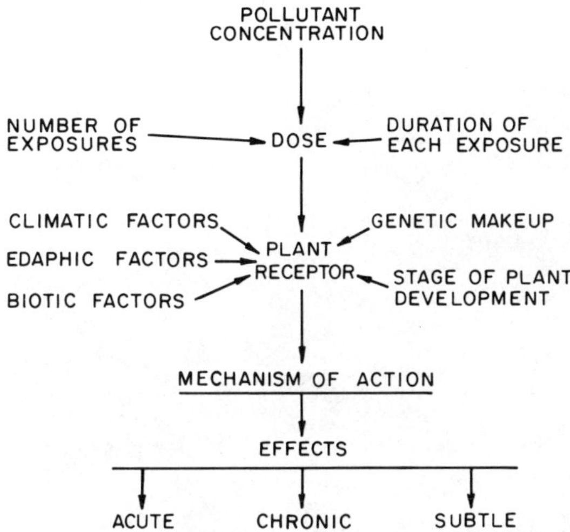

Figure 2.1. Conceptual model of factors involved in air pollution effects on vegetation (adapted by Heck and Brandt[3] from van Haut and Stratman[2]).

coal and oil low in sulfur. On the other hand, large quantities of sulfur-rich ores may be smelted annually for the extraction of minerals. Apart from isolated instances, such as the destruction of the vegetation around Queenstown, Tasmania (Figure 2.2), in the late nineteenth century, Australia has not suffered from sulfur pollution. Indeed, since many Australian soils are deficient in sulfur or at least require the addition of sulfur to maintain soil fertility, it seems likely that the fallout of sulfur dioxide is frequently an advantage.[4] At present there is no serious problem in Australia from sulfur dioxide pollution, nor is there likely to be in the near future. Developed countries in the northern hemisphere, however, are not always as fortunate in this regard.

2.2.1. Damage to Floras

In European countries such as Holland, where a detailed knowledge of the country's flora has been available since before the industrial revolution, it has been possible to point to possible changes in the flora in association with industrialization. For example, le Blanc and de Sloover[5] claim that largely as a result of SO_2 pollution during the last century, the following reductions in the flora of Holland have occurred: flowering plants, 3.8%; terrestrial mosses, 15%; epiphytic mosses, 12%; and lichens,

Figure 2.2. Vegetation returning to the hills around Queenstown, Tasmania, once denuded (late 1800s) through removal of timber for mining operations and by SO_2 damage (reproduced with permission of C.S.I.R.O.).

27%. That the disappearance of species of plants from such areas is the result of man's activity is not in doubt. The extent to which it can be attributed to one pollutant or another is often difficult to establish. Even so, no other air pollutant has been so widespread in Europe for such a long time. Other claims are perhaps more difficult to establish. For example, Pyatt[6] reviews the history of the disappearance of lichens from cities in Europe during the last 120 years. In cities such as Paris it has been attributed to various causes, including the reduction in the numbers of trees, the generally drier city atmosphere, and smoke pollution. Although it is likely that drier atmosphere and possibly generally higher temperatures in cities are of some importance, there is much evidence to support the generally accepted view that pollutants, primarily SO_2, are chiefly responsible for the decline in the lichens.

Several authors refer to the reduction of lichen populations in and around industrial cities (Pyatt,[6] le Blanc and de Sloover,[5] Daly,[7] Ferry et al.,[8] Richardson and Young[9]). Lichens are frequently entirely absent from the central industrial areas. Moving from the unpolluted zone into this dead-heart area, the first forms to disappear are the species with greatly divided thalli (large surface area/volume ratio) and the juvenile forms. The most resistant are the mature forms of species with more-or-less entire (crustose) thalli. By surveying the lichen flora of a region first, it is possible to establish the concentration gradients for a pollutant over the polluted area and to map the dispersal area in relation to the source of the pollutant.

In Australia any changes in the lichen flora have passed unnoticed. However, in cities such as Melbourne, crustose lichens may be found between stones in streets of low traffic density, and a variety of types occur on slate and terra-cotta tile roofs. Mosses have been used in a similar way to lichens to monitor the effects of sulfur dioxide pollution (Gilbert,[10] Daly[7]).

2.2.2. Changes in Plant Communities

Attention was first attracted to the effects of pollution on vegetation by the dramatic destruction around smelters in England, Europe, and North America. Brandt and Heck[11] refer to several areas associated with the smelting of copper, zinc, lead, nickle, and other ores. Typical of these areas was the Duckville Copperhill area of the Tennessee copper basin, where according to Treshow[12] 7000 acres of forest immediately surrounding the smelter was completely denuded as a result of SO_2 damage and a further 17,000 acres of forest was replaced by grass. It is probable that less noticeable changes occurred beyond the grassland area. Associated with these changes were a 10% reduction in rainfall and a 3.5°C increase in mean annual temperature.

Changes in vegetation do not need to be dramatically obvious to cause significant long-term effects in a plant community. The most important aspect of dramatic change, which is usually relatively localized, is that it draws attention to the possibility of potentially more significant chronic effects, which may be much less obvious and more widely spread than the dramatic acute effects. This is demonstrated by a study of a forest in the vicinity of an iron sintering plant at Wawa, Ontario, Canada. The study showed how, in forest where no change was noticeable to a casual observer, changes were in progress which in the long term would significantly change the nature of the plant community. Gordon and Gorham[13] made an ecological survey along a 36-mile transect through the Wawa forest away from the sintering plant along the plume line (direction of prevailing smoke drift). At 36 miles from the works there were 28 tree and shrub species in the forest plant community. This number remained constant until within 10 miles of the sintering works the number of species began to decline until there were none closer than 2 miles from the works. Consequently, it might have been concluded that no change was occurring in the forest more than 10 miles from the works, had the survey ended there. However, when the survey was directed away from adult forms of the 28 species to juvenile forms, it was discovered that no seedlings or juvenile plants of *Pinus strobus* occurred within 30 miles of the sintering plant, while none of *Picea mariana* or *Populus tremuloides* occurred within 15 miles of the plant. Consequently, should smelting continue in the area until the adult forms of the three mentioned species reach the end of their reproductive life, there would be changes in the forest over a much greater area than was at first believed.

Other studies of the effects of sulfur dioxide on plant communities have been made. Van Haut and Stratman[2] provide photographs and diagrammatic interpretations of what can happen in the zone of damage equivalent to the 2- to 10-mile zone referred to above (Gordon and Gorham[13]). Closest to the source of pollutant there is a completely denuded zone. Then there is a zone where some volunteer weed species come in, then zones of grass and scrub, zones of dying trees, and eventually the intact plant community. Guderian[14] discusses some of these same examples and mentions others in giving examples of the differences in sensitivity of individual plants and species of plants to pollutants.

With this example in mind, Treshow[12] pointed out that at that time nothing was known about the effects of pollutants generally on such vital plant processes as photosynthesis, respiration, vigor, growth regulation, or reproduction. In reproduction alone, he pointed out that a pollutant could affect many processes, including pollen release, pollen germination, flower initiation and development, as well as fertilization and fruit and seed set. Pollution could induce flower and/or fruit drop, reduce the numbers of viable ovules, and thereby reduce the number of viable seeds. It could also affect seed viability and seedling survival. Subsequent work has shown that it is likely that most of these processes are affected in various cases, although not all in one species. Much attention has been paid to the effects on photosynthesis, respiration, and changes in cell membranes, but much more work is required on the effects of pollutants on reproductive physiology.

2.2.3. Toxic Levels and Exposure Times

Present-day plants have evolved in an atmosphere of 78% nitrogen, 21% oxygen, and 1% of other gases, including 0.03% carbon dioxide ($\sim$320 ppm) and from between 0.001 and 0.01 ppm of sulfur dioxide. The levels of SO_2 have apparently remained constant over a very long time and, apart from local variations, appear to be remaining constant on a world basis.[4] In localities where atmospheric levels reach five times the ambient level, however, very sensitive plant species sustain injury. Most commonly, however, plants are not affected until atmospheric concentrations reach 0.15 ppm for at least 6 hr day^{-1}.

2.2.4. Action against Plants

Plant health can be influenced by either the direct action of the pollutant on the plant itself or indirectly through its effects on some biological or physical component of the plant's environment. It is likely that both types of action often occur together.

2.2.4.1. Direct Effects

When SO_2 is discharged into a moist atmosphere, a fine mist of sulfuric acid may form and cause the development of numerous small, brown flecks on contaminated leaves. The flecks represent small areas of cells killed through contact with the acid. Such injury is not common and is not regarded as an important aspect of the damage caused by this pollutant.

The more important aspect of SO_2 injury is the internal damage caused by the gas taken into the leaf tissues during respiration (Figure 2.2). When taken in at low concentrations it is oxidized and assimilated, sulfur being an important plant nutrient required for protein synthesis. However, when the rate of uptake exceeds the rate at which it can be assimilated, then toxic effects appear in the leaf tissues. Treshow[15] reviews the various theories put forward to explain how the increased SO_2 uptake causes injury. He concludes that the most likely explanation was proposed by Bleasdale,[16] who attributed the toxic nature of excess SO_2 to its strong reducing capacity. Consequently, its ability to replace the oxygen in cell components with sulfur alters the nature of the components and thereby disrupts the processes in which the particular component was involved. In healthy plants there is an equilibrium between the reduced forms of sulfur, the sulfydrils, and the oxidized forms, the sulfites. When this equilibrium is maintained, as it is in healthy plants, the formation of amino acids and proteins proceeds normally. However, excessive SO_2 causes a buildup of sulfydrils at the expense of sulfites. This interferes with amino acid metabolism and so affects the production of enzymes and various other proteins. This was the explanation for the effects of SO_2 on garden pea (Horsman and Wellburn[17,18]). Changes in various enzyme levels in pea suggested ways in which certain processes were disrupted. A reduction in the activity of ribulose 1,5-bisphosphate carboxylase was related to the reduction in chlorophyll b, which resulted in a corresponding reduction in the rate of photosynthesis. Increased activity of the enzymes glutamate-pyruvate transaminase and glutamate-oxaloacetate transaminase reflected disturbance to amino acid metabolism, while increased peroxidase activity was a direct indication of increased respiration, which commonly occurs in diseased plant tissues.

Keller and Schwager[19] noted a correlation between reduction in ascorbic acid content of plants and increased susceptibility to sulfur dioxide exposures as well as to a number of fungal diseases. Factors such as shading and exposure to oxidizing air pollutants such as ozone or PAN caused reductions in ascorbic acid.

Excessive SO_2 intake impairs photosynthesis and increases respiration rate in most plants. Consequently, on the one hand, SO_2 impairs the plant's ability to manufacture sugars, while on the other hand, it hastens the use of sugars already in store. Sugars produced during photosynthesis are essential to the plant: first, as the basic building material for all plant components, and second, as the energy supply used to drive the many chemical processes which constitute the life of the plant. The net result of SO_2 pollution is therefore a slowing down of the growth rate and eventually a smaller, less productive plant.

An early sign of injury at the cell level is breakdown of cell structure. Chlorophyll leaks from the chloroplasts and becomes dispersed in the cell cytoplasm. Loss of chlorophyll explains, in part, two additional symptoms: first, there is the reduction in photosynthesis in which chlorophyll is essential, and second, the loss of chlorophyll means a loss in color, so that affected plant parts turn yellow (chlorotic) as the chlorophyll content is reduced. Such damage does not occur uniformly throughout the leaf. Consequently, chlorotic patches develop, usually between veins. If disruption is severe enough or prolonged enough, many of the worst affected areas die (become necrotic). The type of necrosis caused by SO_2 is noteworthy because it is initially white and appears scorched. This differs from most other types of disease, in which the necrotic tissue is usually light to dark brown from the beginning. In SO_2 injury, the necrotic tissue usually begins as near white and may remain so or later become light tan to brown (Darley and Middleton[20]).

Detailed descriptions of the types of injury and the symptoms of injury caused to plants by SO_2 are available (Barrett and Benedict,[21] Treshow,[15] Heck and Brandt[3]). Heck and Brandt direct attention to other agents which can cause similar plant injury symptoms. It should always be borne in mind that several agents may be responsible for most symptoms. It should be noted that although SO_2 does affect pollen, ovaries, and seed in some gymnosperm species (Karnosky and Stairs[22]), it does not usually cause damage to the reproductive system of angiosperms except when very high concentrations (100 ppm) prevail (Treshow[15]).

Barrett and Benedict[21] provide a list of plant species sensitive to SO_2, which includes most common vegetable species, all cereals, a number of forage legumes, and some ornamental and fruit trees. Generally, most forest tree species are either moderately affected or resistant, with the exception of some gymnosperms, which are fairly sensitive. This is also the case in Australian tree species as far as is known, including two species of *Callitris* which are native gymnosperms, although no attempt has been made to study the effect of SO_2 on reproduction in the *Callitris* spp. (O'Connor et al.[23]).

2.2.4.2. *Indirect Effects*

At the beginning of this chapter it was stated that many factors interact during the evolution of a plant community and that the resulting interactions eventually define the character of the community itself. The continued interactions and equilibrium between these factors and the species making up the community constitutes an ecosystem. The introduction of any foreign factor into an ecosystem is liable to bring about additional interactions, which may then interfere with the balance and stability of the ecosystem, with the result that the nature of the plant community can change. The reduction in the number of individuals of a particular species through the direct action of a pollutant could cause such a change, but so could the indirect effect of a pollutant on aspects of the environment that control any aspect of the life cycle of one or more species in the community. A change in one species in a community may cause no further change, or it may start a chain reaction leading to considerable change. An example of what can happen is given in a second paper by Gorham and Gordon.[24] On studying the species present in a series of small lakes in the vicinity of the iron sintering plant at Wawa, which released 100,000 tons of SO_2 annually, Gorham and Gordon noted a number of changes in the lakes as they approached the smelter. There were 24 species of aquatic plants and algae in most lakes 15 miles or more distant from the smelter, and the pH of the water was almost neutral. There were numerous species of aquatic animals, including fish and a variety of aquatic and other birds and land animals dependent on the lakes for a living. As they approached the smelter, the pH of the water became progressively more acid and the levels of hydrogen cyanide, calcium, and other minerals increased, while the number of species of both plants and animals associated with each lake became fewer. It was evident that several important interactions in the lakes' ecosystems had been disrupted and that as more interactions were affected greater changes were apparent in the animal and plant communities dependent on the lakes.

There are many examples of how pollutants can cause changes by indirect action. One is the change in the pH of soil. Although many soils have a strong buffering capacity against changes in pH, it is usually possible, by adding enough lime, to raise soil pH. To raise the pH of one soil from 5.5 to 7 may require the application of 5 tons of lime per acre, whereas an application of only 2 tons per acre may achieve the same change in a different soil. This difference can be seen as a crude measure of the difference in the buffering capacity of the two soils. In a similar way the application of SO_2 to soil will lower pH, but the extent of any change will depend on the amount of SO_2 applied and the buffering capacity of the soil. In many cases there may be no change at all.

A change in soil pH is liable to have both direct and indirect effects on plants. For example, under some conditions, especially in acid soils, the lowering of pH increases the availability of heavy metal ions to plants. This is a direct action that may be an advantage or disadvantage, depending on circumstances. Many of the indirect changes, however, may be more important. In general, bacteria are favored by moist to wet alkaline conditions, whereas fungi grow best in dry to moist, slightly acid soils. There are, of course, some exceptions to the rule. Since most, if not all plants derive some benefits from microbes growing in association with their roots and in the adjacent soil, changes in the populations of bacteria and fungi caused by changes in soil pH are likely to affect plants in some way. The effect will depend on the nature of the change and the degree of dependence of the plant on particular organisms. Sometimes there is an obligatory requirement on the part of the plant for the association with particular bacteria or fungi. The importance of nitrogen-fixing bacteria in the genus *Rhizobium* to leguminous plants is generally known. Since 1914 it has been known that SO_2 fallout can lower soil pH sufficiently to reduce the populations of bacteria in soil (Treshow[15]). It is also known that the nodulation of legumes under acid conditions is less effective than under neutral to slightly alkaline conditions. Although the resultant reduction in nitrogen fixation initially affects the legumes, it represents a generally lowered availability to the whole community of plants in which the legumes are a part. Other kinds of association are equally important to plants. The occurrence of mycorrhizal associations with most plants improves their ability to take up minerals from soil. Factors that impair mycorrhiza formation are likely to reduce plant vigor or in some cases (e.g., some *Eucalyptus* and *Leptospermum* species growing in poor soils) can halt growth altogether. According to Slankis,[25] whereas the lowering of soil pH may favor mycorrhizal fungi, increased sulfur levels can be detrimental.

In a less specific way, however, the vast populations of soil microorganisms, including bacteria, fungi, and insects, are responsible for the recycling of nutrients, such as nitrogen, phosphorus, and minerals. Interference with these natural cycles through the action of pollutants is liable to reduce soil fertility and thereby reduce plant vigor. Wainwright[26] has found that leaves, litter, and soil adjacent to areas of high sulfur dioxide pollution are rich in species of sulfur-oxidizing bacteria and fungi. The significance of this is difficult to interpret, but it does indicate a possible change in the composition of the soil microflora.

2.2.4.3. *Interactions*

There are three kinds of interaction, which will be mentioned briefly. First, there is the direct effect of SO_2 on the ability of the plant to resist attack by pathogens and pests (Sinclair,[27] Heagle[28]). On the other hand, SO_2 can also interfere with the ability of fungi to attack plants. For example, SO_2 reduces the ability of rust fungi to attack cereals and lessens the effects of various ascomycetes, including powdery mildews, to attack their host plants. The generally inhibitory effects of SO_2 on fungi are reviewed by Nash.[29] Trees in which vigor has been reduced by exposure to SO_2 are rendered more liable to attack by root-rotting fungi such as *Armillaria mellea* than are fully vigorous trees. Unfortunately, although most fungi are sensitive to SO_2, the levels required to inhibit their growth are generally detrimental to plant health.

The second type of interaction is the synergistic effect of combined pollutants. Sulfur dioxide and ozone, each at concentrations below the levels required to cause injury when either is the only gas present, can cause acute injury when mixed (Sinclair,[27] Heagle[28]). This creates two sets of problems. It makes it difficult to identify the effects of the various pollutants in the field, and it undermines any attempts to set "safe" levels for pollutants in the environment.

The third type of interaction is that between the plant and prevailing weather conditions, which can determine the degree of sensitivity of the plant to a pollutant at a particular time. Factors such as light intensity and relative humidity strongly influence the regulation of stomatal aperture and thereby influence processes such as photosynthesis and respiration rate, which depend on gas exchange. Plants are most susceptible to SO_2 injury when light intensity is increasing and when relative humidity is high. It is unfortunate that conditions most conducive to rapid plant growth are also those which predispose plants to SO_2 injury. Such conditions are most likely to occur early in the day and in spring (Hiblen[30]).

2.3. OZONE

Ozone is a constituent of photochemical smog. It is a powerful oxidizing agent composed of three oxygen atoms. It has been regarded as the most important plant pollutant in some parts of the world, including Northern America, and although this claim is now challenged by the importance of PAN, there is no doubt that in areas of high light intensity and high motor car densities it is a most important hazard to plant health. Ozone is a natural part of the environment; the normal background level at the earth's surface is considered to be about 2 to 2.5 pphm (parts per hundred million). In areas where ozone levels rise above the background level, as they do in many areas, plant health problems arise, crop yields are reduced, and economic losses result (Brandt and Heck,[11] Ormrod[31]). Ozone is also present as a layer in the upper atmosphere, where it is feared that water vapor from high-flying jet aircraft may damage the ozone layer and thereby increase radiation at the earth's surface. Consequently, there is a paradoxical concern about this pollutant, because it is apparent that there is too much in many places on the surface of the earth and it is feared that there may be a reduction of the zone in the stratosphere.

2.3.1. Critical Exposure Levels and Times

Only extremely sensitive species are affected by exposure to ozone levels less than 5 pphm. It usually takes 5 to 12 pphm of ozone from 2 to 4 hr to injure sensitive species such as lucerne, cereals, potatoes, and tobacco (Hill et al.[32]). These levels are common and are often sustained for 12 hr or longer per day in some cities and the surrounding countryside. In Los Angeles the daily maximum level of ozone has reached 100 pphm. The high generation rate and the nature of the surrounding topography has led to damage from the city's ozone affecting vegetation over 100 miles from the city. Feder[33] reports several studies which demonstrate that plant growth is suppressed when plants are grown to maturity in air with an ozone content of 8 to 10 pphm. Consequently, large areas of forest and farmland have been affected by the large-scale generation of ozone in cities.

Miller[34] refers to damage to forests 60 miles inland from Los Angeles being affected by ozone concentrations generated in that city. Such reports are not uncommon in advanced countries with high population densities. Even in countries where populations are generally lower, but where large cities occur, similar findings are being made. Until recently, Australians did not consider that ozone concentrations were sufficient to cause problems outside large metropolitan areas (O'Connor et al.[35]). Ozone concentrations within cities had exceeded 15 pphm on few occasions in Sydney or Melbourne and then for only a few hours. More recent work, however, suggests that ozone may accumulate in areas outside the metropolitan areas. For example, ozone generated in Sydney may be taken out to sea for several hours and then, as the wind direction changes, be blown back on to the city. When this occurs, ozone levels of 38 pphm have been recorded.[36] The effects of these concentrations on vegetation in the city and the farmlands beyond have yet to be assessed.

2.3.2. Direct Action of Ozone

Ozone is taken into the plant in contaminated air and it permeates between the spongy and palisade mesophyll cells (Figure 2.3). It affects different types of cells in different ways. In exposed leaves cells rapidly lose turgor, so that the leaf appears flaccid and wilted (Treshow,[15] Krause and Weidensaul[37]).Stomates close, this being analogous to a wilted plant. An early sign of ozone injury is a discoloration and waxy appearance of the upper leaf surface which may disappear entirely when exposure ceases (Taylor[38]). Another early sign is the closure of stomates. It was considered that the reduction in gas exchange and transpiration resulting from stomate closure was responsible for the reduction in photosynthesis and general depression of growth (Treshow,[15] Duggar and Ting[39]). However, the degree of injury is not correlated with numbers of open stomates (Treshow,[15] Duggar and Ting[39]). Stomatal regulation is associated with the level of potassium in the leaf. In tomatoes, potassium deficiency reduces ozone damage compared with plants grown in sufficient potassium. Exposure to ozone, on the other hand, increases the uptake of potassium in both deficient and healthy plants (Leone[40]). Since potassium uptake is associated with stomatal opening,[40] this work suggests that the initial closing of stomates because of exposure to ozone is counteracted by the stimulated potassium uptake, which then causes an opening of stomata. There is too little work done to show if this is the exception or the rule. Disks of bean leaf respond in the reverse manner to tomato when exposed to ozone (Evans and Ting[41]). It is likely that different species react in different ways according to which of their processes are most sensitive

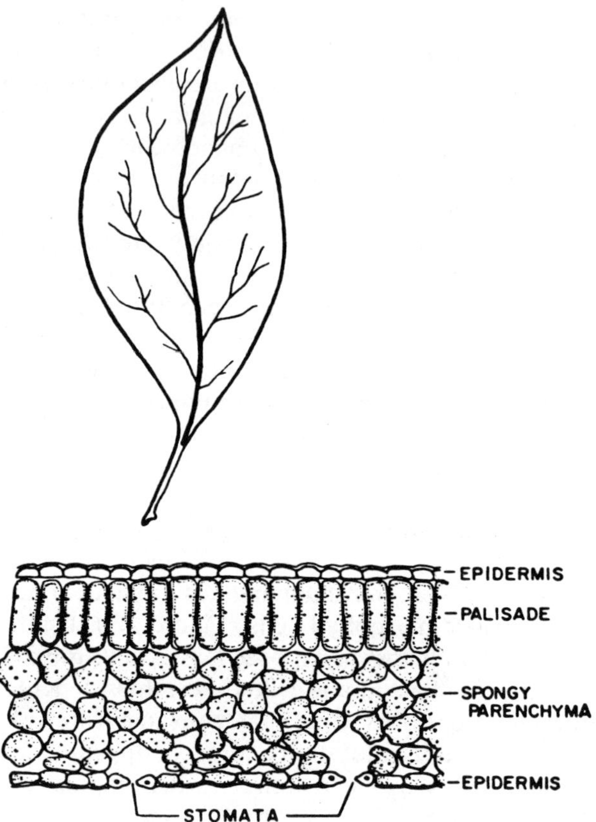

Figure 2.3. Simple diagram of a section through a leaf showing air spaces between cells and the stomates (Brandt and Heck[11]).

to interference by the presence of ozone. It is possible that Leone's[40] results explain in part the results reported by Heagle and Heck[42] that one exposure of tobacco plants predisposes them to injury by a second exposure to oxidants. Heck and Brandt[3] also discuss the question of the interaction of pollutants and stomatal behavior and agree that it is complex but not clearly understood.

A considerable amount of work has been done on the influence of ozone on a range of metabolic pathways in plants. This is reviewed by Mudd.[43] Damage to cell membranes appears to be a universal injury in plants. Spotts et al.[44] reported a 25% reduction in the sterol content of pinto beans exposed to ozone and regarded this as a likely explanation for the changes in structure and function of cell membranes. Ozone damage to membranes causes leakage of electrolytes and water into intercellular spaces and results in a water-soaked appearance of damaged tissue. Ruffner et al.[45] reported similar disruption to the chloroplast membrane, the plasmalemma, and the tonoplast in injured bean. Such changes would impair photosynthesis and other important metabolic processes in cells. Not all tissues in plants are equally sensitive to ozone. For example, palisade mesophyll is more sensitive than is spongy mesophyll (Treshow[15]).

At sublethal levels, short exposure to ozone may cause a temporary decrease in the photosynthetic rate, which may then gradually return to normal as exposure ceases. However, as the suppressing dosage increases or is maintained longer, the plant takes longer to recover, until eventually full recovery does not occur. In reducing photosynthetic rate, ozone lessens the ability of the plant to produce its building material and to store energy to drive its life processes. The energy store is further reduced by the eventual increase in respiration rate in injured tissue. On exposure to ozone, the respiration rate initially drops but as cell injury becomes apparent, the rate increases beyond that of the healthy plant (Treshow[15]).

Ozone injury can also impair growth and reproductive processes. Feder[46] found that in carnation and geranium there was a reduction in side branching, a retardation of flowering, and a reduction in the numbers of flowers produced as a result of daily exposures to 10 pphm of ozone for 5 to 7 hr during a 3-month period. He also noticed a reduction in leaf size, an increase in internode length, progressive leaf destruction, and finally, defoliation. In petunia, flowers were fewer and later developing than in healthy plants. It is likely that there are many subtle effects which add up to a considerable combined reduction in growth and development. This may be best summed up by noting that the value of agricultural crop production losses attributed to ozone damage in California alone exceeds $100 million per year.

Ethylene is a normal plant component with an important role in the regulation of many processes in plants, including fruit ripening, senescence, and leaf fall. Exposure of plants to ozone induces an increase in ethylene production and causes symptoms of accelerated aging (Craker[47]). The symptoms described by Craker attributed to increased ethylene synthesis in ozone treated plants are very similar to those described by Feder[46] in carnation, geranium, and petunia, and they are similar to symptoms reported in plants maintained in flowerboxes in central Melbourne.

2.3.3. Indirect Action of Ozone

While the chemical reactions involved in the generation of ozone may induce a series of further reactions which lead to the formation of peroxyacetyl nitrate, nitrogen dioxide, and so on, thereby changing the aerial environment of the plant (Schuck[48]), there is little, if any, information about changes that ozone might cause in the soil environment. Yet it seems unlikely that changes do not occur in soils exposed for long periods to air highly charged with ozone. The relationships among the soil environment, microbial populations, and plants are important to plant health; they are complex and are difficult to study. They are not generally well understood. Even so, the following discussion suggests the kinds of changes likely to be of significance to plants.

2.3.3.1. Ozone and Saprophytes

Bacteria, fungi, and other microorganisms associated with plant surfaces can influence the success or failure of potential parasites and can influence nutrient uptake by plants. Consequently, by altering the balance between interacting microbes, it is possible to affect the benefits to the plant. Ozone inhibits the growth of some bacteria (Somov et al.[49]) and fungi (Somov et al.,[49] Treshow et al.[50]). Manning et al.[51] found that chronic levels of ozone over 4 weeks increased the numbers of fungi on bean roots. This may be partly due to the death of many roots and partly to the increased leakage of amino acids and sugars from the injured roots. Little consideration has been given to the effects of introducing ozone into these complex systems, although some indication can be found in a review by Emmett and Parbery.[52]

2.3.3.2. Ozone and Pathogens

Pathogens are microorganisms that cause disease. Ozone does often influence the behavior of pathogens by reducing their capacity to invade their hosts. This is true for many rust fungi, powdery mildews, and various other ascomycete fungi (Heagle[28]). Even though a pathogen may be tolerant of it, ozone may still influence the pathogen's capacity to infect its host. The fungus causing powdery mildew of lilac is tolerant of ozone, but the fungus that attacks the mildew fungus, thereby helping to contain its spread in a clean atmosphere, is highly susceptible to ozone, which could lead to increased mildew attack of lilac under polluted conditions (Hiblen and Taylor[53]). Sinclair[27] reports that the presence of photochemicals increased the damage done to ponderosa pine by the mountain pine beetle. Similarly, orchard and forest trees weakened by exposure to photochemical smog become more susceptible to attach by root-rotting fungi such as *Armillaria mellea* (Miller[34]). This is possibly because one effect of ozone on plants is to induce senescence and to reduce vigor. Such changes in plants, regardless of cause, frequently favor the entry of pathogens. This is by no means a general rule. Two papers have independently shown that ozone injury reduces infection in lucerne by *Xanthomonas alfalfae* (Howell and Graham[54]) and in *Pelargonium hortorum* by *Botrytis cinerea* (Krause and Weidensaul[37]). Treshow et al.[50] found that whereas ozone depressed the activity of some pathogenic fungi, it stimulated others. When *Colletotrichum lindemuthianum* and *Alternaria oleracea* were grown for 14 days in an atmosphere of ozone, growth and sporulation were depressed in *C. lindemuthianum*, whereas sporulation was increased in *A. oleracea*. By altering the rate of spore production in each of these fungi, ozone would influence the rate of disease spread in the field.

2.3.3.3. Ozone and Symbionts

Mutualistic symbionts are two organisms that live in close association to their mutual advantage. To a greater or lesser degree, each partner in the relationship depends on the other for its survival. Such relationships are much more common than is generally understood. Two main groups are considered here. The first are the nitrogen-fixing nodules formed on the roots of legumes by bacteria in the genus *Rhizobium*. The bacterium derives an assured supply of carbohydrate from the legume and in return supplies the nitrogen needs of the legume. Assured nodulation of legumes is as important in native plant communities as in agricultural crops. Manning et al.[51] showed that sustained low levels of ozone, 0.1 to 0.15 μl l^{-1}, for 28 days reduced root development and prevented nodulation altogether in pinto bean. Even a single 1-hr exposure to 75 pphm was enough to reduce root growth and to cause a slight reduction in nodulation (Tingey and Blum[55]).

The second group of symbionts are the mycorrhizal fungi, which form close associations with plant roots, which are called mycorrhizas. The fungus gains an assured energy supply from the plant while the plant gains an extremely efficient mechanism for the uptake of minerals, especially phosphorus. This frequently enables plants to survive in habitats that would not be available to them in the absence of the mycorrhiza. Approximately 89% of all plant species form mycorrhizas with fungi. These associations occur in at least some species in all but about four plant families. Many species that are found in poor sandy soils appear to have an obligatory requirement for their mycorrhizal associates. Although mycorrhizas have been known and studied for about 100 years, it is only in recent years that interest has been shown in investigating their potential for improving crop growth. However, such interest has led to only one study (Carney et al.[56]) of the influence of ozone or other oxidants on mycorrhizas. Yet there is good reason to expect that ozone does influence mycorrhiza development and function. First, mycorrhizal fungi are biotrophic in the same way as rusts and powdery mildews which live in living host tissue. With very few exceptions, such fungi are adversely affected by exposure to ozone, either because of the direct effect of ozone on the fungus or because exposure of the host plant to ozone renders it less suitable for colonization by the fungus. Second, ozone interferes with photosynthesis and reduces the availability of sugars in plants. Since a lack of sugar in the root system of a plant reduces its mycorrhiza development, it appears that this and other metabolic changes induced by ozone in the host plant would render it less suitable for the mycorrhizal fungus. Miller[34] implies support for this view. Slankis,[25] in reviewing factors that influence the formation of mycorrhizas, discusses the importance of the presence in and the leakage from roots of amino acids in encouraging mycorrhiza development. Consequently, since changes in amino acid metabolism and/or carbohydrate supply are both important factors in regulating mycorrhiza development and because both ozone and sulfur dioxide can upset each, it seems highly likely that important symbiotic relationships can be upset by exposure to these pollutants.

Carney et al.[56] have studied the effect of the presence of ectomycorrhizal fungi on the reaction of loblolly pine roots to exposure to ozone and sulfur dioxide. They found that the presence of the mycorrhizal fungi afforded some protection to roots against exposure to the two pollutants. On the other hand, they also suggest that exposure to pollutants may reduce the capacity of fungi to form mycorrhizas with roots of trees exposed to pollutants. Thus roots produced after the onset of exposure may be less likely to form mycorrhizas, and consequently would not be protected in any way from pollution effects. These findings support the foregoing view.

Two studies report that some legumes will not develop root nodules in phosphorus-deficient soils until they have first developed satisfactory mycorrhizas: species of *Centrosoma*, *Stylosanthes*, and *Trifolium* (Crush[57]) and *Phaseolus*, *Medicago*, and *Arachis* (Daft and El-Giahmi[58]).

2.3.4.　Interactions

Finally, there are a number of interactions of the host with its environment that can influence its susceptibility to ozone. Saunders[134] introduces an entire number of the journal Environmental Pollution devoted to interactions between various pollutants and between pollutants and microorganisms and considerations of their implications to plant health. First, there is mounting evidence that virus infection protects plants from ozone damage (Brennan,[59] Davis and Smith[60,61]). Bacterial infection may sometimes have a similar effect (Howell and Graham[54]). This has implications in the control of ozone damage in some crops. The deliberate infections of plants with mild strains of virus that produce no symptoms in their host has been advocated as a possible means of protecting some plants from the ravages of severe strains of the protecting viruses (Chamberlain et al.,[62] Stubbs[63]). Should this be acceptable for control of viruses, it may also be considered for protecting against ozone damage. Pinto beans have been protected from ozone by prior infection by bean common mosaic virus (Davis and Smith[61]).

A second notable interaction is the protection of pinto beans from ozone injury by the application of systemic fungicides. Plants sprayed with either benomyl or benzimidazole or related compounds and exposed to ozone were tolerant of ozone when unprotected plants were severely damaged (Moyer et al.,[132] Rufner et al.,[45] Spotts et al.,[44] Pell[64]). Pell found that very young leaves were not protected by benomyl, but that 20-day-old leaves were.

The third set of interactions relate to the physical environment. The time of day, weather conditions, and season of the year may all influence the extent of injury produced by the exposure of plants to similar levels of pollution. Plants kept in the dark immediately prior to exposure are less liable to injury than those at high light intensities (Duggar and Ting[39]). This was considered to be due to stomates being closed and sugar levels being low. Plants also differ in their sensitivity with age (Kress[65]). Individual leaves are most sensitive when they are fully expanded (Treshow[15]). In addition, weather fluctuations, including changes in light intensity, temperature, and relative humidity, all affect sensitivity to ozone. Hiblen[30] has suggested that whereas spring conditions in North America favor SO_2 damage, summer and autumn conditions favor ozone damage. This may not be the case under Australian conditions. White[66] found that ozone injury in the Sydney metropolitan area was apparent on the winter grass, *Poa annua*, between May and September. Suspected ozone damage to poplars in Parramatta Road, Sydney, was observed in October–November (Parbery, unpublished). The highest levels of ozone in Melbourne and Sydney have been recorded in October–November (O'Connor et al[35]). Consequently, in southeastern Australia it appears that the spring is the potential danger time for ozone injury but that it may also occur during autumn and winter. This may be due to the milder, sunnier winters in Australia, which allow active growth of a number of plants.

2.4. PEROXYACETYL NITRATE (PAN)

The formation of ozone is the first step in the formation of photochemical pollutants from the interaction between car exhaust fumes and sunlight. The reaction of ozone with unburned hydrocarbons then causes the formation of a series of compounds known as "peroxyacyl nitrates," of which peroxyacetyl nitrate is highly toxic to plants. It is as widespread as ozone and is an important cause of plant injury in the United States and Japan.

2.4.1. Toxic Levels

Exposure to 1 pphm of PAN for 4 hr is sufficient to cause injury to many plants. PAN concentrations commonly recorded in California and Utah range from 1 to 2 pphm on average but reach a peak of 5 pphm on particularly smoggy days (Treshow[15]). Readings of 5.8 pphm have been recorded in Riverside, California. In Europe levels vary from 0.1 pphm in Milan to 5.7 pphm in Paris (Darley et al.[67]). There have been unconfirmed news reports of levels reaching 7 pphm and 8 pphm in New York and Tokyo, respectively. Even though ozone and PAN have relatively short half-lives and do not accumulate in plant tissue, they are still very widespread in the sunny regions of the world with high densities of motor cars, and together are unchallenged as the most important air pollutants in the United States. PAN levels have not been recorded for Australian cities but are likely to reach toxic levels with about the same frequency as ozone.

2.4.2. Action of PAN

For several years the action of PAN was not distinguished from that of ozone, so it is likely that the silvery leaf surface and other symptoms initially attributed to ozone damage were largely due to PAN injury. In many respects the disruptive action of PAN is similar to that of ozone; however, there are important differences. In the field, while ozone injury causes small flecks or a dark stipple on the upper leaf surface, PAN injury induces a metallic glaze or sheen, ranging from white through silvery to bronze, on the underside of the affected leaf.

Details of the symptoms caused by PAN injury in a range of plants are provided by Treshow[15] and Heck and Brandt.[3] At high concentrations, symptoms of acute injury, including glazing, chlorosis, and necrosis may appear. At lower concentrations, however, chronic injury occurs, which is not so readily detected; PAN causes interference with photosynthesis, respiration, and ion absorption as well as the synthesis of carbohydrates and proteins. Interference with any of these will be reflected in reduced plant growth and possibly reproduction. A measure of chronic injury is again reduced yields in agriculture. The disruptive ability of PAN appears to be related to its interference with DNA synthesis, the constitution of several important enzymes, and the action of the cofactors required to enable enzymes to act on their substrates (Treshow,[15] Mudd[43]).

Although the application of benomyl to pinto beans protected them against the action of ozone alone, it did not protect them from PAN alone (Pell[64]). The two pollutants usually occur in combination in the field and Pell found that benomyl does protect pinto bean against injury from the two gases together. This work[64] emphasizes the need to recognize the different action of these two pollutants and points to the danger of too freely predicting the action of pollutants in the field from experiments done with pure chemicals in the laboratory.

2.5. NITROGEN OXIDES

The status of nitrogen oxides as plant pollutants is not clear. Nitrogen oxides, especially nitrogen dioxide, can cause plant injury when present in high concentrations. However, Treshow[15] stated that up to 1970, no field injury to plants from nitrogen dioxide had been reported in the United States and he regarded it as a minor pollutant. Taylor,[38] on the other hand, regarded it as a major pollutant even though acute injury from exposure to NO_2 was rare. Heck and Brandt[3] note that symptoms of nitrogen dioxide injury are similar to those caused by sulfur dioxide, but that much higher concentrations of nitrogen dioxide are needed to cause acute injury. Lower concentrations for long exposure times may cause early senescence and abscission. Taylor[38] implies that chronic injury, which is so easily overlooked, is much more extensive than realized and attributes this to the symptoms of injury being similar to those caused by ozone damage. However, the conditions under which the two gases act are different. When taken into plants under high-light conditions, ozone is disruptive, even at relatively low concentrations. However, when taken in under high-light conditions, nitrogen dioxide is utilized by the plant provided that it is not present in extremely high concentration. Under low-light conditions, however, it is not assimilated and becomes toxic. A light-dependent enzyme reaction reduces nitrites, formed in the leaf from nitrogen dioxide, to ammonia, which is readily used as a nutrient.

Under low-light conditions this reduction does not occur, and toxic levels of nitrite accumulate in the leaf. A hypothesis to explain the manner in which nitrite accumulation interferes with other processes is proposed by Srivastava et al.[68] There is some evidence that when present in high concentrations with sulfur dioxide, the two gases act synergistically (Bennett et al.[69]).

2.6. DUST

A Senate Select Committee of Enquiry into pollution in Australia in 1969 reported that dust was probably the most important air pollutant in Australia with regard to plant health. The enquiry referred to reductions in fruit production in orchards subjected to high dust loads. The effects of dust on native vegetation does not appear to have been investigated in Australia. However, some work has been done elsewhere on the effects on plant health of dust from roads, extractive and manufacturing industries, and other sources. For example, road dust on leaf surfaces can so alter the absorption of light energy as to lead to increases in leaf temperature (Eller[70]). This would, in turn, affect net photosynthesis.

A comprehensive coverage of dust pollution is provided by Treshow.[15] Even so, there is not a great deal of information available. Dust might be considered as the larger particles in a spectrum of particulates ranging from small suspended aerosols to coarse-grained particles, which rapidly sediment out of air. In determining damage caused by dust, it is important to distinguish between the damage caused by the physical dust particles themselves and any chemicals that may be adsorbed to their surfaces. The damage caused by the physical particles themselves is of various kinds. Particles may clog stomates, thereby reducing transpiration and gas exchange. A thick coating of dust on a leaf can substantially reduce the amount of light available for photosynthesis, in addition to affecting the physics of light absorption (Eller[70]). The presence of dust can alter a plant's reaction to some diseases (Manning[131]). The overall effect is a reduction in growth. Treshow[15] refers to work that has recorded up to 40% yield loss in crops that were heavily dusted and the death of pine trees that accumulated cement dust. Some of the main sources of dust are soot, lime and cement works, other mining operations, and dirt roads.

2.7. FLUORIDE

The harmful effects of atmospheric fluoride have been known and studied for 100 years. The fluoride most damaging to plant health is hydrogen fluoride, HF. Atmospheric fluorides are taken up and accumulated by leaves. The most important source of fluoride is aluminium smelting, although other industries, such as titanium and selenium production and brick manufacture, can also be important (Treshow,[15] Ferry et al.[8]).

Although the level of fluoride in the atmosphere is important in determining the degree of injury caused to plants, fluoride is different from the other pollutants so far considered in that it accumulates in plant tissue. Consequently, provided that fluorides are present long enough for plants to accumulate toxic levels, injury will occur in the zone of accumulation. The concentrations of F in plant foliage can be many thousands of times greater than in the atmosphere (Table 2.1). Fluorides in soil are taken up and concentrated in plants but do not generally cause problems.

Fluoride taken into leaves accumulates toward the leaf tip and around the leaf margins, where it may be from 2 to 100 times more concentrated than in tissue adjacent to veins or toward the base of the leaf. Although the physiological interference caused by fluoride in leaves may be expressed by the appearance of symptoms in fruit and other plant parts, the fluoride itself is not transported from the foliage. Consequently, fruit trees damaged by fluoride may produce fruit with symptoms of fluoride injury, but the levels of fluoride in the fruit are not increased and the fruit is safe to eat. On the other hand, the intake of contaminated foliage by animals or man can lead to severe toxicological problems.

The metabolic disturbances caused in plants by fluoride and the symptoms that they induce are described in detail by several authors (Treshow,[15] Treshow and Pack,[72] Feder,[33] Nash[29]). In dicotyledons the margins of leaves become necrotic and turn brown. In monocotyledons leaf apices die and turn an ivory color. In each group of plants there is usually a reddish-brown zone of tissue separating the necrotic from the living tissue. Many monocots are particularly sensitive to fluoride injury and are useful indicator plants for monitoring toxic levels or fluoride. These include *Crocus*, *Freezia*, *Gladiolus*, *Ixia*, and *Tulipa*. Species of *Gladiolus* and *Freezia* are particularly valuable indicator species.

Table 2.1.   Effect of Airborne Fluorides on Yield of Valencia Oranges (Treshow[15])

	Foliar F concentrations (ppm)	Ambient F concentrations (ppb)[a]	Fruit yield (lb)	Fruit yield (%)
Filtered air	10	2.1	377	100
Unfiltered air	71	6.3	297	79
Outdoor checks	197	6.9	107	28

[a] Parts per billion (1000 million) is expressed as ppb.

Fluorides affect plants by combining with the metal components of proteins or by other inhibitory interactions with them. As protein inhibitors they may interfere with such metabolic processes as respiration, photosynthesis, carbohydrate synthesis, cell wall composition, energy balance, and the synthesis of nucleotides and nucleic acids (Treshow[15]). The biochemistry of this interference is reviewed by Nash.[29] Not all exposures to fluoride are detrimental to plants since several reports (Treshow[15]) refer to the stimulation of growth in the presence of atmospheric HF, in one case even when leaf damage was apparent. Nevertheless, most commonly, fluorides cause injury accompanied by reductions in growth. They can also depress yields of fruit. McLaughlin and Barnes[73] found that fluoride depressed dark respiration and photosynthesis in various South American trees and that the effects were greater in softwood than in hardwood species. The uptake of fluoride causes decreased uptake of manganese and magnesium in *Abies alba* while increasing calcium uptake (Garree et al.[74]). This could explain the disruption of enzyme and protein synthesis (Nash[29]).

Toxic levels of fluoride are not as readily determined as for some other pollutants. In addition to the influences of variation in species, period of exposure, environmental conditions, and atmospheric concentrations of fluorides, there is the ability of each species to accumulate fluoride. The information in Table 2.1 gives some appreciation of the complexities involved. These data suggest three things: (1) that plant foliage has a marked capacity to take up fluoride from air; (2) that increasing the atmospheric fluoride concentration increases the plant's capacity to take it up, but that (3) the rate of uptake by plants grown under the particular glasshouse conditions was less than, although proportional to, the rate of uptake in the open atmosphere. In this case it appears less by a factor of approximately 2.5.

Many lichens are sensitive to fluoride and may be useful as indicators of fluoride pollution (Gilbert,[10] Richardson and Young[9]). It seems that fluoride has similar physiological effects on the algal partners in lichens as it has on higher green plants. Little research has been done so far and there does not seem to be any information regarding its effect on the fungal partner in lichens. It is known, however, that must fungi are sensitive to fluoride salts, which are frequently used as fungicides for preserving wood. There is some evidence (Heagle,[28] Manning[71]) that plants exposed to fluoride are less susceptible to some diseases than plants which are not.

One of the main problems with fluoride contamination of plants is that animals fed on contaminated fodder or pasture develop toxicity symptoms which can be very serious and sometimes lethal if large volumes are eaten.

Man is more fortunate than the grazing animal since for the most part he eats nonfoliar parts of plants. In common with the heavy metal pollutants, fluoride enters the human and animal food chain through plant uptake (Treshow[15]).

2.8.   HEAVY METALS

Many of the heavy metals, such as cadmium, copper, iron, mercury, nickel, lead, and zinc, become toxic to plants and animals once they exceed the levels the species have evolved with and can cope with. Several are essential to plant and animal health in trace amounts but become toxic at higher levels. Many plants appear to be less susceptible to heavy metal toxicity than the animals, including man, that feed upon them. Consequently, although some of the metals referred to here are not generally toxic to plants, they will be considered because of their contamination of plants, by which means they find their way into various animal food chains.

Plants have evolved to cope with the levels of various heavy metals that have occurred in their environment during evolution. In some cases the metals, such as iron, copper, cobalt, and zinc, have become essential for incorporation into enzymes and proteins. Thus at normal concentrations they are essential to plant and animal life. However, plants and animals have evolved mechanisms for excluding unwanted metals and for regulating the supply of the essential ones. Consequently, there has developed a concept of the "biological half-life" of metals in animals. In man the half-life of the three metals to be considered here are methyl mercury, 0.2 year; other organic forms of mercury, 0.8 year; lead, 4 years; cadmium, 16 to 33 years. It might be expected that the half-life of the metal is inversely correlated with the natural levels of the element in the environment during evolutionary time. Organisms have evolved mechanisms for coping with or tolerating the levels of heavy metals in their environment during their evolution. When environmental levels rise above these levels and exceed the threshold for tolerance in the organism as a result of accumulation, toxic effects become apparent.

2.8.1.   Lead

"The body burden of lead is now much greater in man than in his modern ancestors and as one source of lead is food, its acquisition, mobility and effects in plants and animals, as well as its partitioning among air, soil, water and biotic fractions must be understood" (Rains[75]).

During the 70-year history of the motor car, the lead content of the average American is claimed to have risen 125-fold. Today it is said to be almost at the critical level of tolerance (Martyn,[76] Huisingh[77]). Lead is a severe poison. Chisolm[78] states that lead poisoning in children is a major source of brain damage, leading to mental deficiency and behavioral problems. Lead poisoning is difficult to diagnose. High amounts of lead can be stored in tissues, and damage done during such storage is not understood. Lead poisoning in children is often followed by kidney disease in later life. Lead poisoning causes various disorders of the general nervous system, including general irritability. Gout was common in America during the prohibition era, when much of the "moonshine" was distilled in lead equipment (Chisolm[78]). Ahlberg et al.[79] reported evidence of organolead compounds causing genetic deficiencies in onions and fruit flies. Shaar et al.[80] have reported lead accumulation in the nuclei of mosses. Schubert[81] reports chromosome breakage in victims of lead poisoning. There is much evidence that lead does affect behavior in plants and animals.

A high proportion of lead pollution comes from the use of tetramethyl- and tetraethyllead, used in gasoline. In the United States, 80% of all lead pollution comes from that source and it is introduced to the environment as an air pollutant (Huisingh[77]). For every 30 m from the source of emission, there is a 30% reduction in the atmospheric load of lead deposited on the surrounding soil and vegetation. This pattern of distribution is fairly typical of that for other heavy metals emitted from a point source. Many studies of roadsides have been made. Fairly typical of them is that made by Chow.[82] He drew the following conclusions from his study:

1. The greatest amounts of lead occurred nearest the roadsides of the oldest and busiest highways.
2. The greatest accumulation occurs in the upper few centimeters of soil.
3. Most of the 300 to 800 ppm of lead found in the roadside grasses was there not merely as deposit, but as accumulation within the grasses.

This last point, however, has not been universally accepted. From the relatively small amount of work done it is evident that plants vary greatly in their ability to take up lead. Fungi, lichens, and mosses appear to take it up fairly readily, whereas higher plants vary greatly both in their uptake and in the way they accumulate it. Quinche et al.[83] and Zuber et al.[84] made extensive studies of roadside plants in Switzerland. They showed that plants which produced leaves with a large surface area/volume ratio trapped more lead than those with a smaller surface area/volume ratio.

However, they also demonstrated uptake. They reported greater accumulation of lead on fruit and leaves than in roots but concluded that while the bulk of lead accumulated on plant surfaces, some was taken into the plants and accumulated in the tissues. Baumhardt and Walsh[85] reported that in maize there was accumulation in the foliage but not in the grain. In radish, most lead remains in the root cortex, and of the little that passes the endodermis, most remains in the vascular tissues (Lane and Martin[86]). Huisingh,[77] in reviewing the implications of heavy metals in agriculture, states that dicotyledons take up lead more readily than monocotyledons and reports that vegetation in the vicinity of a smelter contained 20 to 200 ppm of lead, 120 to 150 ppm being considered toxic to horses and cattle. Lead is taken up and accumulates in the outer tissues of at least one edible fungus when growing along roadsides (McCreight and Schroeder[87]). Man stores lead very efficiently (Chisolm[78]). The ability of cattle, sheep, and other flesh animals to store lead is not clear. Zuber et al.[84] found that cattle fed on lead-laden hay excreted most of the lead in feces and urine.

The intake of lead in contaminated food is a potential problem for man. Some food species do accumulate lead in edible parts (McCreight and Schroeder,[87] Lane and Martin[86]), although many do not appear to do so. Leaf vegetables could present a health problem since aerial plant parts can accumulate lead as a surface contaminant. It is of concern in fodder grown near roadsides and consumed by animals. Rains[75] investigated the deaths of 13 horses in the Benicia Vellejo, where they were fed oats containing 700 to 1000 ppm of lead. Rains concluded that at certain times of year the oats took up large amounts of lead and that this explained the only abnormal finding regarding the dead horses—they contained very high lead levels. Whether the lead was taken into the fodder or was merely a surface contaminant (which it was not) would seem unlikely to matter to the horses! Small native mammals trapped close to roadsides, whose diets include plant material, have been found to have a body burden of lead two to four times that of animals living away from busy roadsides (Lowell et al.[88]). Some plants, such as tobacco, do develop symptoms of leaf narrowing as a result of excessive lead uptake (Huisingh[77]). Lead, cadmium, and other metals, taken up in fairly low concentrations, interfere with stomate function and thereby influence gas exchange and photosynthesis in sunflower (Bazzar et al.[89]). However, many plants appear to accumulate lead without severe disruption.

One encouraging report comes from Denmark (Salgaard et al.[90]), where the lead content of cereals has dropped progressively from 1962 to 1977 because of a change in the process of producing nitrogenous fertilizers with lower lead content. Lead content in cereals grown by roadsides, however, was still found to be high.

2.8.2. Mercury

A wide range of inorganic and organic salts of mercury has been used in agriculture for the control of plant diseases for many years. Substitutes for mercury to control some diseases, such as bunts, are difficult to find. However, because of the extremely toxic nature of mercury and the apparent buildup of mercury in the general environment, it is now widely restricted in agricultural use.

The main effects of mercury and other heavy metals taken up by plants in excessive amounts is that they interfere with the metabolic processes involving iron and other essential metals. However, the overriding effects are that mercury enters the food chain of animals, including man, with potentially serious consequences. Its effects on seed-eating birds (Mellanby[91]) are well known, as are its even more serious effects on birds of prey (Fimeite,[92] Borg et al.[93]). That mercury-contaminated plant material fed to animals can seriously affect humans who eat the contaminated flesh is well established (Caldwell[94]).

Plants differ in their ability to take up mercury. Rice treated with mercury as a seed dressing to control diseases contains 50 times the amount of mercury as that contained in untreated rice (Caldwell[94]). On the other hand, wheat does not take up appreciable amounts of mercury applied as a seed dressing. Methyl mercury hydroxide taken up by maize causes severe stunting, accompanied by disruption of mitochondria and interference with respiration (Lipsey[95]). Several mercury salts are toxic to a wide range of plants.

2.8.3. Cadmium

When cadmium is present in low concentrations in soil, even though it may be taken up by some species of plants, it rarely appears to cause detrimental effects in the plant itself. However, since cadmium is a severe, accumulative toxic metal in animal health, its accumulation in food plants is of concern. Some plants take up cadmium more effectively than do others. Haghiri[96] noted, when investigating the ability of various vegetables to accumulate cadmium from soil, that the following accumulated the most, in order of greatest amount to the least: lettuce, radish tops, celery stalks, celery leaves, green pepper, radish roots. He also found that increasing applications of cadmium to soil caused increased uptake of cadmium by soybeans and wheat accompanied by decreased yields. Jones et al.[97] obtained similar results with soybeans, and Turner[98] found that increased applications of cadmium to soil increased the uptake of zinc as well as cadmium. Lowering soil pH increases the uptake of the cadmium as well as that of other heavy metals by plants (Tyler,[99] McIlveen and Cole[100]).

Coughtrey and Martin[101] have found that clones of the grass *Holcus lanatus* L. growing in the vicinity of an area receiving high concentrations of cadmium fallout had been selected for tolerance and would grow better in cadmium-contaminated soils than clones from an area free from cadmium pollution. They discuss other work, which indicated that heavy metal tolerance, which is known in several grasses and microorganisms, requires large amounts of energy to maintain and that even under normal conditions of growth, tolerant clones are slow-growing and produce less biomass than do nontolerant clones.

2.8.4.   Indirect Effects of Heavy Metals on Plant Health

Apart from having serious implications in animal and human health, by their entry into food chains through the plants we eat, heavy metal pollution can affect plant health indirectly by interacting with microorganisms. It is known that cadmium, nickel, and lead all affect fungal growth and reduce mycorrhiza development in plants exposed to them (McCreight and Schroeder[102]). Similarly, the undue uptake of cadmium, cobalt, copper, and zinc by red clover plants reduces nodulation (McIlveen and Cole[100]). Consequently, by reducing the benefits derived from symbiotic organisms, heavy metals may have a more marked effect on plant health than by their direct effects on plants.

One of the most worrying aspects of heavy metal pollution is associated with the liberation of methylated and ethylated forms of lead and mercury into the environment. Because of biotransformation of these alkyl forms by microbes in soil and water, the metals are kept mobile in natural cycles, whereas nonalkyl forms are relatively fixed. The alkyl forms, not normally present in soil and water, are much more readily taken up by plants and animals than other forms of these metals (Huisingh,[77] Jernelov and Martin[103]). This change of the form of heavy metals in the environment is regarded as one of the most serious threats to plants and animals resulting from heavy metal pollution.

2.9.   AGROCHEMICALS

The ever-increasing demands on agriculture to produce more food and raw materials for textiles and other uses place increasing pressure on agricultural industries to increase their productivity. Since it is frequently inappropriate to achieve this through increasing land areas in use, increased productivity is most effectively achieved with the aid of chemical inputs in the form of fertilizers, herbicides, pesticides, and fungicides. Before considering the pollution problems created by the use, or abuse, of agrochemicals, it is necessary to remind ourselves of the benefits of agrochemicals, to keep the importance of these problems in perspective.

The large scale of crop production needed to support the world population can be maintained only by the replacement of chemicals removed from the system at harvest, by the addition of fertilizers prior to or at sowing. Much research is at present directed toward improved efficiency in fertilizer use and better exploitation of natural means of increasing fertility. It is never likely, however, that world agriculture will be independent of some form of chemical input.

Increased production is gained at the expense of removing various natural buffers and barriers against pests and diseases. Thus the most productive systems of agriculture are generally the most prone to destruction by pests and diseases. Consequently, crop protection has to be managed by various systems of biological and chemical control, and increasingly by integrated biological and chemical control. It is highly unlikely that chemicals can ever be dispensed with for plant protection, although in many cases their use may be diminished.

There are many advantages to be gained by the increased use of herbicides in agriculture. Even though Roman farmers were aware of it, few today appreciate that in many soils, the only reason for plowing is for the control of weeds. The long-term plowing of many soils, however, destroys their structure, leads to considerable erosion, and reduces their productivity. Current investigations show that just as good a crop of wheat and other crop plants can be obtained on unplowed land provided that weeds are controlled. Weed control by use of chemicals therefore has several important advantages. It maintains good soil structure, reduces erosion, can slow the decline in fertility, and allows farmers to sow sooner after rain than they otherwise can. As pointed out by Gunn,[104,105] both sides of the question of agrochemical use have to be carefully considered when public concern is aroused. The banning of agrochemicals when there are no sound scientific reasons for doing so is as irresponsible as the careless use or abuse of them. The problems of nonuse frequently outweigh the problems of use. The complete ban on the use of mercurial fungicides and of DDT are cases in point, where regulated use could be beneficial.

2.9.1. Fertilizers

Of greatest concern at present are nitrogenous fertilizers. In some parts of the world, such as Australia and New Zealand, the widespread use of nitrogen-fixing legumes in most agricultural systems has minimized the use of nitrogenous fertilizers. In such systems, there does not appear to be significant loss of nitrate from soil into water courses.

In other parts of the world, different practices have been adopted and highly soluble forms of nitrogen applied as fertilizers are used extensively to promote high crop yields. In areas where such fertilizers are used widely, it is not uncommon to find increased nitrate levels in streams and pondages. Where such water is used for domestic purposes, there is evidence that dependent communities suffer an earlier onset of hypertension than is normal (Malberg et al.[106]).

High levels of nitrogen in soil do not generally affect plant health adversely unless the nitrogen level is out of balance with potassium and phosphorus. When this occurs, plants can become etiolated, prone to lodging, and may become increasingly susceptible to a number of diseases, such as powdery mildews and rusts.

2.9.2. Herbicides

These are also called "weedicides" in agriculture and commerce and "defoliants" in relation to military use. Their mode of action is commonly through the hormonal control of plant development and function, and when used for their intended purpose, they kill unwanted plant species. At low concentrations, however, they may not kill but can impair plant function and development, thereby causing partial or complete crop loss if applied at a critical stage of development. This has commonly caused problems in Australia in crops such as grapes, tomatoes, mangos, bananas, and papaws, by either delaying or preventing flowering, or by causing young flowers and fruits to abciss (fall off). Such direct damage has usually been caused by careless or thoughtless use, and as a result of the threat of litigation is generally not now common. The application of small doses of 2,4-D and similar chemicals can reduce photosynthesis in plants (Wolf[107]). Ormrod[31] also refers to spray-drift problems.

There are, however, a number of possible indirect effects which may sometimes be important and need to be considered. The herbicide (defoliant) 2,4,5-T (2,4,5-trichlorophenoxyacetic acid) contains trace amounts of an impurity, TCDD (2,3,7,8-tetrachlorodibenzo-p-dioxin, also known as dioxin), which is an extremely toxic substance. Workers in factories where dioxin is made or where 2,4,5-T is made can develop a skin condition known as chloracne, which is attributed to the TCDD (Hussain et al.[108]). The explosion of a dioxin factory in Italy in 1976 led to great concern for the welfare of unborn children, to the extent where abortions were advised for women in the early stages of pregnancy, since TCDD is mutagenic (Hussain et al.[108]). Such concentrations, however, were manyfold greater than the amounts occurring in 2,4,5-T sprayed into the environment.

The levels of TCDD in 2,4,5-T are in trace amounts, but even so, there is great public concern regarding the possible effects of this mutagen released into the atmosphere, especially during spraying and later burning of weeds, such as blackberry. There is no substantial evidence that the field use of 2,4,5-T causes problems in man, but there is great public concern.

Of more immediate concern here, however, is the possible indirect effect of herbicides on soil microorganisms and thereby on soil fertility and subsequent plant growth. Lundkvist[109] found that 2,4-D (2,4-dichlorophenoxyacetic acid) and MCPA (4-chloro-2-methyphenoxyacetic acid) each depressed the ability of blue-green algae to fix atmospheric nitrogen in soil. The degree of suppression was increased by increased amounts of the herbicides. Since microbes fix 30 million tonnes of atmospheric nitrogen annually on a worldwide basis, the effects of widespread herbicide use could be significant. Dubey[110] has reported that 2,4-D and other herbicides affect the viability of pollen in some plant species, with up to 70% of pollen grains killed 3 days after spraying with 1000 ppm of spray. Ormrod[31] discusses the importance of the timing of herbicidal sprays so as not to influence the crop being protected from weeds. He refers to work which suggests that fungi which normally occur in the root zone of plants and which may be antagonistic to root pathogens can be suppressed by the use of herbicides, which can increase disease problems. This kind of finding was predictable since Sadasivan[111] drew attention to work showing that a wide range of sprays applied to the foliage of plants can have a marked effect on populations of organisms on root surfaces. This work indicates that any aerial or soil pollutant might upset ecological interactions at the soil/root interface and thereby change the incidence of certain diseases.

2.9.3.   Pesticides

The inorganic pesticides (including fungicides) have been dealt with in Section 2.8. This is important because most other pesticides either do not accumulate in soil (DDT is an exception) or do not affect plant health. Mellanby[91] has observed the polluting effects of copper used as an orchard spray.

Two important detrimental effects of insecticides are described by Ormrod.[31] He refers to the reduction in the numbers of pollinating insects and its effects in reducing fruit set and thereby yield of fruit. Similarly, by reducing the populations of insect hyperparasites, some insecticides can actually lead to increases in numbers of some insect pests. Apart from these two effects, there can be detrimental effects on crop production if pesticides are irresponsibly used (Dubey[112]), but where care is exercised, pesticides seldom cause significant direct damage to plants.

The extent to which detrimental indirect effects may result from pesticide use is unknown. The following example, however, demonstrates how complex such effects can be. Bays[113] describes the case of eutrofication of the Chew Valley Lake, which supplies potable water to several communities. Sheep dip, used several miles from the lake, found its way, via natural drainage channels, into the lake, where its concentration was sufficient to kill a small crustacean that fed on a species of alga in the lake. It eventually proved that the population of alga in the lake was held in check by the crustacean, so that without it, the alga grew profusely. As the alga population aged and plants began to die, the demand for oxygen by the heterotrophic organisms that broke down the alga was so great that the oxygen levels in the lake dropped so low that all life in the lake perished and the water became unusable.

A further example of the need to consider indirect effects of pesticides on plant growth is provided by Mosse.[114] She has shown that the amount of root infection of onion by the mycorrhizal fungus, *Glomus mossae*, can be substantially reduced by applications of some insecticides and fungicides. The effect of this on plant growth may be significant or not, depending on soil conditions.

2.10. PLANT MONITORS OF POLLUTION

When it is known that there is a possibility of pollution occurring in a given area where valuable plant communities exist, it becomes important to monitor the environment to detect early warnings of potential widespread damage in time to avoid serious problems. For this reason much attention has been given to the selection and use of biological monitors. Such monitors may be natural components of the flora or they might be plants specifically placed throughout the area as monitors. Perhaps the oldest form of biological monitor was the canary, taken into mines and caves to warn of toxic levels of carbon monoxide.

The great genetic variation between and within species of plants enables particular plant groups, such as lichens, or species or particular cultivars of species to be selected which are especially valuable in detecting very low levels of particular pollutants. One of the earlier reports of useful variation referred to onions (Engle and Gabelman[115]), where a line was selected that was resistant to tip burn caused by ozone. In the resistant line, stomatal closure occurred very rapidly and the plants suffered considerably less damage than did other onion lines. The character was genetically dominant and easily bred for. In breeding for monitors, however, it is not resistance or tolerance that is sought but extreme sensitivity. Such a program was initiated by the United States Forest Service in an effort to monitor pollution in forest areas. Eventually, they selected three specific lines of eastern white pine to detect each of the pollutants sulfur dioxide, fluoride, and ozone (and PAN). At the same time they ran a second program to select tolerant lines. This work is discussed in detail by Sinclair.[27] The value of plant monitors over chemical analysis is that they take account of the various interactions of pollutant, plant physiology, and environment, which no chemical analytical procedure can do.

A second approach to finding plant monitors is to select components of the existing flora rather than to select and breed special lines. This approach is explained by Weinstein and McCune.[117] They set out the following guidelines for the selection of suitable monitors:

1. The monitor species should be sensitive to the pollutant at levels below the sensitivity of the species of economic or aesthetic value that it is desired to protect.

2. It should be widely distributed throughout the area of concern.

3. The symptoms produced should be distinctive for the pollutant and readily seen.

4. The species should be present throughout the growing season.

5. The species should grow from a terminal shoot throughout the growing season.

Weinstein and McCune[117] draw attention to the desirability of selecting the least valuable plants in a community as indicators, and it is interesting to note that they advocated the use of various weed species as well as mosses and lichens.

Much can be learned about the distribution patterns of pollutants by studying the moss (Gilbert[10]) and lichen (Pyatt[6]) flora surrounding the source of a pollutant. Algae and other life forms are also useful (Phillips,[118] Melhuus et al.[119]). It is likely, however, that the lichens will prove most useful monitors of pollution among the plants, because as well as there being changes in the species distribution as described by Gilbert[10] and Pyatt,[6] the lichens have an almost spongelike capacity to take up fluoride and heavy metals. This appears to be associated with the fungal partner in the alga/fungus symbiosis that makes up the lichens. The fungi take up the metals and deposit them in their cell walls outside the plasmalemma, which separates the living cell contents from the nonliving cell wall. Consequently, the chemical analysis of lichens in a given area provides most reliable data on the quantities of various metals falling out over their habitat. They have been used to monitor radioactive fallout as well as the concentrations of lead and other heavy metals from industrial fallout. There are a number of key references to the considerable literature on the use of lichens as monitors of pollution (Hawksworth,[120] Hawksworth and Rose,[121] Nieboer et al.,[122] Seaward,[123] Tomassini et al.,[124] Richardson and Young[9]). A leaf-spotting fungus of sycamore, *Rhytisma acerinum*, has also been considered as a monitor species (Vick and Bevan[125]).

Thomas et al.[126] prepared a useful bibliography of abstracts of literature dealing with biological indicators of environmental quality. Their bibliography covers a very large field, including the detection of pesticides and other pollutants by using cell-free extracts, cell culture, tissue culture, organisms, and communities of organisms.

2.11 CONCLUSIONS

There is no doubt that several pollutants cause severe disease and loss in economic crop species (Heggestad[127]). Similarly, damage and change occur in forests and native plant communities, where measurement of damage and its economic assessment are difficult to obtain. Studies such as those by Gordon and Gorham[13] and use of plant monitors help evaluate what happens in the field. Some studies indicate that species sometimes have the capacity to adapt to the polluted environment through the development of mechanisms for tolerating the presence of the pollutant. This is a process of natural selection. For example, Horseman and Wellburn[18] found that when exposed to sulfur dioxide, clones of plants growing in highly polluted areas suffered less disruption to enzyme activity than plants obtained from areas free from or sustaining low levels of pollution with sulfur dioxide. Similarly, tolerance to various heavy metals has been found in grasses and other organisms following several generations of heavy metal pollution (Coughtrey and Martin[101]).

This does not mean that pollution problems will eventually take care of themselves through natural selection. Although the genetic capacity to adapt to changes in most aspects of the environment is inherent in most species of plants and other organisms, it will only operate effectively when change is not extreme or rapid. Rapid and extreme increases in pollutants that cause acute injury are unlikely to induce adaptation but are likely to cause death. In addition to enabling natural selection for adaptation to changing environment, the genetic reserves of plants provide means by which man can adapt plants, through breeding and selection programs for use as sensitive monitors of particular pollutants in the field as well as providing tolerant cultivars for growth in highly polluted areas. It seems, however, that there is a price to be paid for using cultivars that are tolerant to some pollutants, since there is evidence that plants tolerant of heavy metals are less efficient in terms of biomass production (Coughtrey and Martin[101]). Exclusion mechanisms which may be termed "avoidance" rather than "tolerance" (Garsed and Read[128]) may be preferred as a means of overcoming pollution effects.

There is an increasing body of information which supports the view that the indirect effects of some pollutants (e.g., heavy metals) on plant health may be more serious than their direct effects. There are various life-support systems provided by microorganisms associated with higher plants which are often important, if not essential, to the growth of the higher plant. Evidence suggests that the influence of pollutants on such systems may often have a more detrimental effect on plant welfare by depriving them of the benefits of the supporting microorganisms than by interfering directly with the plants' metabolism. Mycorrhiza development in plants is depressed by the presence of heavy metals (McCreight and Schroeder[102]) and by sulfur dioxide in soil (Slankis[25]) and possibly by other pollutants, including agrochemicals. Various agrochemicals applied to plants as foliar sprays can influence the composition of the populations of microflora associated with plant roots (Sadasivan[111]). Heavy metal pollution can also reduce the effectiveness of nitrogen fixation in legumes by *Rhizobium* bacteria (McIlveen and Cole[100]), and the agrochemical herbicide 2,4-D and others reduce the ability of nitrogen-fixing soil algae to fix atmospheric nitrogen (Lundkvist[109]) as well as reducing the viability of pollen from a number of species (Dubey[110]).

Many factors govern the magnitude of pollution problems. Ozone damage is a serious problem in countries where there are high densities of automobiles and extended periods of sunshine. Thus ozone pollution problems on an economic scale are found only in certain areas of the northern hemisphere. Generalizations about pollution damage can be misleading, however, because there is still a great deal of information required about most problems and because each species of plant has its own range of sensitivities to each pollutant, which cannot be predicted. In addition, plant reactions to pollutants will differ in response to changes in atmospheric and plant physiological changes. In general, it appears that gymnosperms are more sensitive to air pollutants than are angiosperms (Jacobson and Hill[129]). Similarly, there is some evidence that among the angiosperms the dicotyledons accumulate heavy metals more readily than do monocotyledons (Huisingh,[77] Beavington[130]).

Finally, although heavy metal pollutants do not always induce stress in the plants in which they accumulate, they are of concern because they may enter animal food chains through such plants and may cause serious problems as a result.

ADDENDUM

Recently, numerous papers have been published which support the various findings already reported. Several provide additional evidence that the indirect effects of a number of pollutants on bacteriorhizas and mycorrhizas can affect plant growth.

One, however, is particularly noteworthy in drawing attention to the likelihood that new problems resulting from the actions of society will continue to be diagnosed. The death of Norfolk Island Pine (*Araucaria heterophylla*) on the eastern seaboard of New South Wales (Australia) has been a problem recognized for several years and attributed to various causes, including the combined action of ozone and seaspray (Hartigan, personal communication, 1972). Dowden and Lambert[133] now report that the death of *A. heterophylla* is caused by high levels of detergent in the seaspray around large cities such as Sydney and Wollongong, which affects foliage, causing it to take up larger amounts of salt than it can tolerate.

REFERENCES

1. Goodman, G. T., and Roberts, R. M., *Nature*, 245, 287 (1971).
2. Haut, H. van, and Stratman, H., *Colour-Plate Atlas of the Effects of Sulphur Dioxide on Plants*, Girardet, Essen, 1970, p. 206.
3. Heck, W. W., and Brandt, C. S., "Effects on Vegetation: Native, Crops, Forests," in A. C. Stern (Ed.), *Air Pollution*, Vol. 2, Academic Press, New York, 1977, pp. 157–229.
4. Ahlberg, J., Ramel, C., and Wachtmeister, C. A., *Ambio*, 1, (1972).
5. le Blanc, F., and de Sloover, J., *Can. J. Bot.*, 48, 1485 (1970).
6. Pyatt, F. B., *Environ. Pollut.*, 1, 45 (1970).
7. Daly, B. T., *Proc. N.Z. Ecol. Soc.*, 17, 70 (1970).
8. Ferry, R. W., Baddeley, M. S., and Hawksworth, D. L. (Eds.), *Air Pollution and Lichens*, Athlone Press, London, 1973.
9. Richardson, D. H. S., and Young, C. M., "Lichens and Vertebrates," in M. R. D. Seaward (Ed.), *Lichen Ecology and Biogeography*, Academic Press, London, 1977.
10. Gilbert, O. L., *New Phytol.*, 67, 15 (1968).
11. Brandt, C. S., and Heck, W. W., "Effects of Air Pollutants on Vegetation," in A. C. Stern (Ed.), *Air Pollution*, 2nd ed., Vol. 1, Academic Press, New York, 1968.

12. Treshow, M., *Phytopathology*, 58, 1108 (1968).

13. Gordon, A. G., and Gorham, E., *Can. J. Bot.*, 41, 1063 (1963).

14. Guderian, R., *Air Pollution* (English Ed.), Springer-Verlag, Berlin, 1977.

15. Treshow, M., *Environment and Plant Response*, McGraw-Hill, New York, 1970.

16. Bleasdale, J. K. A., *J. Park Admin.*, 18, 300 (1952).

17. Horsman, D. C., and Wellburn, A. R., *Environ. Pollut.*, 8, 123 (1975).

18. Horsman, D. C., and Wellburn, A. R., *Environ. Pollut.*, 13, 33 (1977).

19. Keller, T., and Schwager, H., *Eur. J. For. Pathol.*, 7, 321 (1977).

20. Darley, E. F., and Middleton, J. T., *Annu. Rev. Phytopathol.*, 4, 103 (1966).

21. Barrett, T. W., and Benedict, H. M., "Sulphur dioxide," in J. S. Jacobson and A. C. Hill (Eds.), *Recognition of Air Pollution Injury to Plants: A Pictorial Atlas*, Informative Report No. 1, Air Pollution Control Association, Pittsburgh, 1970.

22. Karnosky, D. F., and Stairs, G. R., *J. Environ. Qual.*, 3, 406 (1974).

23. O'Connor, J. A., Parbery, D. G., and Strauss, W., *Environ. Pollut.*, 7, 7 (1974).

24. Gorham, E., and Gordon, A. G., (1963) Some effects of a smelter upon aquatic vegetation near Sunbury, Ontario. Can. J. Bot., 41, 371 (1963), 371–378.

25. Slankis, V., *Annu. Rev. Phytopathol.*, 12, 437 (1974).

26. Wainwright, M., *Environ. Pollut.*, 17, 167 (1978).

27. Sinclair, W. A., *J. For.*, 67, 204 (1969).

28. Heagle, A. S., *Annu. Rev. Phytopathol.*, 11, 365 (1973).

29. Nash, T. H., "The Effects of Air Pollution on Other Plants, Particularly Vascular Plants," in Ref. 8, pp. 192–223.

30. Hiblen, C. R., *Plant Dis. Rep.*, 53, 544 (1969).

31. Ormrod, D. P., *Pollution in Horticulture*, Elsevier, Amsterdam, 1978.

32. Hill, A. C., Heggestad, H. E., and Linzon, S. N., "Ozone," in Jacobson and Hill, Ref. 21.

33. Feder, W. A., "Cumulative Effects of Chronic Exposure of Plants to Low Levels of Air Pollutants," in J. A. Naegele, *Air Pollution Damage to Vegetation*, Advances in Chemistry Series 122, American Chemical Society, Washington, D.C., 1973, pp. 21–30.

34. Miller, P. L., "Oxidant-Induced Community Change in a Mixed Conifer Forest," in Naegele, Ref. 33, pp. 101–117.

35. O'Connor, J. A., Parbery, D. G., and Strauss, W., *Environ. Pollut.*, 9, 181 (1974).

36. Ahlberg, J., Ramel, C., and Wachtmeister, C. A., Ecos, No. 16, 3 (1978).

37. Krause, C. R., and Weidensaul, T. C., *Phytopathology*, 68, 301 (1978).

38. Taylor, O. C., in Naegele, Ref. 33, pp. 9–20.

39. Duggar, W. M., and Ting, I. P., *Annu. Rev. Plant Physiol.*, 21, 215 (1970).

40. Leone, I. A., *Phytopathology*, 66, 734 (1976).

41. Evans, L. S., and Ting, I. P., *Atmos. Environ.*, 8, 855 (1974).

42. Heagle, A. S., and Heck, W. W., *Environ. Pollut.*, 7, 147 (1974).

43. Mudd, J. B., (1973) Biochemical effects of some air pollutants on plants. in Naegele, Ref. 33, pp. 31–47.

44. Spotts, R. A., Ludezic, F. L., and Hamilton, R. H., *Phytopathology*, 65, 39 (1975).

45. Rufner, R., Witham, F. H., and Cole, H., *Phytopathology*, 65, 345 (1975).

46. Feder, W. A., *Environ. Pollut.*, 1, 73 (1970).

47. Craker, L. E., *Environ. Pollut.*, 1, 299 (1971).

48. Schuck, E. A., in Naegele, Ref. 33, pp. 1–8.

49. Somov, V. S., Chekunova, L. N., and Slezkina, V. F., *Mikol. Fitopatol.*, 6, 260 (1972).

50. Treshow, M., Harner, F. M., Price, H. E., and Kormelink, J. R., *Phytopathology*, 59, 1223 (1969).

51. Manning, W. J., Feder, W. A., Papia, P. M., and Perkins, I., *Environ. Pollut.*, 1, 305 (1971).

52. Emmett, R. W., and Parbery, D. G., *Annu. Rev. Phytopathol.*, 13 (1975).

53. Hiblen, C. R., and Taylor, M. P., *Environ. Pollut.*, 9, 107 (1975).

54. Howell, R. K., and Graham, J. H., *Plant Dis. Rep.*, 61, 564 (1977).

55. Tingey, D. T., and Blum, U., *J. Environ. Qual.*, 2, 341 (1973).

56. Carney, J. L., Garrett, H. E., and Hedrick, H. C., *Phytopathology*, 68, 1160 (1978).

57. Crush, J. R., *New Phytol.*, 73, 743 (1974).

58. Daft, M. J., and El-Giahmi, A. A., "Effects of *Glomus* Infection on Three Legumes," in F. E. Sanders, B. Mosse, and B. Tinker (Eds.), *Endomycorrhizas*, Academic Press, London, 1975.

59. Brennan, E., *Phytopathology*, 65, 1054 (1975).

60. Davis, D. D., and Smith, S. H., *Phytopathology*, 64, 383 (1974).

61. Davis, D. D., and Smith, S. H., *Environ. Pollut.*, 9, 97 (1975).

62. Chamberlain, E. E., Atkinson, J. D., and Hunter, J. A., *N.Z. J. Agric. Res.*, 7, 480 (1964).

63. Stubbs, L. L., *Aust. J. Agric. Res.*, 15, 752 (1964).

64. Pell, E. J., *Phytopathology*, 66, 731 (1976).

65. Kress, L. W., *Phytopathology*, 61, 130 (1971).

66. White, N. H., "Observations on Air-Oxidant Injuries on Plants in the Sydney Metropolitan Area," Clean Air Conf., Sydney, 1962.

67. Darley, E. F., Taylor, O. C., and Middleton, J. T., "Photochemical and Exhaust Fume Damage to Plants," Abstracts, 1st Int. Cong. Plant Pathol., Lond., 1968.

68. Srivastava, H. S., Jolliffe, P. A., and Runeckles, V. C., *Environ. Pollut.*, 9, 35 (1975).

69. Bennett, J. H., Hill, A. C., Soleimani, A., and Edwards, W. H., *Environ. Pollut.*, 9, 127 (1975).

70. Eller, B. M., *Environ. Pollut.*, 13, 99 (1977).

71. Manning, W. J., *Environ. Pollut.*, 9, 87 (1975).

72. Treshow, M., and Pack, M. R., "Fluoride," in Jacobson and Hill, Ref. 21

73. McLaughlin, S. B., and Barnes, R. L., *Environ. Pollut.*, 8, 91 (1975).

74. Garree, T. P., Plebin, R., and Lhoste, A. M., *Environ. Pollut.*, 13, 159 (1977).

75. Rains, D. W., *Nature*, 233, 210 (1971).

76. Martyn, D. F., *Environmental Pollution*, Spec. Publ. No. 6, Australian Conservation Foundation, Canberra, 1970, p. 5.

77. Huisingh, D., *Annu. Rev. Phytopathol.*, 12, 375 (1974).

78. Chisolm, J. J., *Sci. Am.*, 224, 15 (1971).

79. Ahlberg, J., Ramel, C., and Wachtmeister, C. A., *Ambio*, 1, 29 (1972).

80. Shaar, H., Ophus, E., and Golluog, B. M., *Nature*, 241, 215 (1973).

81. Schubert, J., *Ambio*, 1, 79 (1972).

82. Chow, T. J., *Nature*, 225, 295 (1970).

83. Quinche, J. P., Zuber, R., and Bovay, E., *Phytopathol. Z.*, 66, 259 (1969).

84. Zuber, R., Bovay, E., Tschannen, W., and Quinche, J. P., *Rech. Agron. Suisse*, 9, 83 (1970).

85. Baumhardt, G. R., and Walsh, L. F., *J. Environ. Qual.*, 1, 92 (1972).

86. Lane, S. D., and Martin, E. S., *New Phytol.*, 79, 273 (1977).

87. McCreight, J. D., and Schroeder, D. B., *Environ. Pollut.*, 13, 265 (1977).

88. Lowell, L., Getz, L. V., and Prather, M., *Environ. Pollut.*, 13, 151 (1977).

89. Bazzar, R. A., Carlson, R. W., and Rolfe, G. L., *Environ. Pollut.*, 7, 241 (1974).

90. Salgaard, P., Aarkrog, A., Fenger, J., Blyger, H., and Graaback, A. M., *Nature*, 272, 346 (1978).

91. Mellanby, K., *Pesticides and Pollution*, Collins, London, 1971.

92. Fimriete, N., *Environ. Pollut.*, 1, 119 (1970).

93. Borg, K., Erne, K., Hanko, E., and Wanntorp, H., *Environ. Pollut.*, 1, 91 (1970).

94. Caldwell, R. L., "Mercury in Fungicides," 11th Annu. Beltwide Cotton Prod. Res. Conf., Atlanta, Jan. 11–13, 1971.

95. Lipsey, R. L., *Environ. Pollut.*, 8, 149 (1975).

96. Haghiri, F., *J. Environ. Qual.*, 2, 93 (1973).

97. Jones, R. L., Hingsly, T. D., and Ziegler, R. L., *J. Environ. Qual.*, 2, 351 (1973).

98. Turner, M. A., *J. Environ. Qual.*, 2, 118 (1973).

99. Tyler, G., *Ambio*, 1, 52 (1972).

100. McIlveen, W. D., and Cole, H., *Phytopathology*, 64, 583 (1974).

101. Coughtrey, P. J., and Martin, M. H., *New Phytol.*, 19, 273 (1977).

102. McCreight, J. D., and Schroeder, D. B., *Phytopathology*, 64, 583 (1974).

103. Jernelov, A., and Martin, A. L., *Annu. Rev. Microbiol.*, 29, 61 (1975).

104. Gunn, D. L., Flnn, Appl. Biol. **72,** 105 (1972) in D. E. Hathaway (Ed.), *Foreign Compound Metabolism in Mammals*, Vol. 3, Chemical Society, London, 1975.

105. Gunn, D. L., Ann. Appl. Biol., 72, 105 (1972).

106. Malberg, J. W., Savage, E. P., and Osteryoung, J., *Environ. Pollut.*, 15, 155 (1978).

107. Wolf, F. T., *Bot. Rev.*, 43, 395 (1977).

108. Hussain, S., Ehrenberg, L., Lofroth, G., and Gejvall, T., *Ambio*, 1, 32 (1972).

109. Lundkvist, I., *Sven. Bot. Tidskr.*, 64, 460 (1970).

110. Dubey, J. A., *Environ. Pollut.*, 13, 169 (1977).

111. Sadasivan, A., "Rhizosphere Microfloras," Proc. Summer Sch. Bot., Darjeeling, 1960, pp. 363–370.

112. Dubey, J. A., *Phytopathology*, 60, 485 (1970).

113. Bays, L. R., *Environ. Pollut.*, 1, 205 (1971).

114. Mosse, B., *The Effects of Pesticides on VA Mycorrhiza*, Report for 1977, Rothamstead Exp. Sta., Pt. I. Adlard and Son Ltd./Bartholomew Press, U.K., 1978, pp. 238–239.

115. Engle, R. L., and Gabelmann, W. H., *Am. Hort. Sci. Proc.*, 89, 423 (1966).

116. McCune, D. C., "Summary and Synthesis of Plant Toxicology," in Naegele, Ref. 33.

117. Weinstein, L. H., and McCune, D. C., "Field Surveys, Vegetation Sampling and Air and Vegetation Monitoring," in Jacobson and Hill, Ref. 21.

118. Phillips, D. J. H., *Environ. Pollut.*, 13, 281 (1977).

119. Melhuus, A., Seip, K. L., Seip, H. M., and Myklestsad, S. *Environ. Pollut.*, 15, 101 (1978).

120. Hawksworth, D. L., *Lichenologist*, 8, 87 (1976).

121. Hawksworth, D. L., and Rose, R., *Lichens as Pollution Monitors*, Studies in Biology No. 66, Edward Arnold, London, 1976.

122. Nieboer, E., Richardson, D. H. S., Puckett, J. K., and Tomassini, F. D., "Lichen Response Studies," in T. A. Mansfield (Ed.), *Effects of Air Pollutants on Plants*, Cambridge University Press, Cambridge, England, 1975.

123. Seaward, M. R. D., "Performance of *Lecanora muralis* in an Urban Environment," in D. H. Brown, D. L. Hawksworth and R. H. Bailey (Eds.), *Lichenology: Progress and Problems*, Academic Press, London, 1976, pp. 323–358.

124. Tomassini, F. D., Puckett, K. J., Nieboer, E., and Richardson, D. H. S., *Can. J. Bot.*, 54, 1591 (1976).

125. Vick, C. M., and Bevan, R., *Environ. Pollut.*, 11, 203 (1976).

126. Thomas, W. A., Goldstein, G., and Wilcox, W. H., *Biological Indicators of Environmental Quality*, Ann Arbor Science Publishers, Ann Arbor, Mich., 1973 (546 abstracts).

127. Heggestad, H. E., *Phytopathology*, 58, 1089 (1968).

128. Garsed, S. G., and Read, D. J., *Environ. Pollut.*, 13 (1977), 173.

129. Jacobson, J. S., and Hill, A. C. (Eds.), *Recognition of Air Pollution Injury and Vegetation: A Pictorial Atlas*, Informative Report No. 1, Air Pollution Control Association, Pittsburgh, 1970.

130. Beavington, F.,*Environ. Pollut.*, 8, 65 (1975).

131. Manning, W. J., *Environ. Pollut.*, 2, 69 (1971).

132. Moyer, J. W., Cole, H., and Lacasse, E., *Phytopathology*, 64, 584 (1974).

133. Dowden, H. G. M., and Lambert, M. J., *Environ. Pollut.*, 19, 71 (1979).

134.. Saunders, P. J. W., *Eviron. Pollut.*, 9, 85 (1975)

3

EFFECTS OF ATMOSPHERIC POLLUTION ON METALS AND PROTECTIVE COATINGS

K. Bartoň

Prague, Czechoslovakia

3.1. INTRODUCTION

The atmosphere is the most common corrosive environment. Metallic materials, metallic technical equipment, and other goods are exposed to this environment and have to be protected against deteriorative effects that shorten the effective lifetime, cause a decrease of reliability, and increase the probability of failure. Protection against atmospheric corrosion is a technical and economic necessity. Without adequate protection, losses by rapid deterioration of metallic products would cause economic difficulties for all industrially developed countries. Even though different protective measures are applied, some of them being very effective, losses due to atmospheric corrosion are very high. The cost of protecting metals against atmospheric corrosion, as well as associated maintenance, renewal of corroded parts, and repairs is very high. This cost amounts to 1 to 2% of the national income of industrialized countries.[1,2]

The corrosion aggressiveness of different atmospheres depends not only on their climatic characteristics, but also, and in many cases predominantly, on their content of stimulating chemical compounds. Besides salt particles in the vicinity of the sea, industrial pollution plays a dominant role. In areas with high concentrations of industry and dense population, the air is intensely polluted by compounds of sulfur, chlorine, ammonia, and so on. Most of these pollutants are active stimulants of atmospheric metallic corrosion.

Air pollution is therefore not only a danger from the biological viewpoint, but its influence on the technical basis of modern human society is very severe. It might be said that modern technical processes are producing highly detrimental environmental conditions for their own existence.

This chapter's aim is to summarize the present state of knowledge about the stimulating effect of industrial air pollutants on atmospheric corrosion of technical metals used in modern society.

3.2. ATMOSPHERIC CORROSION–A SPECIAL CASE OF ELECTROCHEMICAL CORROSION

3.2.1. Definition of Corrosion

Corrosion of metals is usually defined as the deterioration of technically important properties of metal due to their chemical (electrochemical) reactions with the environment. Simultaneously, active physical effects such as different kinds of stressing and erosion, may change the forms of corrosion and accentuate its technical danger. For example, stress corrosion cracking as the result of static tensile stresses, corrosion for specific metal–environment systems, and corrosion fatigue induced by dynamic loading of certain metals in aggressive environments limit the technical applicability of many metallic materials. The term "deteriorative effect of corrosion" should be interpreted rather broadly. Because of different features of corrosion, the following changes may be of importance:

• Breaking of metallic parts if corrosion has lowered their mechanical strength by loss of the *effective cross section* or by changes of structure, by cracks, and so on.

- Perforation of walls of tubes, reactors, and car bodies by nonuniform corrosion.
- Loss of important surface properties (electrical, optical, thermal, etc.) by changes in surface roughness or formation of solid layers of corrosion products.

3.2.2.   Principles of Corrosion in Aqueous Electrolytes

Research on the general principles of electrochemical corrosion (i.e., corrosion in ionic conductive environments, mainly in solutions of electrolytes in water) has been a topic of intense work during this century. Its results–the principal mechanisms and kinetic laws—are described in a number of monographs.[3-9]

It is not intended to discuss here the general mechanisms and kinetics of corrosion in solutions of electrolyte. For a better understanding of the special topic of this chapter, it is, however, necessary to summarize the main rules of this very broad topic. Only one special part of the broad field of corrosion science will be discussed here—the corrosion of metals in nearly neutral solutions of electrolytes in water in contact with the surrounding atmosphere. This topic is the basis for understanding the principles of atmospheric corrosion

The corrosion process of metals in an electrolytic solution includes a sequence of steps of physical, chemical, and electrochemical nature taking place at or near the metal/solution interface and in the bulk solution. The single reaction steps form a system in which each following step depends on the preceding. Figure 3.1 schematically describes the sequence of reactions of metallic corrosion in nearly neutral electrolytic solution (pH~4 to 8). Such conditions are typical for the electrolyte formed by condensed water from the atmosphere.

The mechanism of step 1 the dissolving of gaseous, chemically active components of the atmosphere, is purely physical. If surface effects on small droplets (fog, rain) are neglected, the concentration of dissolved gases obeys Henry's law, the Henry constants for the single gases being negatively dependent on temperature. The content of oxygen in the air is constant, so that the amount dissolved in water (or a diluted electrolytic solution) is controlled only by temperature. The CO_2 content in the air changes in broader limits from the background value to 0.3% in strongly polluted or other special atmospheres (in tropical forests, in indoor microclimates, animal sheds, etc.). The concentration of gaseous atmospheric pollutants due to human activity, such as sulfur compounds (mainly SO_2, H_2S), nitrogen compounds (NO_x, NH_3), and other compounds (or-

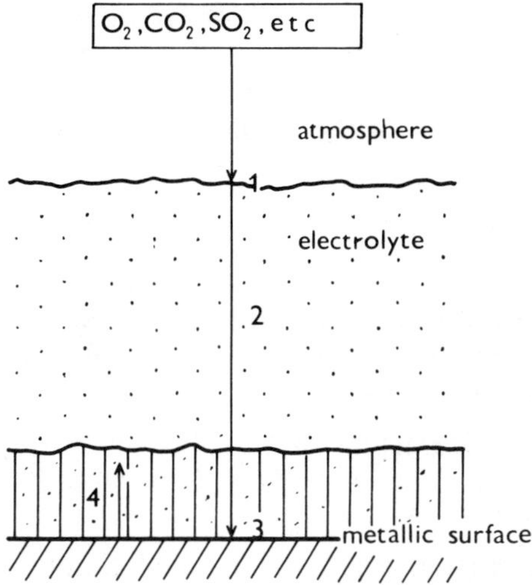

Figure 3.1. The system metal/corrosion products/electrolyte/atmosphere.

ganic acids) is usually much lower if compared with the "natural" components of the air important for the kinetics of the corrosion process. Most of them are readily soluble in water and extremely reactive with regard to metal. Very low amounts of sulfur dioxide (e.g., ~0.1 ppm) or of hydrogen sulfide (<0.01 ppm) dissolved in a film of water on the metallic surface (either by dissolution of the gas in the surrounding atmosphere or primarily adsorbed on the surface of the metal) are therefore very effective stimulants of the corrosion process.

Step 2, transport of the reactive species to the metallic surface, in the schematic description if Figure 3.1 is a purely physical one. In the case of a system contacting the atmosphere at constant temperature (and at partial pressures of the components of the atmosphere O_2, CO_2, SO_2, etc.), the transport of dissolved gases to the metallic surface is controlled by the gradient of their concentration. As the system is an open one (due to the contact of the electrolyte surface with the surrounding atmosphere of constant composition), the first law of Fick is valid. The solution may contain components able to react with the dissolved gases or to catalyze their reactions. It is well known that SO_2 dissolved in a surface solution on metal covered with corrosion product is rapidly converted to SO_4^{2-}:

$$SO_2 + O_2 + 2e \rightarrow SO_4^{2-} \tag{1}$$

by the catalytic effect of Fe^{3+} and Cu^{2+} ions in solution or solid corrosion products of the same metals.[10,11]

Step 3 is the most important, as it causes the transition of the metal into its ionic products and thus the harmful changes, which fall under the definition of corrosion. Under the assumption that the corrosion is thermodynamically possible, the electrochemical reactions have to be treated as a redox system in which the dissolution of the metal is generally an oxidation process.

$$Me \rightarrow me^{n+} + n(e) \tag{2}$$

The electric balance has to be maintained by the reducing action of (e) reacting with some reducible component in the system. For the case limited by neutrality and abundant oxygen dissolved in the surface solution, the following reactions of reduction predominate:

$$H_2O + O_2 + 4e \rightarrow 4OH^- \tag{3}$$

$$Me^{n+} + m(e) \rightarrow Me^{(n-m)+} \tag{4}$$

Both parts of the electrochemical corrosion reactions, the oxidation (anodic) and reduction (cathodic), are a closed system: The oxidation (the actual corrosion) cannot proceed without simultaneous reduction.

If the preceding and following steps are neglected, the kinetics of the electrochemical reactions can be derived from generally valid laws of electrochemical kinetics. The velocity of any electrode process is described by the equation

$$j = k \exp \left(\frac{\alpha n F \zeta}{RT} \right) \tag{5}$$

where j is the current density (i.e., the velocity of the reaction), k a rate constant, α the transition coefficient, n the number of electrons transferred, α the overpotential, F the Faraday constant, R the gas constant, and T the absolute temperature. This law is generally valid except for cases of passivation at higher current densities (e.g., stainless steels, aluminum, chromium, etc.) for the anodic partial reaction, or insufficient reducible matter in the solution leading to the limiting current density of the cathodic partial process. Figure 3.2 is a schematic picture of the simultaneous electrochemical reactions, showing all possibilities of arriving at the final current density, the process velocity. In the Figure $+j_a$ is the current density for active dissolution of a metal not controlled by a limited supply of reducible species in the system, $-j_l$ the velocity if the cathodic reaction is limited, and $+j_p$ the corrosion velocity of passive metal. In the latter case $+j_p$ need not be equal to $-j_p$ because other steps (e.g., step 4 in Figure 3.1) control the velocity.

If the solution contains anions at high concentrations (e.g., SO_4^{2-}), the primary anodic step may lead to a solid salt. As will be shown later, this

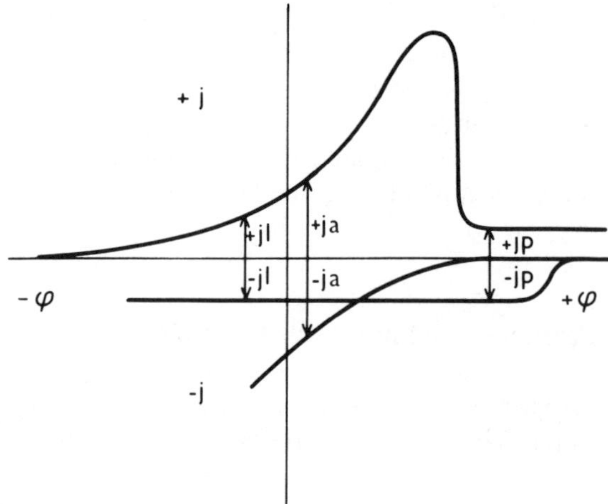

Figure 3.2. Schematized electrochemical principle of corrosion; $+j$, anodic (dissolution) current; $-j$, cathodic current; index a, active corrosion, p, passive corrosion; l, limiting current.

is typical for atmospheric corrosion of iron (carbon steels) in atmospheres with higher SO_2 contents, in which solid $FeSO_4$ is a very common component of the corrosion product.[12–15]

In neutral, well-aerated solutions, step 4 (Figure 3.1) is very complex. Primary products of both elementary electrochemical reaction interact and form solid corrision products:

$$Me^{n+} + nOH^- \rightarrow Me(OH)_n \tag{6}$$

Other components of the solution may also take part in the formation of corrosion products. So SO_4^{2-} and Cl^- participate in the formation of hydroxide (oxide) salts with low solubility. The corrosion products of zinc, lead, copper, and other metals mostly contain such secondary reaction products. If the metal forms more-or-less stable insoluble compounds, differing in the degree of oxidation, solid-phase reactions then change the composition of the corrosion products. The first solid reaction product may be of a colloidal nature. This changes to crystalline compounds. All these processes affect the corrosion velocity as they change the conditions at the surface of the metal. Solid corrosion products may adhere and hinder transport steps 2 and 4 or change the conditions at different surface areas, so that the corrosion velocity is not uniform. Further, they may have catalytic influence on important reactions of the system. Oxidation of dissolved (adsorbed) SO_2 to SO_4^{2-} is actively promoted by rust; corrosion products on copper and zinc are less active and on aluminum are practically inert.[16] The formation of solid corrosion products tends to reach steady-state conditions. In some real systems, as in atmospheric corrosion of steel, steady properties of the corrosion products are reached after very long periods (months or years).

To summarize, in corrosion systems with formation of layers of more-or-less adhering corrosion products, their properties are a very important factor determining the corrosion velocity.

3.2.3.   Special Features of Atmospheric Corrosion

Atmospheric corrosion is a special case of metallic corrosion in aqueous electrolyte solutions. The following characteristics define this case:

- The volume of the solution is very small; that is, the solution forms a very thin film on the large surface of the corroding metal.
- The corrosion process is discontinuous; only if atmospheric conditions are favorable for the formation and existence of the surface electrolyte are electrochemical corrosion reactions possible.
- Due to the chemical constituents of the atmosphere, the water formed by condensation processes from atmospheric moisture is a nearly neutral solution, containing reactive dissolved components of the air, such as O_2, CO_2, chlorides (in maritime regions), and pollutants arising from human activities. The water also contains compounds of sulfur (sulfides, sulfur dioxide, sulfates, etc.), nitrogen (ammonia, various nitrogen oxides), and chlorine (Cl_2, chlorides, etc.).

3.2.3.1.   Formation and Destruction of the Surface Solution

The surface electrolyte forms by:

- Falling liquid precipitates (rain, mist, melting snow, or ice).
- Condensation of water vapor on surfaces cooler than the ambient dew-point temperature.
- Sorption of water vapor, resulting in liquid-phase water.

Its action ends by evaporation or, at low temperatures, by freezing.

In an outdoor atmosphere the liquid precipitates falling and wetting the surface have a predominant influence. It should be stressed, however, that it is not the total amount of precipitation but the duration of rain or fog periods that is important for the atmospheric corrosion process. The agressivity of regions with frequent long rainy periods, but with moderate total amounts precipitation per year may be much higher than those with very short but violent rainfall such as subtropical and tropical regions.

Dew formation on cool surfaces is a very important factor of corrosion in partially or completely closed spaces, such as under shelter, in storehouses, and in packages, where the surfaces are protected against the direct action of the weather. If the atmosphere in such spaces is not conditioned and communicates with the outdoor air either directly or indirectly through walls or windows with more or less insulating capacity, the following process takes place. During the cooler part of the day, the temperature of metallic equipment decreases. As the absolute humidity of the air does not change rapidly, the relative value rises. The cooling of the metal is necessarily slower than the decrease of the air temperature, so no surface dewing takes place. The rise of the environmental temperature has an opposite effect, especially if the mass and hence the thermal capacity of the metallic object is high. Its temperature remains lower than the ambient one, so that in the microclimate near the surface, the dewpoint temperature is reached and liquid water is formed on the surface. Conditions for the existence of liquid water film formed by surface dew formation may last relatively long if the atmosphere in the closed space is humid, as often occurs in packages. Another well-known case of long-lasting, and from the corrosion viewpoint dangerous, dew formation is the transfer of cool metallic objects into warm humid environments (e.g., of a motor car in winter from the street into a warm, but humid garage).

Formation of surface electrolytes by sorportion is mostly induced by the presence of hydroscopic matter on the surface. Another possibility—multilayer sorption of water vapor and capillary condensation—is highly improbable and can take place only at humidities very near the saturation value at the ambient temperature. Hygroscopic particles are contained in airborne dust.

3.2.3.2. *Properties of the Surface Electrolyte*

Another specific characteristic feature of atmospheric corrosion is the limited volume of the surface solution. The general laws of corrosion are derived from systems in which the ratio of the volume of solution to the corroding surface is high. In atmospheric corrosion this ratio is on the contrary extremely low. The thickness of the solution is never greater then 0.1 to 0.2 mm. Thicker layers are observed only on specially shaped surfaces not allowing the surplus water to flow off. Only during periods of rain or melting snow can layers of such high thickness be reached. If the surface electrolyte solution forms by condensation of water vapor, its amount is much lower (about 10 μm), not ignoring the possibility of an uneven dropwise condensation. Sorption of vapor on hygroscopic substances on the corroding surface yields even thinner layers of the electrolyte, estimated to be about 1 μm or less.

The low volume/surface ratio typical for atmospheric corrosion allows us to deduce some special properties of the corroding system:

- If water-soluble matter is present on the surface, the solution attains high concentration and may even be saturated.
- The thin layer of the solution is ineffective as a barrier to the transport of oxygen and other components of the atmosphere through it. Moreover, very frequent but small fluctuations of the atmospheric temperature induce a constant "inner" movement and thus mixing of the layer, so that the transport processes of active components toward the metal/solution interface are not diffusion-controlled, the convective transport predominates. Atmospheric corrosion is therefore very seldom controlled by the cathodic part of the electrochemical reaction system. The electrochemical reduction of dissolved oxygen or of another reducible compound reoxidized by it is very fast, and owing to rapid replenishment of the system by the convective supply of oxygen, the effect of the "limiting" current density of the cathodic reaction (see Figure 3.2) is small.

Exceptionally, if the form of a horizontal surface leads to allowing thicker layers of solution which cannot flow off, cathodic control may be of importance.

The high concentration of all soluble compounds in the reaction system and the periodic drying of the surface supports the formation of solid precipitates or corrosion products. If there is covering of the surface by a more-or-less uniform layer of corrosion products, a new and very important factor in the reaction system arises.

The chemical properties of the surface electrolytic solution can be derived from general knowledge of the characteristic composition of atmospheric precipitates. As the solution contacts the corroding metal, covered by a more-or-less developed, but always present layer of solid corrosion products, reactions between the components of both parts of the system must take place, so that the original composition of the surface solution rapidly changes, attaining steady-state dynamic equilibrium conditions. Even if the primary precipitate is acidic, as in highly polluted atmospheres containing oxides of sulfur, chlorine or its compounds, and so on, the pH value of the solution isolated from corroding metallic surface will be close to the neutral point. The system shows a high buffering capacity, probably as the result of rapid interactions of H_3O^+ and OH^- ions and other anions with the oxides and hydroxides of the corrosion products, which are often present in reactive amorphous form. The interactions may be physical adsorption on the surface of solid particles, or chemical, as in the transformation of hydroxides into hydroxide salts with very low solubility.

3.3. EFFECTS OF POLLUTION ON THE MECHANISM AND KINETICS OF ATMOSPHERIC CORROSION

Atmospheric pollutants affect the corrosion process by taking part at all steps of the process, thus influencing its mechanisms and kinetics. The mechanism of their action is best demonstrated with help of the schematic description of the stepwise processes described in Sections 3.2.2 and 3.2.3.

3.3.1. Reaction System of Atmospheric Corrosion in the Presence of Stimulating Pollutants

3.3.1.1. Entrance of the Emission into the System

Chemically reactive pollutants contact the system from the outside, as schematically shown on Figure 3.1. The first step of their action is necessarily settling or sorption on the surface of the corroding metal.

If the surface solution layer of water is present (during wet periods), they dissolve in it and may react with its other components. The formal application of Henry's law is of no practical significance here, because no equilibrium can be reached in the system discussed, in which the dissolving species continually reacts with other components. A typical example, already mentioned, is that of the dissolution and rapid transformation of atmospheric sulfur dioxide in the surface electrolyte. In the presence of oxidizing agents, as dissolved oxygen or ferric ions, the gas, which is only very slightly dissociated, is oxidized to sulfate:

$$SO_2 + H_2O + O_2 \rightarrow H_2SO_4 \tag{7}$$

Other anionic component entering into the system, such as chlorides, sulfates, or nitrates, may remain unchanged, but their reaction with the metal or corrosion products already present may entirely change the system.

If the pollutant contacts the surface during a dry period, it cannot dissolve, but if gaseous, it may be adsorbed on the solid surface of the metal or the corrosion product covering it. Also, solid particles of salt or salt containing dust may settle on the surface.

There is some information on the adsorption process of SO_2 on different metals.[16] The amount of adsorbed matter and the adsorption velocity much depend on the kind of metal, the presence of corrosion products on its surface, and the relative humidity of the atmosphere (Figure 3.3). There are metals where the adsorption of the pollutant is generally very low and not much influenced by the presence of corrosion products or by relative humidity (e.g., SO_2 on aluminum).

Other metals, such as zinc and lead, adsorb the gas in limited amounts, and the dependence of the adsorbed amount on humidity is not very strong. These metals show a slightly increased adsorption of corrosion products over their surface.

Iron behaves quite differently. The dependence of adsorbed amounts of SO_2 on humidity (Figure 3.4) is very steep. If rusty iron is the adsorbent, at relative humidities above 95% nearly all SO_2 molecules contacting the surface are "caught" by it. The shapes of the curves show the same features as have been shown in the chemical deductions of the "critical humidity" value of Vernon[17] and Patterson and Hebbs.[18] The different sorption of SO_2 on metals is valid for very low concentrations of the gas in real atmospheric conditions (<1 ppm). At higher concentrations the mechanism of SO_2 adsorption is different as is its secondary transformation by chemical and electrochemical reactions. This is the principal reason why attempts to "accelerate" atmospheric corrosion in the laboratory by raising the SO_2 content of simulated environment necessarily lead to failures.[19,20]

If the reason for the different behavior of the metals toward SO_2 is

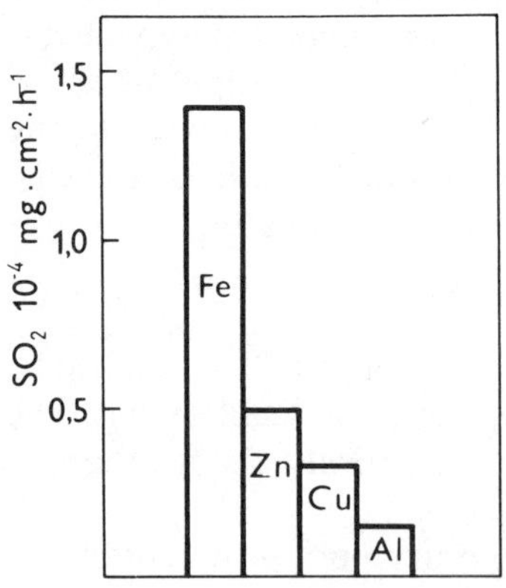

Figure 3.3. Steady-state adsorption rate of SO^2 on metals at 95% relative humidity and 0.1 ppm SO_2 (Sydberger and Vannerberg[16]).

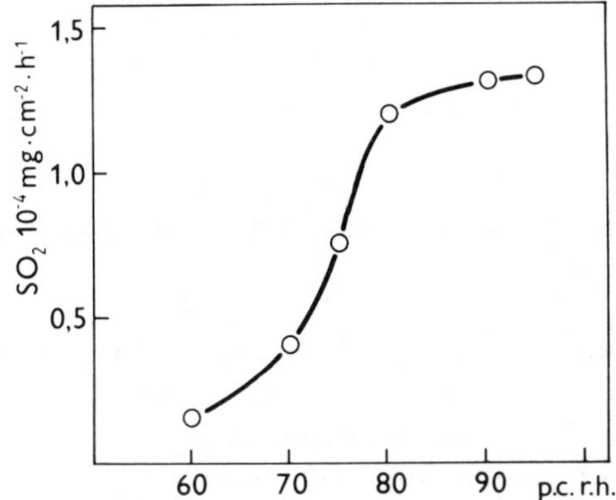

Figure 3.4. Steady-state adsorption rate of SO_2 (0.1 ppm) on rusted iron in SO_2 dependence on relative humidity (Sydberger and Vannerberg[16]).

studied, the following deductions may be made. The sorption is probably higher if chemisorption follows or if the primarily adsorbed gas is rapidly transformed by chemical reactions. This may take place on rusted iron as

$$SO_2 \xrightarrow{\ O_2,\ Fe^{3+},\ Fe^{2+}\ } SO_4^{2-} \tag{8}$$

or on copper.

Less adsorption is to be expected on metals with not very pronounced redox catalyzing properties, as on zinc and lead. Here the formation of basic salts may be of importance, and even the primary existence of sulfites cannot be excluded.

Passive metals such as aluminum do not adsorb higher amounts of SO_2, as there is no subsequent reaction disturbing the quickly achieved sorption equilibrium. Even if adsorbed SO_2 is slowly oxidized to sulfate, the passive layer is able to resist its action, and only after rinsing the acid by rain may new SO_2 be adsorbed. Analogous behavior can be expected with chromium, high-alloy stainless steels, and other metals keeping up passivity at atmospheric conditions (Ti, Zr).

An interesting piece of information about primary adsorption of SO_2 on silver has been published by Michailovski,[21] who assumes the formation of highly stable clathrates, $SO_2 \cdot 6H_2O$, that ''survive'' dry periods and after wetting induce the corrosion process even if no new SO_2 contacts the surface (Section 3.2.1).

Hydrogen sulfide is readily adsorbed by many metals. Its action on lead and other reactive metals proceeds analogously to that of SO_2 starting on discrete sites, the character of which has not yet been clearly explained.

There is very little information about the primary adsorption of other gaseous contaminants of the atmosphere, such as nitrogen oxides, chlorine, and so on, although their kinetic effect has been studied.

3.3.2. Mechanism of Corrosion Reactions Induced by Pollutants and Their Reaction Products

As shown in Section 3.3.1, most information on the mechanisms of action of pollutants in the process of atmospheric corrosion of metals is available for sulfur dioxide. This is due to the predominating importance of this gas on metals used for engineering and structural purposes. Therefore, it seems necessary to treat this part preferably on the basis of knowledge gained by thorough research of the reaction mechanism of atmospheric corrosion of metals in the presence of this gas. Attention will be paid, however, to other polluting species in the air.

3.3.2.1. *Reactions of Atmospheric Rusting*

It has been shown (Section 3.3.1) that iron (plain carbon steels, cast iron, some low-alloy steels) is highly reactive with SO_2 and that at higher relative humidities and in the presence of corrosion products, the primary step of the reaction system—the sorption of SO_2 on the surface—is very effective. If "metallic" iron surfaces are contacted with moist atmospheres containing SO_2 at relatively low concentrations (<10 mg m^{+3}), the gas gets primarily adsorbed on "active" sites. The velocity of sorption depends on the relative humidity, as was shown in the preceding section. Because of the simultaneous action of oxidizing agents in the reaction system (oxygen), the primary attack takes place very quickly, and corrosion products (ferrous, ferric) accelerate both the adsorption process and the catalytic oxidation of SO_2 to SO_4^{2-}.

From the technical viewpoint the mechanisms of the primary corrosion reaction are of limited importance. Their knowledge may only characterize borders of corrosion or no corrosion. Only one factor controls this—the presence or absence of a surface solution. If the metal is dry, no corrosion begins; if it is wet, corrosion may start.

The condition for the formation of a surface electrolyte have been treated in Section 3.2.3.1. It has to be stressed that no functional dependence of surface wetting and the starting of corrosion exists. Many additional factors are involved, such as (1) the presence of stimulating substances on the "metallic" surface (adsorbed pollutant, dust particles containing soluble and aggressive components), (2) the "activity" of the metallic surface (e.g., freshly grit- or sand-blasted surfaces show a higher tendency for primary corrosion than if they are kept for several hours in noncorrosive conditions), (3) the chemical and structural composition of the surface (active sulfide inclusions may positively induce the corrosion process), and so on. The beginning of atmospheric corrosion can be better described by a probability function, even though the existence, frequency, and duration of surface wetting periods play the most important role. In open atmospheres with frequent, long wetting periods, rusting starts within a few hours or days. In drier indoor environments where wetting is rare, surfaces may remain without rust for weeks.

Atmospheric rusting of iron starts on discrete points of the surface. It has been shown that in the absence of stimulating agents settled on the surface from the environment (salts, etc.), the starting point coincides with active sulfide inclusions:[22]

$$(Mn, Fe)S + H_2O + 3O_2 \rightarrow (Mn, Fe)_2O_3 + H_2SO_4 \qquad (9)$$

[through the intermediate formation and hydrolysis of $(Mn, Fe)SO_4$]. The first local "rust" therefore contains high amounts of Mn_2O_3.

The voluminous primary rust preferably adsorbs SO_2 from the environment and this is catalytically oxidized to H_2SO_4. The sulfuric acid formed attacks new points of the surface and the "technical" rusting process, covering the surface completely with corrosion products, starts and proceeds further. At this stage the very important and unique feature of atmospheric corrosion of iron—the formation and spreading of sulfate nests—take place.[12-15] Schwarz[12] first described the occurrence of discrete salt agglomerations at the steel/rust interface. These contain predominantly solid $FeSO_4 \cdot 4H_2O$ (with lower amounts of other $FeSO_4$ hydrates). The nests show an extraordinary stability and can be destroyed only with great difficulty. Their formation, spread, and extreme stability have been thoroughly studied and the following summarized theory seems to be valid.

In the presence of a limited volume of the surface solution (e.g., at conditions without rain but with high humidity), the sulfate formed by previous reaction dissolves, and a highly concentrated solution is in contact with the iron surface. Taking into account the electrochemical mechanism of anodic dissolution of iron in concentrated solutions containing sulfates,[23] the following reaction steps may be expected:

$$Fe + H_2O \rightarrow Fe(OH^-)_{ads} + H^+ \qquad (10)$$

$$Fe(OH^-)_{ads} \rightarrow Fe(OH)_{ads} + e \qquad (11)$$

$$Fe(OH)_{ads} + SO_4^{2-} \rightarrow FeSO_4 + OH^- + e^- \qquad (12)$$

After Ref. 24, even solid $FeSO_4$ may be the primary reaction product. If $FeSO_4$ first appears dissolved in the very limited amount of water present at the surface, it remains concentrated at the point of its origin, and during the next drying period, cystalline $FeSO_4 \cdot 4H_2O$ fills a shallow point on the surface. The unexpected stability of these sulfate nests, once formed (in the presence of water and oxygen they should be rapidly transformed to hydroxide: $2FeSO_4 + 3H_2O + \frac{1}{2}O_2 \rightarrow 2FeOH + 2H_2SO_4$), can be explained with the help of the mechanism shown in Figure 3.5.

By the action of humidity, the surface on the ferrous sulfate becomes covered by a thin, colloidal hydroxyl membrane.[25] This membrane is supposed to be permeable for anions and neutral water molecules only. During moderately wet periods, the membrane forms and shields the nest from direct atmospheric attack.

As the iron surface in contact with iron sulfate is an anodic site, new SO_4^{2-} can be transported through the membrane and a mildly acidic reaction (pH $\cong$ 3 to 4) is maintained. Drying (e.g., during dry and hot weather) destroys the colloidal character of the membrane by aging and crystallization. At the next contact with highly humid air or liquid water, the membrane is not able to follow the expansive forces because of the high osmotic pressure within the nest. Its bursting leads to an outflow of a drop of highly saturated acidic $FeSO_4$ solution. This can initiate the formation of a new satellite sulfate nest[15] (Figure 3.6). Prolonged contact with liquid can lead to the same result. The osmotic pressure forces water through the permeable membrane into the nest, the volume of which

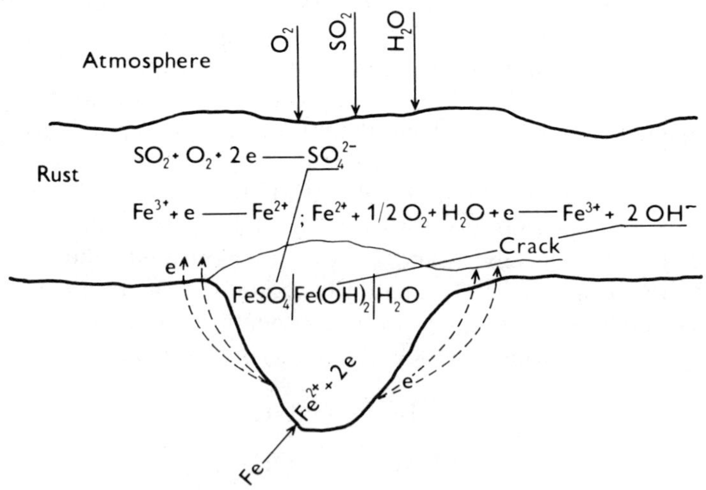

Figure 3.5. Schematic picture of a "sulfate nest."

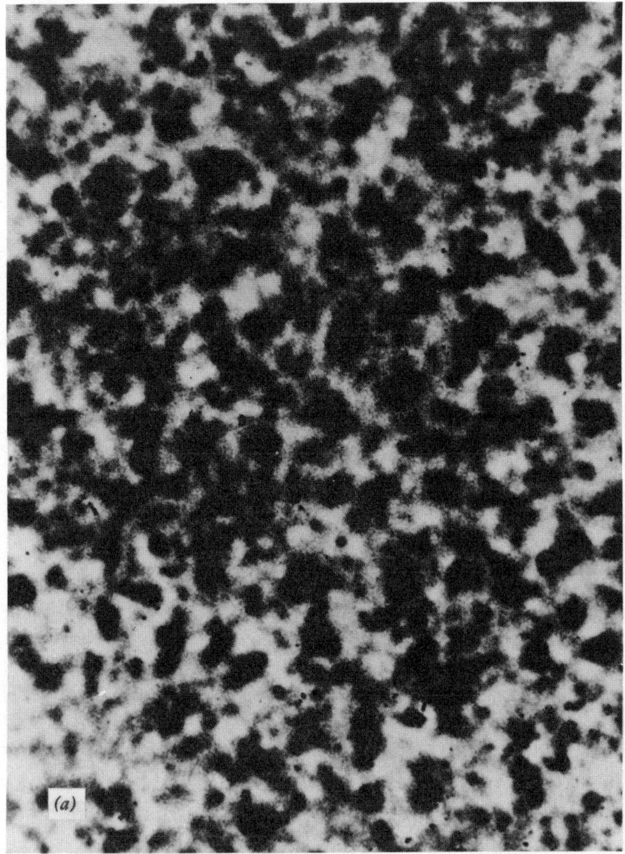

Figure 3.6. Autoradiographic print of (*a*) newly originated sulfate nest; (*b*) surface spreading of sulfate nest.

grows and breaks the membrane. In laboratory experiments it has been shown that the expulsion of the solution takes place primarily at the edge of the nest, rarely at its top. If the change of humidity is very abrupt (e.g., if a dried sample is exposed to a spray of water), the expulsion of the drop may be very vigorous, so that it settles on the surface at a certain distance from the original boundary of the nest. In most cases it flows out only. The trace of a spreading sulfate nest shows a typical chainlike form similar to the filiform corrosion found to take place on steel surfaces polluted by different corrosion-stimulating salts.[26]

This mechanism leads to a final more-or-less uniform distribution of sulfate in the rust layer. Although the relative stability of sulfate nests has

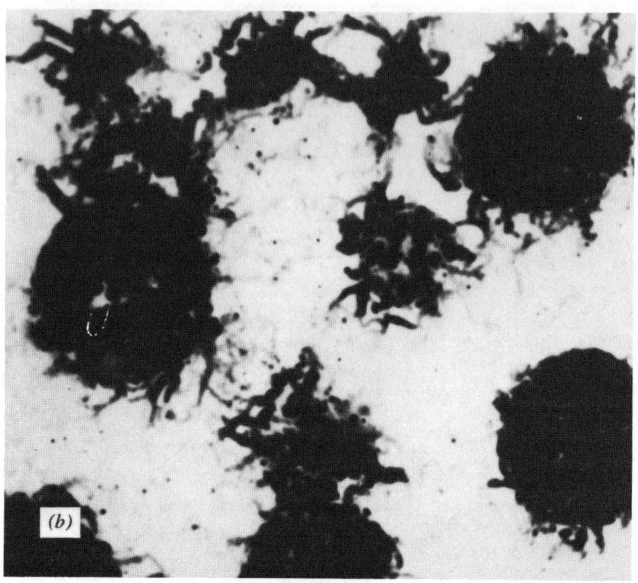

Figure 3.6. (continued)

been postulated, the mechanism leading to the formation of satellite nests results in uniform covering of the surface by sulfate nests. The original or primary sulfate agglomerations may "die" or their relative importance becomes comparable with that of their satellites. There seems to exist a relation between the SO_2 content of the air and the final steady-state density of sulfate nests covering the iron rust interface.

It is necessary to assume a limited life of the sulfate nest even though we speak of its surprising stability. If a once-formed nest should not "die," it would cause a typical craterlike nonuniform corrosion pattern. Practical experience shows, however, that old, rusted surfaces are to a high degree uniformly corroded even though existing nests can be traced by chemical prints or isotope tracers. It can be deduced that sulfate nests "die" and reappear at random, the pattern of this mechanism depending on the properties of the steel and the environment. Concluding, the mechanism of the stimulating action of sulfurous pollution on the corrosion of iron differs from that on other metals. Every molecule of atmospheric SO_2 stepping into the system has the ability to initiate corrosion of several atoms of iron (according to Schikorr, up to 40 or 50). This special feature of rusting is of outstanding technical importance if iron (steel) has to be successfully protected against atmospheric attack.

Salt agglomerations form on steel by the action of other stimulants, also. Chlorides have been found to initiate the growth of typical forms of nests, the behavior of which slightly differs from that of sulfate nests. Their formation probably starts by the solution of a soluble chloride containing particles settled on the surface. At high humidity, a concentrated chloride solution forms and the inherent passivity is destroyed. Chloride concentrates in a pit forming the center of the nest, whereas OH^- ions are formed concentrically. This special mechanism has been demonstrated by tracer methods with $^{35}Cl^-$ and $^{23}Na^+$ (Figure 3.7). The mechanism is typically electrochemical; the center of the nest is anodic and Cl^- ions are transported to it in the electric field. OH^- ions are distributed around the center. The existing pH gradient helps to stabilize the nest. It shows a less evident tendency to spreading than do sulfate agglomerations, most probably because the steep pH gradient leads to a colloidal ring-form membrane of rust hydroxides between the anodic and cathodic surface areas, with higher resistance to changes of humidity.

The information gained by studying the formation and behavior of sulfate nests leads to the conclusion that hydrolysis of primary-formed $FeSO_4$ is the main source of rust formation, and this controls the kinetics of atmospheric corrosion of plain carbon steels and similar materials, such as cast iron. Hence higher corrosion velocities should be expected in environments favoring the formation of sulfate nests and their hydrolytic destruction [i.e., in humid (but not too rainy) and highly polluted atmospheres].

3.3.2.2. *Long-Term Course of Atmospheric Corrosion of Plain Carbon Steel in Environments Polluted by SO₂*

The knowledge of the long-term course of atmospheric corrosion provides most important technical information and is the basis for choosing proper measures of protecting equipment against the destructive action of the atmospheric environment.

Diverse properties of atmospheres resulting from macro-, mezzo-, and microclimatic influence lead to highly different time dependencies of corrosion, as can be seen on Figure 3.8. The common feature of all corrosion–time curves of steels is a rather high corrosion velocity in the starting period of the process (the short induction periods cannot be seen; their duration in open atmospheres with a higher aggressivity is a few hours or days), with a later progressive decrease of the corrosion velocity. Within several months (from 12 to 60 and more) the process seems to attain steady-state conditions. The corrosion velocity becomes formally constant.

For technical purposes it is very important to get a thorough knowledge of the kinetic parameters of the process as well as of the main reactions of steel corrosion in atmospheric environments. This knowledge is useful for design engineers. Kinetic data and a description of the chemical mechanism help in selecting correct protective measures and their application so as to obtain their optimal possible properties.

Studies organized to clarify the dependence of the long-term course of atmospheric steel corrosion on characteristic properties of the environ-

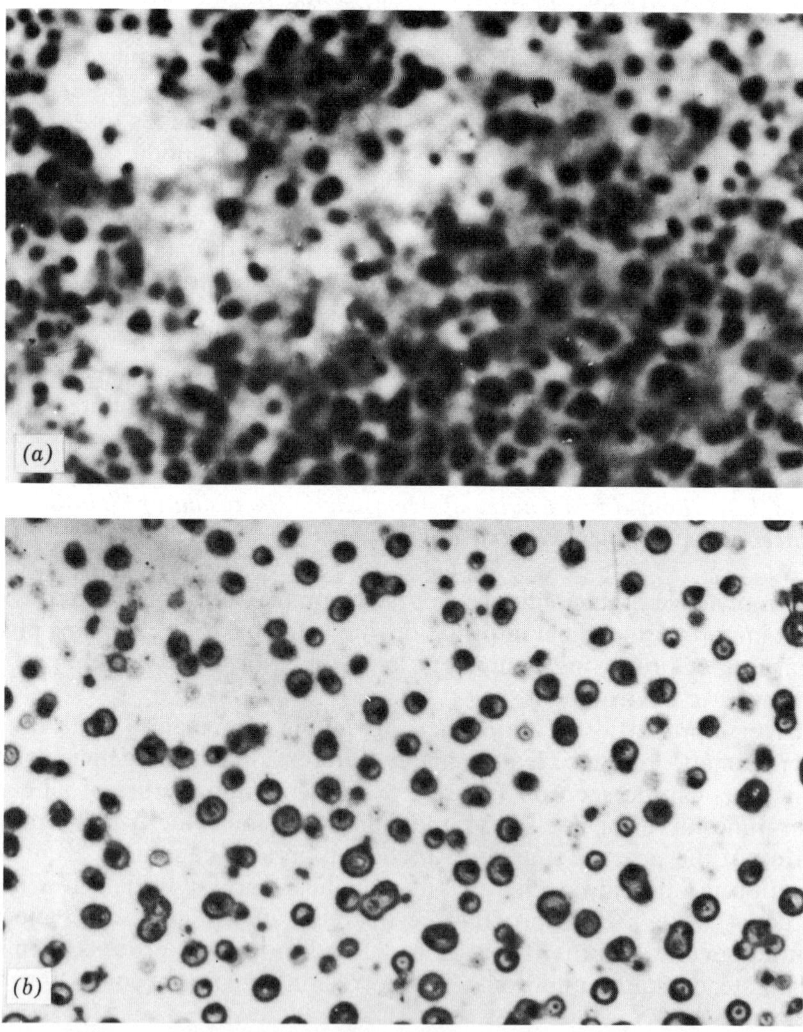

Figure 3.7. Autoradiographic print of chloride nests: (a) $^{35}Cl^-$; (b) ^{23}Na.

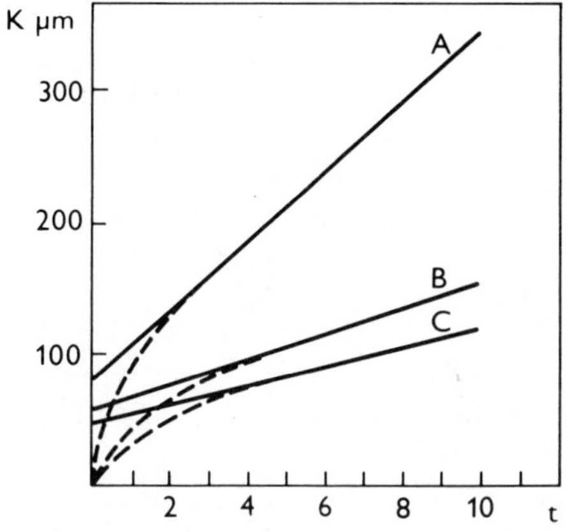

Figure 3.8. Corrosion–time curves of plain carbon steel on three atmospheric sites (K, corrosion; t, years).

ment have led to the statement that two factors control the kinetics of the process[27–30]: the time of surface wetness and the degree of contamination of the atmosphere with stimulating matter.

In most types of atmospheres, sulfur dioxide and other sulfur compounds are the most important stimulative species. Other factors are of less importance. Temperature plays a restricted role, as it affects the time of surface wetness. Below the freezing point, corrosion stops and even at temperatures close to it, the corrosion process is so slow that it loses technical importance.[31] Within the temperature range of existing surface electrolytes (i.e., $>0°C$ to about $20°C$), the temperature dependence of atmospheric corrosion is practically negligible. This can be explained by the contrary effect of increasing reaction velocity and decreasing solubility of gaseous active components of the atmosphere (O_2, SO_2) in the electrolyte with increasing temperature. The controlling action of "time of surface wetting"[30] on the amount of corrosion (K) can be expressed by

$$K = \sum \tau V_K \tag{13}$$

in which V_K is the corrosion velocity at any moment of active corrosion (i.e., in the course of each period of the wetting). As shown on Figure 3.8, the corrosion velocity (V_K) slows down with time (under the assumption of constant average values of surface wetting).

The corrosion velocity V_K is, if temperature is neglected, mainly affected by the composition of the surface electrolyte—the activities of the stimulating anion (e.g., SO_4^{2-}) and of the water. This can be seen in Figures 3.9 and 3.10, which show the results of laboratory measurements of

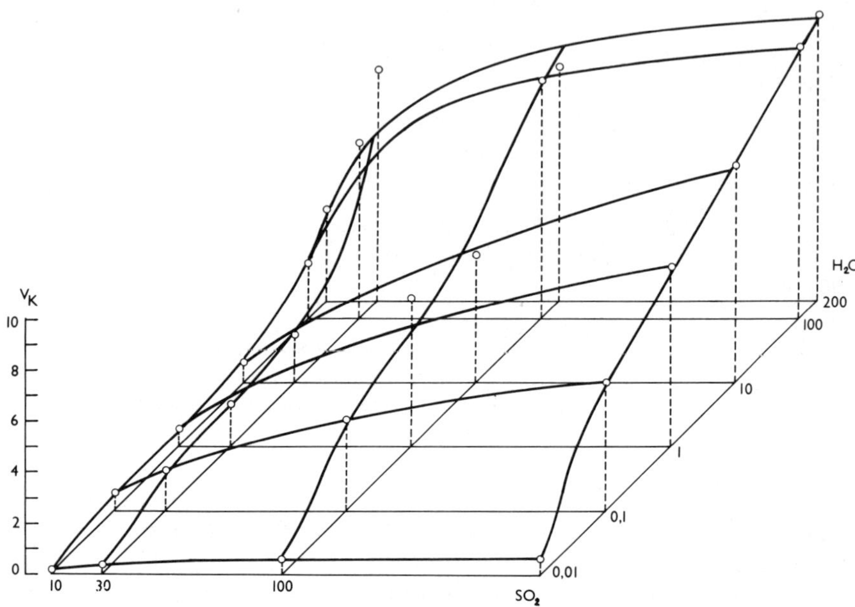

Figure 3.9. Dependence of corrosion rate on SO_2 pollution (mg m^{-2} day^{-1}) and activity of water expressed in thickness of electrolyte film in μm; 5°C (Bartoň et al.[31]).

the corrosion velocity of plain carbon steel and its dependence on the amount of the water and SO_4^{2-} ions in the electrolyte.

The previously mentioned long-term atmospheric corrosion experiments were planned to reveal the relations between atmospheric corrosion and the properties of different atmospheres and allowed the determination of several important principles describing the agressivity of the environment toward plain carbon steel and other metals. For explaining the final results, it seems necessary to describe the methodical principles of this program.

Meteorological values are measured by standard techniques: relative humidity and temperature by thermohygrographs and separate daily readings at 7 A.M. and 2 and 9 P.M., with interpolations between values at 9 P.M. and 7 A.M. of the next day. The amount and duration of precipitation is registered daily. Air pollution is evaluated monthly by cummulative sorption of acidic sulfur compounds on a geometrically defined alkaline surface (a porous filtering cardboard impregnated with sodium carbonate[32]). In addition, collection and analysis of dust, energy of solar radiation, and wind velocity and direction measurements are performed.

The time of surface wetting is determined with the help of a computer

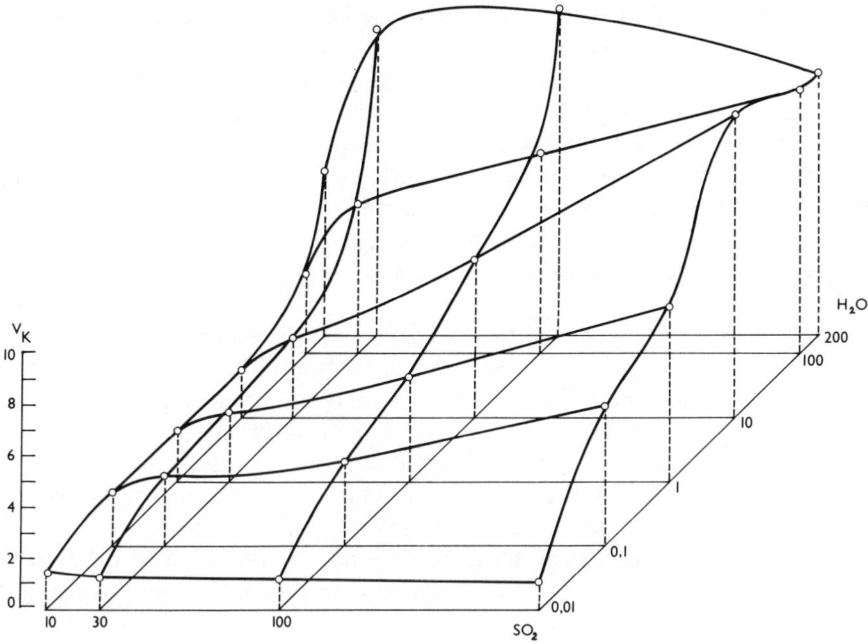

Figure 3.10. As Figure 3.7, at 15°C.

on the following assumption. Based on laboratory tests, atmospheric corrosion of steel can be expected to proceed with technically significant velocities at temperatures $>0°C$ and relative humidities $>80\%$. For each site and month of given period, wetting periods are computed and cumulative values are analyzed by linear regression:

$$\tau_m = A + \tau t$$

where τ_m = monthly sums of wetting times (hours)
    t = total time (hours)
    A = a constant
    τ = significant "average" wetting time of the site (hr day^{-1})

This "average" τ - value includes periods with rain, falling mist, and so on, during which relative humidity is generally above 80%. Only the duration of situations with melting snow or frost are not expressed by the criterion mentioned. These situations are relatively scarce and short-termed and should not significantly distort the necessary information. The duration and amount of liquid precipitation is neglected in the experiment described. Higher amounts of stagnant water during rain can accelerate the process, but flowing water dilutes the surface electrolyte, the effect of which is to lower the aggressivity of the surface electrolyte.

Duration of surface wetness can, however, be measured directly. Two types of instruments are available. The first is based on the function of a battery of couples designed to yield measurable currents if their surface is wetted. The metals used are Cu–Fe[33] or Pt–Zn.[27] The electrodes of each metal (noble, nonnoble) are connected. Each couple is separated by a very thin layer of insulating material. The distance between dissimilar metal is <0.5 mm. The time of wetness is registered with the help of a mA meter.

Another device[34] works with a passive pair of electrodes of the same mostly passive (nickel- or chromium-plated brass, stainless steel) or immune metal (gold, platinum). The system is continually polarized to a constant potential difference (e.g., 1 V). During dry periods no current flows. If the surface of the sensing element gets wetted, the resistivity falls abruptly and current is registered. The evaluation unit registers the frequency of wetting and its duration and summarizes both.

The amount of adsorbed sulfur compounds on the alkaline cardboard discs is estimated as SO_4^{2-} (after extraction and oxidation) and expressed in mg SO_2 m^{-2} day^{-1}.

The characteristic values of the testing sites are summarized in Table 3.1. For expressing the relations between these characteristics of the environment and the course of corrosion, the corrosion–time curve should be described mathematically. The simplest method to express the long-term corrosion course follows from the obvious setting up of steady-state conditions described by a linear time–corrosion relation. If the first part of the corrosion–time dependence (with decreasing corrosion velocity) is not taken into account (e.g., the corrosion up to 6 or 12 months), the

Table 3.1.   Long-Term Climatic and Aerochemical Values of Testing Sites

	A: Urban	B: Heavy industrial	C: Rural
Mean temperature (°C)	8.8	9.65	10.45
Mean relative humidity (%)	79.5	74.0	75.6
Precipitation (mm a^{-1})	584	449	526
Wetness[a] (hr a^{-1})	3890	3050	3700
SO_2 (mg m^{-2} day^{-1})	83	131	32

[a] Hours with relative humidity $\geq$ 80% at $\geq$ 0°C.

remaining part of the relation can be expressed by

$$K = A + Bt \qquad (14)$$

where K is the corrosion, A a constant cutting the corrosion axis and describing the amount of corrosion during the nonsteady-state period, B the linear corrosion velocity, and t the time.

If linear regression is applied to the curves of Figure 3.8, the results shown in Table 3.2 are obtained. The regression coefficients show that the general linear relation expresses the corrosion–time dependence quite satisfactorily.

It is, of course, possible to find high regression coefficients for such relations as

$$K = At^B \qquad (15)$$

or for hyperbolic functions with the slope of the asymptote describing the steady-state velocity of corrosion. The simplicity of linear regression and the easier handling of constants A and B in further analyses of the influences affecting the long-term kinetics of atmospheric steel corrosion outweighs its lower precision.

The results obtained by computing results of 9 years of exposure are summarized in Table 3.2. It is obvious that the main influence on corrosion velocity has to be attributed to the degree of SO_2 pollution of the environment. To find a mathematical expression of this effect as well as of the previously mentioned "time of wetness" principle, it was tried to fit time of wetness (hr day^{-1} mean value from Table 3.1) and SO_2 pollution X (mg m^{-2} day^{-1}) with the "steady-state" corrosion velocity (constant B from Table 3.2) according to

$$B = M\zeta^{n_1}X^{n_2} \qquad (16)$$

The results, comprising the values of n_1, n_2, and the coefficients of correlation, are shown in Table 3.3. Equation 16 can easily be expressed by a nomograph (Figure 3.11).

To show the isolated influence of SO_2 pollution, the results were transformed by computing

$$\frac{B}{M\tau^{n_1}} = f(X) \qquad (17)$$

that is, the course of changes of the steady-state velocity per hour of wetness with increasing SO_2 pollution. Figure 3.12 shows the result. The highest relative influence of growing SO_2 air pollution is observed in the

Table 3.2. Linear Regression Constants A amd B and Correlation (R) of Corrosion–Time Courses for 10 Unalloyed Steels

	Site (Table 3.1)								
	A			B			C		
Steel number[a]	A (μm)	B (μm a^{-1})	R	A (μm)	B (μm a^{-1})	R	A (μm)	B (μm a^{-1})	R
1	60.3	7.8	0.85	53.4	36.1	0.96	44.0	6.5	0.89
2	58.3	9.6	0.92	80.1	26.4	0.85	47.0	7.3	0.96
3	65.6	10.2	0.97	73.52	38.0	0.99	52.3	7.3	0.94
4	60.1	10.3	0.92	68.7	32.4	0.98	45.3	7.1	0.97
5	62.0	10.0	0.94	58.1	34.7	0.99	43.3	6.7	0.96
6	66.8	5.7	0.88	68.0	31.9	0.99	47.3	5.5	0.97
7	60.1	5.8	0.84	67.2	26.2	0.99	48.3	4.3	0.91
8	59.4	7.9	0.83	72.5	27.0	0.99	45.4	4.8	0.89
9	69.0	9.8	0.96	74.8	35.3	0.94	58.0	6.3	0.84
10	57.3	4.3	0.92	73.5	24.4	0.98	44.5	2.9	0.92

[a] Chemical composition, Table 3.4.

Table 3.3. Regression of Steady-State Corrosion Velocity (B) with Time of Wettness (τ) and SO_2 Pollution (X) according to $B = M\tau^{n_1}X^{n_2}$ for Low-Carbon Steel, Zinc, Copper, and Aluminum

	M	n_1	n_2	R
Steel	0.015	0.43	0.57	0.87
Zinc	0.0003	0.50	0.72	0.90
Copper	0.00003	0.55	0.50	0.90
Aluminum	8×10^{-9}	0.68	1.16	0.79

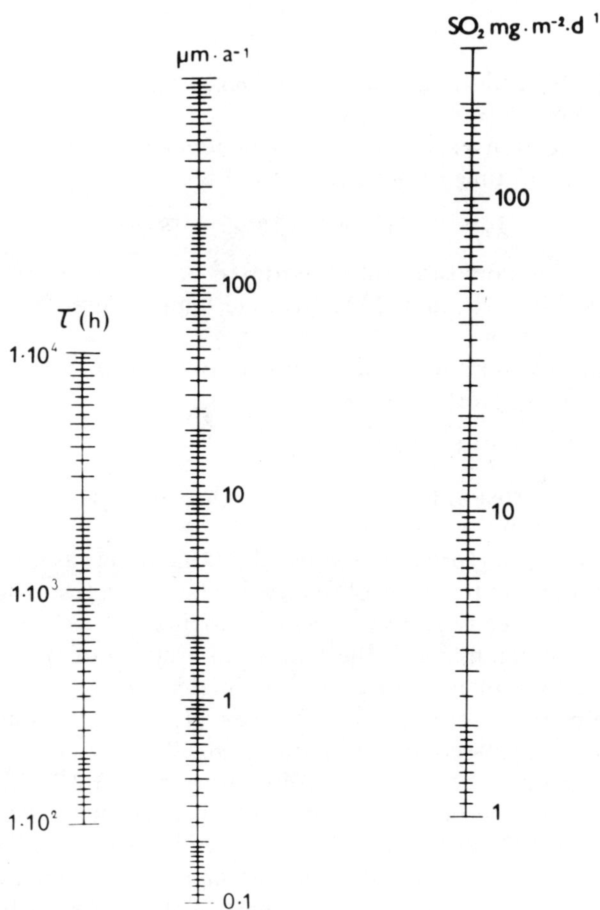

Figure 3.11. Nomograph, expressing the dependence of steady-state corrosion rate of low-carbon steel (in $\mu m\ a^{-1}$) on SO_2 pollution (in $mg\ m^{-2}\ day^{-1}$) and time of wetness (τ, hours per year).

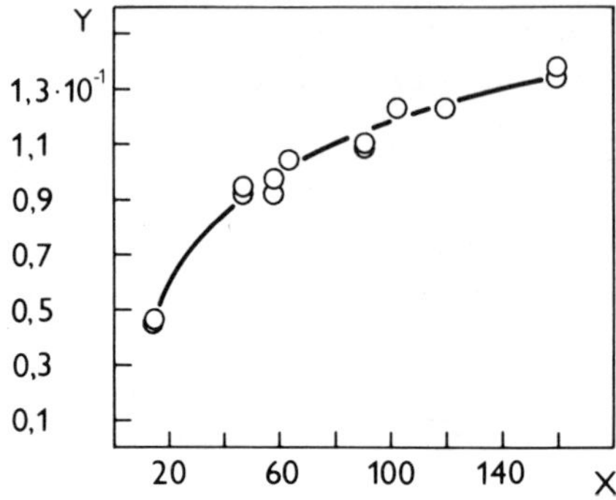

Figure 3.12. Dependence of steady-state corrosion rate of steel per hour of surface wetness on SO_2 pollution, X (mg m^{-2} day^{-1}).

region of low SO_2 pollution values; with increasing pollution, the relative influence becomes lower.

Sereda's[27] equation expressing the dependence of steel corrosion on sulfur pollution and time of surface wetting is as

$$Y = 0.131X + 0.0180Z + 0.787$$

where Y is the log corrosion rate (mg dm^{-2} per day of wetness), X the SO_2 (SO_3 dm^{-2} day^{-1}), and Z the average temperature (°F) during wet days.

Haynie and Upham[28] give the following expression based on results obtained on sites in North America:

$$Y = 9.013e^{\,0.00161SO_2}(4.768t)^{\,0.7512\,-\,0.00520Ox}$$

where Y is the corrosion loss (μg SO_2 m^{-3}), t the time, and Ox the oxidizing matter (μg m^{-3}).

Both these relations formally express the same things as ours does, but they are computed on the basis of relatively short time tests, not allowing the expression of steady-state corrosion velocities.

For better understanding of the long-term kinetics of steel corrosion, it is necessary to explain the reason for the decreasing and later steady corrosion velocity observed in most types of open atmospheres. This phenomenon is closely connected with the mechanism of rust formation under the stimulating effects of sulfur dioxide (Section 3.2.1). Sulfate nests play an important role in the growth mechanism of rust. The higher the concentration of SO_2 in the air, the more densely the surface becomes covered. During later periods thicker rust layers are formed (mainly by hydrolysis of primary $FeSO_4$) and the part of the surface where sulfate nest are active decreases. Old nests "die" and new ones arise; their amount becomes constant eventually, so that steady-state conditions are attained. Even then the summarizing principle of

$$K = \sum \tau V_K \qquad (13)$$

remains valid (Figure 3.8), but from the technical viewpoint the long-term course of corrosion becomes linear and steeper in more polluted atmospheres, where the steady-state coverage by sulfate nests is higher.

3.3.2.3. Influence of Chemical Composition of Steels on Corrosion in Polluted Atmospheres

Steel is a general denomination for iron alloys with metals and nonmetallic elements. Many metals are common constituents of steels—mainly Mn, Cr, Ni, Mo, V, Cu, and others. All steels contain nonmetals as carbon (the most important constituent of steel), P, S, and so on. Even the so-called group of plain carbon steels contains changing amounts of C and Mn (as desirable components) and Cu and other metals (e.g., P and S) as indifferent or harmful contaminants.

It is well known that high-alloy steels containing >12% Cr or >18% Cr and >8% Ni (and other metals) become passivated in many environments, including atmospheric. These "stainless" steels do not rust in atmospheres not containing higher amounts of sulfurous or other activating pollutants. Only in highly aggressive atmospheres do stainless steels lose their passivity and become covered by a thin adherent layer of corrosion products, containing compounds of iron and the alloying elements. Their corrosion velocity is, however, extremely low, particularly as far as Cr18Ni9 or Cr18Ni9Mo steels are concerned. The formation of corrosion products leads to aesthetic surface deterioration only and does not significantly affect the strength of stainless steel parts. In carbon and low-alloy steels the content of Cr (or Cr and Ni) is not sufficient to build up stable passivity in polluted atmospheres. The chemical composition of such steels, however, strongly affects the long-term kinetics of atmospheric corrosion.

By multiple correlation analyses the effect of alloying elements and impuruties on the kinetic parameters of 10 sorts of carbon steels (Table 3.4) tested for about 9 years on three sites with different aggressive atmospheres (Table 3.1) was analyzed. First the corrosion–time dependence was computed according to

$$K = A + Bt \qquad (14)$$

where K is the corrosion (g m^{-2}), t the time (days) and A and B are constants. Terms of evaluation of 1, 2, 3, 5, and 9 years were used and constants A and B as well as correlation coefficients were computed

Table 3.4.   Chemical Analyses of Tested Nonalloyed Steels (percent)

Steel Number	C	P	S	Mn	Si	Cr	Cu
1	0.07	0.020	0.027	0.28	0.10	0.04	0.21
2	0.10	0.013	0.033	0.21	0.08	0.05	0.12
3	0.03	0.025	0.013	0.25	0.09	0.05	0.12
4	0.04	0.025	0.050	0.30	0.10	0.04	0.10
5	0.045	0.011	0.025	0.25	0.08	0.03	0.13
6	0.21	0.015	0.015	0.28	0.10	0.03	0.07
7	0.13	0.011	0.021	0.39	0.20	0.05	0.18
8	0.06	0.006	0.023	0.30	0.05	0.04	0.11
9	0.12	0.015	0.017	0.71	0.10	0.05	0.08
10	0.58	0.015	0.020	0.61	0.25	0.03	0.10

(Table 3.5). The next step was to correlate the chemical composition with the corrosion velocities B on the single sites. Thus the coefficients of regression listed in Table 3.5 were obtained.

An analysis of the results shows that there exist significant differences in the corrosion behavior of carbon steels due to their chemical composition. It is not surprising that growing sulfur content leads to a steep increase of corrosion velocity; the "inhibitive" action of carbon is, however, unexpected.

By including nine sorts of low-alloy steels (Table 3.6) into the analysis, the sense of the results does not change (Table 3.7).

Table 3.5.   Multiple Linear Regression Coefficients of Steady-State Corrosion Rate (Constant B, Table 3.2) with Chemical Composition of Carbon Steels (Table 3.4)

	Site (Table 3.1)		
	A	B	C
R	0.85	0.86	0.89
C_0	9.2	33.5	6.6
C_1 (C)	−9.6	−8.6	−6.4
C_2 (P)	−1.1	−3.1	−0.9
C_3 (S)	30.2	20.6	21.1
C_4 (mn)	−0.4	−0.8	−0.5
C_5 (Si)	−1.7	−1.0	−0.9
C_6 (Cr)	−1.5	−18.6	−1.2
C_7 (Cu)	−1.5	7.2	−0.24

Table 3.6.   Chemical Analysis of Low-Alloy Steels (percent)

Steel Number	C	P	S	Mn	Si	Cr	Cu (+ Ni)
11	0.34	0.003	0.014	1.05	1.10	0.04	0.11
12	0.49	0.015	0.010	0.80	1.60	0.05	0.10
13	0.43	0.017	0.015	0.85	0.40	0.42	0.16
14	0.43	0.016	0.014	0.70	1.40	0.50	0.23
15	0.30	0.019	0.010	0.83	1.20	1.30	0.30
16	0.27	0.011	0.030	0.50	0.37	1.30	0.24
17	0.28	0.014	0.031	0.53	0.25	2.50	0.15
18	0.19	0.012	0.020	1.10	0.25	0.70	0.16
19	0.44	0.016	0.017	0.78	0.25	1.50	0.21

It can be concluded that the differences between corrosion velocities as affected by the chemical composition of nonpassive steels are more pronounced in severely polluted atmospheres than in rural ones.

The empirical experience that some alloying elements such as Cu and P raise the corrosion resistance of steel in open atmospheres has led to the production of special low-alloy "slow-rusting" steels. These contain small additions of Cu, Ni, Cr, Si, Mn, and P. Their corrosion resistance is due to special protective properties of the rust "patina" forming on them in the course of atmospheric exposure. After several months steady-state conditions are reached, characterized by very slow corrosion velocities, <1 to $2 \ \mu m \ a^{-1}$. Only in extremely polluted atmospheres are the

Table 3.7.   Multiple Linear Regression Coefficients of Steady-State Corrosion Rate of 19 Carbon and Low-Alloy Steels with Chemical Composition

	Site (Table 3.1)		
	A	B	C
R	0.82	0.69	0.82
C_0	7.60	16.57	3.88
C_1 (C)	-4.63	-4.22	-1.17
C_2 (Mn)	-0.88	-4.94	-0.97
C_3 (Si)	-1.32	-6.19	-1.44
C_4 (Cr)	0.40	-7.73	-0.07
C_5 (Cu)	-2.79	-14.23	-2.22
C_6 (P)	-1.53	-20.02	-2.75
C_7 (S)	94.05	777.45	131.07

corrosion velocities somewhat higher, 4 to 5 μm a^{-1}, the comparable corrosion rates of plain carbon steels being about 30 to 40 μm a^{-1}. It should be stressed that favorable properties of the rust layer can be expected only in open-air exposure with free access of all atmospheric factors. Under sheltered conditions with excluded rainfall and sunshine, the formed rust does not afford protection, and these steels corrode at comparable rates with plain carbon sorts.

Low-alloy slow-rusting steels are a very useful material for steel structures exposed to the open air, as they need not be painted or protected by other coatings.

The protective nature of rust on low-alloy slow-rusting steels is probably connected with the stability of amorphous FeOOH, formed by hydrolysis of primary $FeSO_4$. The protective amorphous state of rust is due to the presence of ions of the alloying elements in the corrosion products inhibiting the transformation of amorphous FeOOH to coarse crystalline, nonprotective αFeOOH.[35] Amorphous FeOOH covers to sulfate nests and does not allow their spreading by the previously described mechanism of a periodically bursting membrane (Section 3.2.1), but does not hinder the outward diffusion of SO_4^{2-} ions and their leaching by rain.

3.3.3.   Influence of Atmospheric Pollution on the Long-Term Corrosion Course of Nonferrous and Passive Metals

3.3.3.1.   *Metals Bonding Stimulants into Insoluble Products (Zinc, Copper, Cadmium, Lead, etc.)*

The accelerating mechanism of corrosion-stimulating air pollution components on these metals differs from that described for iron (Section 3.3.2). In the system of factors influencing the rate of atmospheric corrosion, the duration of wetness and the access of pollutants to the corroding surface remain dominating.

The first reaction step on zinc (a typical representative of the group) is the formation of a hydroxide product layer:

$$Zn + H_2O + \tfrac{1}{2}O_2 \rightarrow Zn(OH)_2 \tag{18}$$

This reaction is an electrochemical one with the typical mutually dependent anodic and cathodic part. Although zinc is a highly electronegative metal, oxygen consumption controls the cathodic step. Owing to the highly insoluble, adherent, and dense layer of $Zn(OH)_2$, the reaction is strongly polarized. Once the layer has been formed, the corrosion rate becomes very low. The kinetic principle of atmospheric corrosion of zinc is the same as that of passive metals. The controlling step is the destruction of the previously formed "steady-state" layer of the protective product. The higher the rate of its destruction, the higher is the overall corrosion velocity, as new $Zn(OH)_2$ has to be formed to hold up steady-state conditions. [The renewal of $Zn(OH)_2$ is fast if compared with its destruction and therefore has no influence on the corrosion velocity.]

The mechanisms destroying the protective $Zn(OH)_2$ layer include dissolution, abrasion, erosion, and other physical effects. If no or very low concentrations of stimulants are present, the highly insoluble layer of $ZN(OH)_2$ can react with atmospheric CO_2, forming an additional very protective hydroxide carbonate, $3Zn(OH)_2 \cdot ZnCO_3$. With growing sulfurous pollution, the reaction on the corrosion product/atmosphere interface changes to

$$n Zn(OH)_2 + m SO_2 + \frac{m}{2} O_2 \rightarrow m ZnSO_4 \cdot (n - m) Zn(OH)_2 \qquad (19)$$

The resulting hydroxide sulfate is more soluble than the primary $Zn(OH)_2$. Its dissolution and washing off is easier, so that more $Zn(OH)_2$ has to be formed by new corrosion of zinc to renew steady-state conditions. The higher the SO_2 content in the atmosphere, the more $ZnSO_4$ is contained in the secondary corrosion product and the easier the products are dissolved. In extremely polluted atmospheres, conditions for direct formation of $ZnSO_4$ can arise.[36] Unlike iron (Section 3.3.2.1), every SO_2 molecule taken up by the system is able to destroy only one zinc atom. Also taking into account that the sorption of SO_2 by zinc covered with corrosion products is less than on rusted iron, a different dependence of the corrosion velocity on the pollution grade of the environment should be expected. Figure 3.13, derived on the basis of long-term experiments on sites with differentiated pollution, shows this dependance, computed with

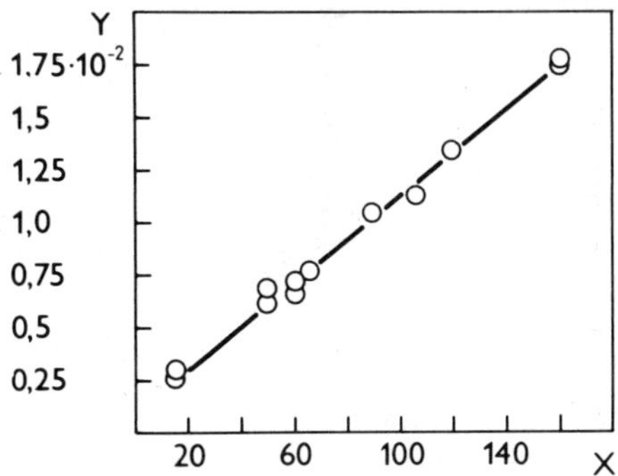

Figure 3.13. As Figure 3.10, for zinc.

help of the equation

$$\frac{V_K}{M\tau^{n_1}} = f(X) \tag{17}$$

where V_K is the steady-state corrosion rate of zinc in an open atmosphere ($g\ m^{-2}\ d^{-1}$), τ the duration of surface wetness (hr day^{-1}), and M and n_1 are constants.

Another form of eq. 17,

$$V_K = M\zeta^{n_1}X^{n_2} \qquad \text{(see also Section 3.3.2.2)}$$

expresses the dependence of the steady-state corrosion rate (V_K) on surface wetness and the grade of SO_2 pollution (SO_2 mg m^{-2} day^{-1}) measured by cumulative sorption on an alkaline surface. Its nomographic form (Figure 3.14) allows a rapid estimation of the corrosion velocity of zinc if time of wetness and sulfur pollution are known. Another expression

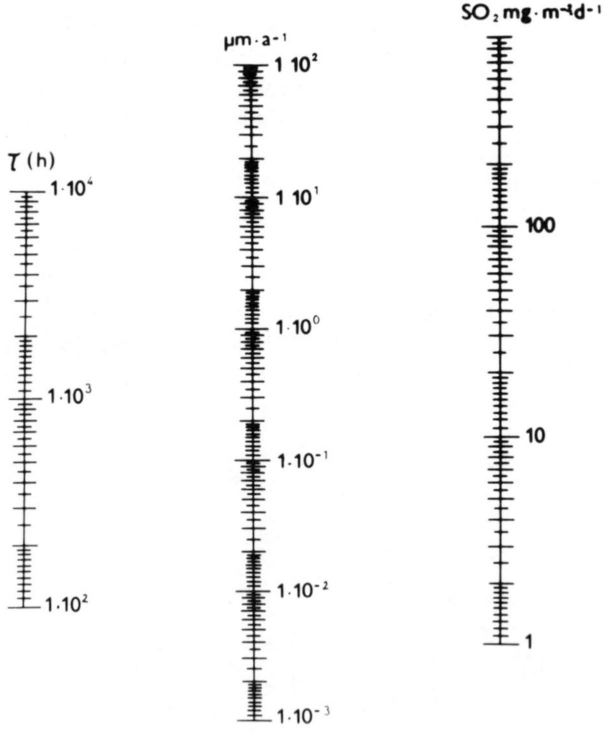

Figure 3.14. As Figure 3.9, for zinc.

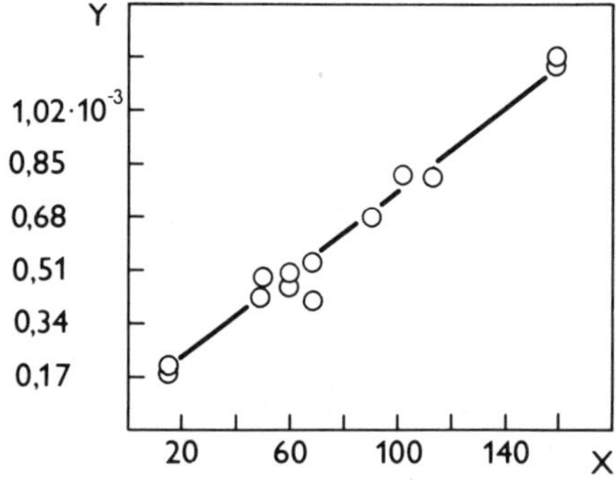

Figure 3.15. As Figure 3.10, for copper.

relating the atmospheric corrosion rate of zinc to time of wetness and sulfurous pollution has been described by Haynie and Upham.[28]

Although no experimental results are available, a similar dependence of the corrosion rate of cadmium on pollution (or pollution and time of surface wetness) can be expected.

The behavior of copper is slightly different. The primary corrosion products are oxides (Cu_2O, CuO) with high protective capacity. These interact rather slowly with atmospheric SO_2 and humidity to form the well-known green patina, mostly composed of $CuSO_4 \cdot 3Cu(OH)_2$. This product is, however, more soluble than the primary oxides, so that the corrosion steady-state velocity is not much less than that of zinc. Figure 3.15 shows the dependence of the steady-state corrosion velocity of copper (per hour of surface wetness) on the pollution grade of the environment. The nomograph in Figure 3.16 is based on the same equation as those for iron (steel; Figure 3.11) and zinc (Figure 3.14).

Copper and its alloys readily react with H_2S and other sulfur compounds polluting the air. Corrosion rates may reach quite high values near an oil hydrogenation plant, reaching values as high as >5 μm a^{-1}.

No experimental results are available for expressing the corrosion pollution dependence for lead, although this metal should obey the foregoing general principles, explaining the accelerating action of SO_2 on metals forming corrosion products with SO_4^{2-} bound in scarcely soluble compounds.

Also, lead is attacked by atmospheric H_2S, forming sulfide with low protective properties.

In slightly polluted (rural) atmospheres nickel keeps up its passivity. If the concentration of sulfur dioxide is higher (urban, industrial), nickel is very slowly attacked, forming hydroxide sulfates as the corrosion product.

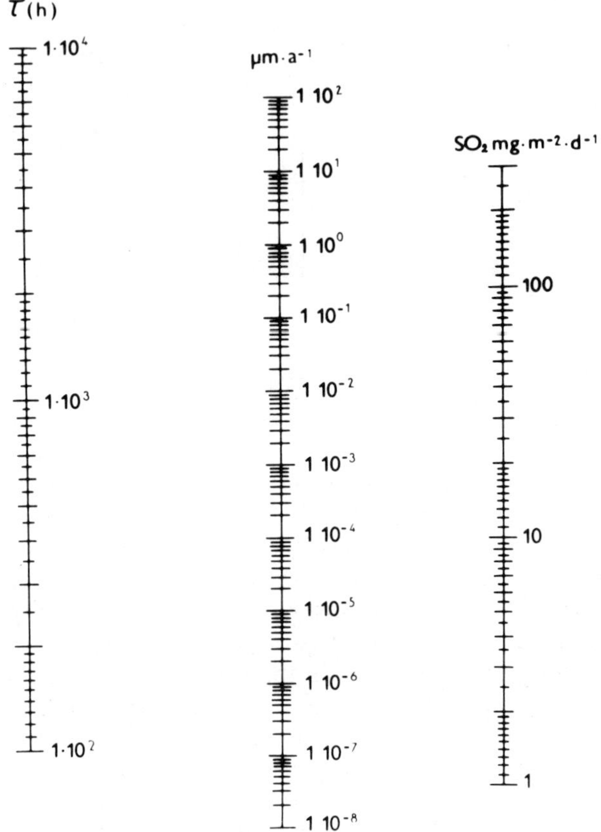

Figure 3.16. As Figure 3.9, for copper.

Magnesium and its technical alloys corrode in polluted atmospheres according to the same principles as zinc, but faster. The protective layer of $Mg(OH)_2$ and $MgCO_3$ formed in relatively clean environments becomes proportionally changed to $MgSO_4$ as the pollution grade increases. This soluble salt does not, of course, provide any protection.

3.3.3.2. *Nonferrous Passive Metals*

The main representatives of this group are aluminum and its alloys and alloys based on titanium and zinconium. Chromium is applied for coatings only. As an alloying constituent, it renders high passivity to NiCr alloys

($>$20% Cr) that do not corrode even in highly polluted atmospheres, and to the previously discussed stainless steels.

Pure technical aluminum is highly resistant to the action of polluted open air conditions. Schikorr[37] has proclaimed that aluminum declines to react with atmospheric SO_2. This empirically based rule has been verified by laboratory experiments,[16] proving that different from iron alloys, zinc, or copper; a very small proportion of SO_2 molecules contacting passive aluminum surfaces become adsorbed and are able to take part in the corrosion reaction.

Long-term experiments in different atmospheres have proved the extremely low corrosion rate of passive aluminum and its alloys. Alloying aluminum with nonnoble components (Zn, Mg, Si, Mn) increases the corrosion rate to a certain but low degree. Copper as an alloying component is much more detrimental. Corrosion velocities grow through its action up to 4- to 20-fold. It is interesting to note that the relative influence of alloying on decreasing the corrosion resistance of aluminum is more distinct in relatively clean (less polluted) atmospheres, whereas in industrial environments with extreme SO_2 concentrations the relative differences are suppressed even though the absolute corrosion velocities are by one to two orders of magnitude higher. These conclusions can be deduced from Table 3.8 and Figure 3.17

The relation between corrosion velocity and the SO_2 pollution grade (Figure 3.18) differs strongly from that of iron (plain carbon steels) or

Table 3.8.   Steady-State Atmospheric Corrosion Rate (μm a^{-1}) of Aluminum and Its Alloys

	Site (Table 3.1)		
	A	B	C
Al 99.85%	0.114	2.057	0.034
Al 99.5%	0.139	2.215	0.040
Al 99.95%	0.145	2.215	0.033
AlMg2%	0.188	2.191	0.056
AlMg3%	0.228	2.169	0.050
AlMg5%	0.223	3.419	0.056
AlMn1%	0.272	2.584	0.046
AlMgSi	0.214	2.055	0.040
AlMgSiCu	0.317	3.984	0.076
AlCuMg	0.490	2.950	0.116
AlCu4%-Mg1%	0.354	4.389	0.158
AlCu4%-Mg	0.413	3.63	0.145

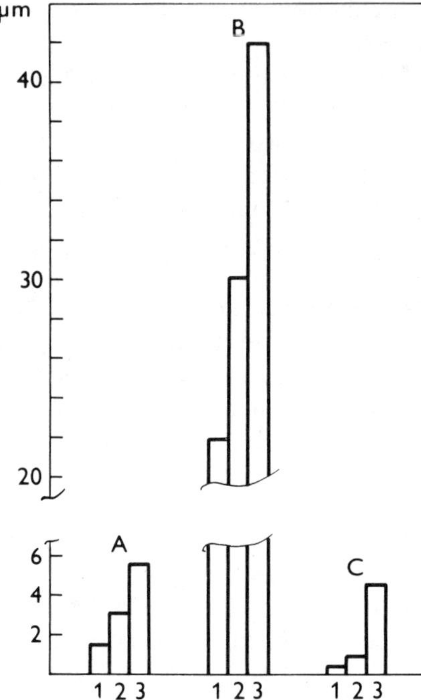

Figure 3.17. Corrosion of aluminum alloys in different polluted atmospheres: 1, Al (99.95; 99.85; 99.5%); 2, Al alloys not containing copper; 3, Al alloys with copper; A, B, C, sites from Table 3.1 (A, urban; B, heavy industrial; C, rural).

nonferrous metals of the Zn, Cd, Cu mechanism group (Section 3.3.3.1). Within the range of low and middle grades of SO_2 pollution (rural, urban, light industrial types), the corrosion velocity grows very slightly. A steep increase of the corrosion rate is observed between light industrial and heavily polluted atmospheres in regions with highly concentrated mining,

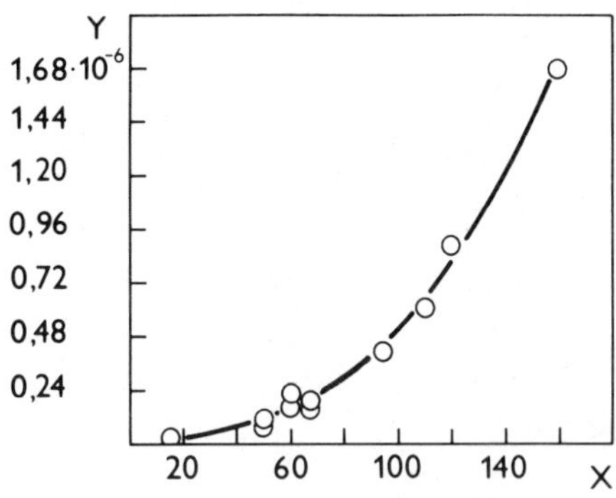

Figure 3.18. As Figure 3.10, for aluminum.

energy, or chemical and metallurgical industries producing extreme sulfurous pollution.

It seems very probable that the rapid rise in the corrosion rate of aluminum in these environments is due not only to gaseous contamination of the air. Settling dust particles contain appreciable amounts of active stimulants. If these are not washed off by rain, any period of wetting will produce a concentrated solution of components, destroying the inherent passivity of the metal. This mechanism can be clearly shown by comparing time–corrosion curves of aluminum exposed in the open air and under sheltered conditions, excluding the direct action of rain but allowing the access of all other corrosive factors of the atmosphere (Figure 3.19).

Whereas in open-air exposure, the time–corrosion dependence shows the typical shape of steady-state conditions with low corrosion velocity, the picture obtained under shelter is completely different. After an induction period characterized by undisturbed inherent passivity of the metal, the corrosion rate rapidly rises and eventually attains distinctly higher steady-state values than in the open air. This phenomenon is observed not only in the most aggressive, highly polluted atmospheres, but also in semirural and urban environments.

Few results of atmospheric corrosion measurements with Ti and Zn are available. It is to be expected that even in the most polluted atmospheres, these metals will not lose their inherent passivity and from the technical viewpoint no corrosion will take place.

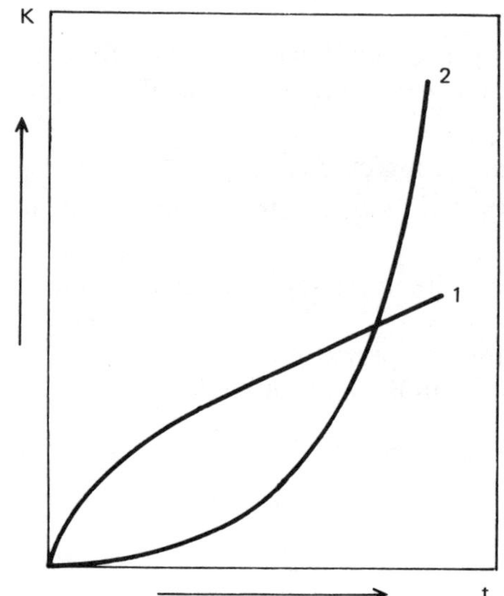

Figure 3.19. Corrosion–time curves of aluminum in heavy industrial atmosphere: (1) open exposure; (2) under shelter.

3.3.3.3. *Noble Metals*

Atmospheric pollutants do not induce corrosive processes on thermo-dynamically noble metal such as gold or the platinum group. Silver is readily attacked by small traces of hydrogen sulfide, forming Ag_2S as the corrosion product. The reaction rate is slow. For electrotechnical applications of silver, even very thin layers of this product may be detrimental. SO_2 is another stimulant of its formation, probably by the reaction

$$4Ag + 2SO_2 \rightarrow Ag_2S + Ag_2SO_4 \tag{20}$$

The interesting theory of SO_2 clathrates formed in atmospheric conditions on silver[21] has already been mentioned.

3.4. DETERIORATION OF SYSTEMS USED FOR PROTECTING METALS AGAINST CORROSION IN POLLUTED ATMOSPHERES

3.4.1. General Comments

Metallic equipment has to be protected against atmospheric corrosion to prevent or postpone deteriorative effects of the environment. The term "deterioration" comprises all effects due to corrosion, preventing rational use of the equipment. From the technical view the following classification of undesired changes by corrosion may be suitable:

Loss of mechanical strength by uniform, nonuniform, or structural corrosion (by intercrystalline or stress corrosion cracking), so that break hazards arise.
Perforation of walls by nonuniform or pitting corrosion.
Change of important properties of the surface (electrical, optical, heat transfer, friction, etc).
Loss of aesthetic properties (e.g., in architectural applications of metals).

The designer and engineer must choose optimal protective systems to obtain the desired technical effects—useful life and economic parameters of the equipment and of its parts.

3.4.2.   Survey of Systems Used for Protecting Metals

For practical reasons the aims of protection may be divided into:

- Systems used for the preservation of the necessary properties of metallic equipment during its technical use.
- Systems protecting metals and metallic goods between completion of one step of the manufacturing process and the next, or of the complete equipment before its practical use; these methods are described as temporary protection.

From the viewpoint of the mechanisms by which protective systems act, the following division seems reasonable:

- Changes of the microclimate surrounding the metal so that the environment becomes less aggressive or does not induce any corrosion at all.
- Selection of materials resisting the corrosive action of a defined grade of corrosion aggressivity. This is done with a view to the type of undesired effects of corrosion deduced from the function of the equipment and of its component as described in Section 3.4.1.
- Use of protective coatings—metallic, inorganic nonmetallic, organic, or duplex (metallic plus organic or inorganic).

Other types of protective methods involve technological principles during production and will not be discussed here.

3.4.3.   Optimal Choice of Protective Systems

The optimal choice of protective systems against atmospheric corrosion is the result of analyzing a system of the following components:

- The aggressivity (corrosivity) of the environment affecting the given metallic equipment (or its component).
- The technical purpose of protection (Section 3.4.1).
- The desired life of the protective method needed for the economical function of the equipment.

The relations describing the reasoning necessary to derive an optimal protective systems (PS) are shown in Figure 3.20. The single components deserve a more detailed explanation.

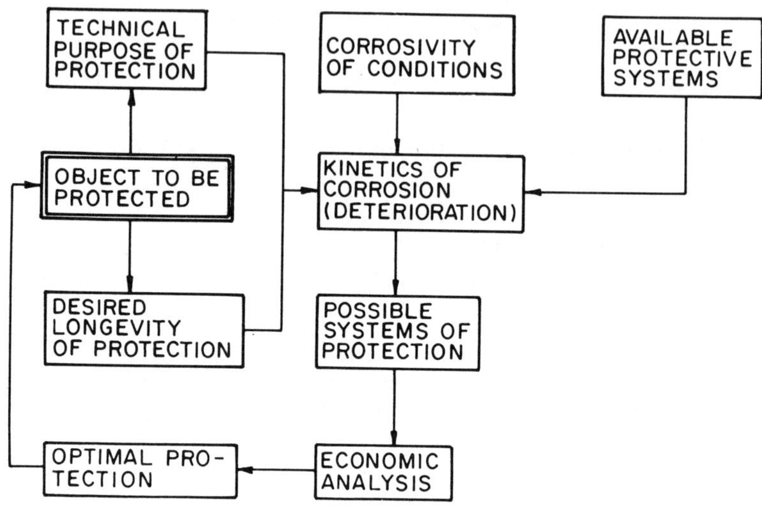

Figure 3.20. Choice of optimal protection against corrosion.

3.4.3.1 *Corrosivity of Atmospheres*

A metallic component often serves a well-defined purpose; that is, it is expected to be used in an environment the corrosive aggressivity of which may be described and technically specified. In many cases the metallic component will be erected in a certain place and will be exposed to the outdoor atmosphere without affecting it, in return. Steel structures such as bridges and masts are an example of this. Other types of equipment produce their own microclimate as a result of interactions of the outdoor atmosphere with the function of the equipment, as in heat production or movement. Moving equipment, such as automobiles, railway engines, and wagons are of this type, as well as many other metallic products producing microclimate effects due to their function. Heat effects are common with pipelines transporting liquid or gaseous media and with electrical equipment.

As has been shown in Section 3, the kinetics of atmospheric corrosion are generally controlled by two simultaneously acting factors: the summed duration of surface wetness periods and the composition of the surface electrolyte during these periods. If the influence of the specific reaction mechanisms of different metals is taken into account (Sections 3.3.1 to 3.3.3), it is possible to derive Figure 3.21, showing atmospheric corrosion as a system of several components with a defined corrosion velocity that is the result of their interactions. This general system may be used for

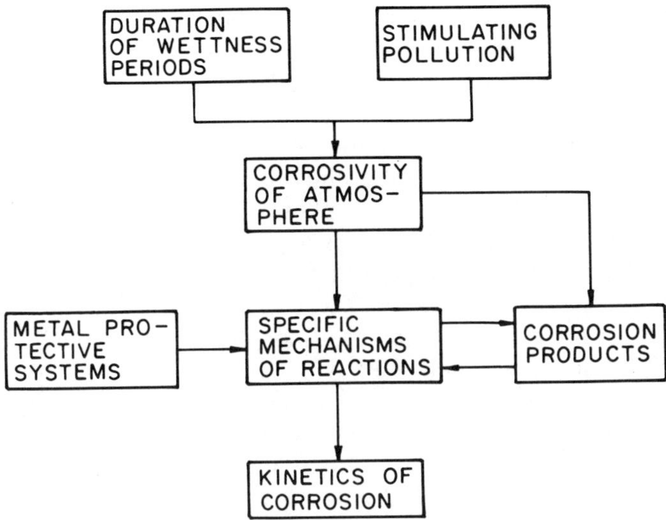

Figure 3.21. Factors influencing the kinetics of atmospheric corrosion.

the definition of a specific corrosivity of the atmosphere surrounding a given metallic product.

Corrosion is possible only if a surface solution is present (the time-of-wetness principle) and if the properties of the electrolyte are able to induce and propagate the process. It is not reasonable to speak of a general corrosivity of the surface electrolyte, as each group of metals is attacked by different stimulating ions contained in it. The same generally stimulating ions act toward the single groups of metals at different critical concentrations. (For more detailed information, Section 3.3 should be consulted.) For the most common metallic materials used for engineering purposes, such as plain and low-alloy steels, copper and its alloys, and zinc (e.g., for coating purposes), it is useful to derive a "corrosivity" classification system based on the ideas expressed in the following grading of both controlling factors: *surface wetting,* where the frequency is from zero to many times daily and duration from zero to thousands of hours per year—and an *excess of stimulating components of the atmosphere*—SO_2 and other sulfur compounds, chlorides, chlorine, NO_x, dust-containing soluble matter, and so on, from the "background" value found in very clean atmospheres to extreme ones typical for surfaces of equipment exposed in severely polluted atmospheres in chemical works or frequently splashed by corrosivity liquids. For practical reasons the pollution grade is expressed by the SO_2 access to the surface (in mg m^{-2} day^{-1}). Other accompanying stimulating pollutants generally follow the concentration of the SO_2 component, with the exception of special cases (e.g., Cl_2 in firms producing chlorine compounds or NO_x in the nitric acid industry). The following grades of both factors have been found useful, as their combination permits the derivation of types of "general" corrosivity (Tables 3.9 to 3.11). These tables show clearly that without "wetting," no or very "light" types of atmosphere predominate even in strongly polluted environments, With increasing frequency and duration of wetness periods, the degree of pollution becomes more important. "No" pollution (i.e., pollution at the background level) does not mean "no corrosion," even in very dry types of atmospheres; the probability of corrosion initiation grows even here with growing time of exposure. The reason for this is that stimulating solid pollutants contain hygroscopic compounds, so that the probability of wetting grows with increasing concentrations of pollutants, even at humidities much below the saturation point of the air. For example, if NaCl is the soluble component of dust, as in maritime regions or near highways after de-icing periods, wetting becomes possible at 76% relative humidity. Gaseous aggressive contaminants act similarly, their primary reaction products with most metals being hygroscopic compounds.

For practical purposes it is useful to define the general corrosivity grades of atmospheres (Table 3.11) by corrosion velocity data for the commonest metallic materials of construction. An example of such definitions is given in Table 3.12

A general classification of the corrosivity of atmospheres may be used for the optimization process for such objects as steel structures. For other types of equipment, the grading problem is more complicated and the general principle serves only as a guidance for detailed description of atmospheric corrosion influences acting on the equipment and its single parts.

The intensity of atmospheric pollution by corrosion-stimulating compounds may be, as shown, classified into grades, which together with a rationally graduation of "surface wetness" conditions, serve as the basis for technical classifications of the corrosivity of atmospheres.

3.4.3.2. *Purposes of Protection*

The general aims of protecting metals were mentioned briefly in Section 3.4.1. Some remarks on the main failure types due to atmospheric corrosion seem to be necessary:

1. The protective system has to ensure that corrosion does not endanger the mechanical strength of the protected part. This demand is very

Table 3.9. Grading of Time of Wetness

| Grade | Wetting | | Example |
	Frequency	Duration (hr yr^{-1})	
1	Nil to very scarce	<10	Indoor conditioned atmospheres
2	Scarce to low frequency seasonal	10–100	Indoor unconditioned atmospheres
3	Frequent periodical	100–1000	Shelters, indoor unconditioned atmospheres with sources of humidity
4	Very frequent with continuous wetness periods of >100 hr	1000–5000	Outdoor atmospheres
5	Nearly continuous	>5000	Indoor and outdoor atmospheres with intense sources of humidity

common and concerns steel structures, parts of transport vehicles, and so on. If the protective system has no decorative function, the following possibilities arise:

- Choose dimensions of the part so as to eliminate the expected loss of strength due to corrosion during use (until the end of the life of the part); this method is seldom used, although the undesirable mass increase of the part is in most cases not so high as to be of much significance.

- Use another material with higher corrosion resistance (lower corrosion velocity); this alternative is limited primarily by the physical

Table 3.10. Grading of Pollution

Grade	SO$_2$ (mg m^{-2} d^{-1})	Example of Atmosphere
1	<20	Very clean rural atmospheres; clean (conditioned) indoor environments
2	>20–40	Clean rural and semirural
3	>40–60	Urban
4	>60–130	Heavy urban and industrial
5	>130–(>400)	Extreme pollution near emission centers

Table 3.11.   Corrosivity Grades of Atmospheres[a]

Grade of wetting (Table 3.9)	Grade of pollution (Table 3.10)				
	1	2	3	4	5
1	I	I	I	II	II
2	II	II	II	II	III
3	III	III	III	IV	IV
4	III	III	IV	IV	V
5	IV	IV	V	V	V

[a] I—Very low corrosivity—corrosion velocities not measurable, low probability of corrosion initiation. II—low corrosivity—corrosion rate very low, probability of corrosion initiation to be considered. III—Medium corrosivity—corrosion rates to be taken into account; probable initiation of corrosion within short periods. IV—High corrosivity—fast corrosion, protection for products with long life expectancy should be chosen very carefully. V—Extreme corrosivity—very fast corrosion, special protective methods necessary.

and technological properties of such materials and by their price (e.g., if carbon steel should be replaced by a stainless type by a nonferrous metal or by plastic). For some types of equipment exploited in mildly polluted and hence mildly aggressive atmospheres, it is possible to replace common steel by slow/rusting types or by a suitable aluminum alloy. The higher price of such a material is easily compensated for by savings, as no protective coatings are needed for the new equipment or during its use.

• Apply a coating with a desired protective longevity. In this case, if no decorative aims are claimed, the coating type can be selected from the broad choice of metallic, organic, or ceramic coatings which are known.

2. The aim of protection is to ensure the necessary longevity of the physical surface properties of the object. Parts have to keep up defined roughness, electrical, or optical characteristics in order to serve successfully. In other words, the protective system has to keep up the original properties of the surface. In such cases (e.g., with parts of instrumentation), the possibilities of choosing systems are limited—primarily to metals or metallic coatings keeping up passivity (stainless steels, electrodeposited chromium coatings) or metal coatings such as gold, platinum, or their alloys, which do not become covered by corrosion products even in highly polluted and hence aggressive atmospheres.

Table 3.12. Corrosion of Common Metallic Materials in Atmospheres Defined by Corrosivity Grades (A, Steady-State Corrosion Velocity, μm yr^{-1}; B, Corrosion Loss after 10 Years' Exposure, μm)

Grade of Corrosivity (Table 4.3)	Plain carbon steel		Stainless steel (18% Cr, 9% Ni)		Zinc		Copper		Aluminum (99.5%)	
	A	B	A	B	A	B	A	B	A	B
I	0	<0.1	0	0	0	0	0	0	0	0
II	<1	<10	0	0	<1	<10	<1	<10	0	0
III	1–5	>10–80	0	0	>1–2	10–20	>1–2	>10–20	<0.01	0.1
IV	>5–10	>80–150	<0.01	<0.1	>2–4	>20–40	>2–4	>20–40	<0.1	<1
V	>10->50	>150->500	>0.01	>0.1	>4->6	>40->60	>4	>40	>0.1->2	>1->20

3. The system has to protect the metallic part against loss of strength and simultaneously keep up its decorative properties. This is a very common technical requirement if metallic equipment such as structures parts of buildings, car bodies, and so on, have to be protected against atmospheric corrosion. The longevity of the decorative and protective function of the system should be defined very carefully. An automobile body may serve as an example: The useful average technical life of a car may be about 6 to 8 years. The protective function of the system has to ensure that the critical parts of the body do not lose their mechanical strength to such a degree that they could endanger its safe use. It seems reasonable to define the longevity of the decorative function of the protective system as 2 to 3 years. The decorative changes due to the deterioration of the system should be scarcely visible after 2 years use, after 3 years only very slight. Later the car is an "old" one, and the decorative part of the surface protection become less important, so that the safety factor has to be regarded as predominant.

The protective system can be chosen from the very broad set of materials and coatings available. The choice of the most effective system is a result of a detailed analysis (Figure 3.20), in that the limits of tolerable decorative and protective deterioration of the system have to be defined very precisely.

Many other examples of combined aesthetic and protective requirements on protective systems can be cited, especially metallic parts of buildings, and metallic goods in broad public use, such as household goods and sporting equipment.

3.4.3.3. *Longevity of Protection*

Economic viewpoints determine the controlling factors for defining the optimal life of the protective system. Different aspects are considered if a protective system for a steel bridge has to be chosen than for sporting equipment such as skates or ski bindings. In the first case, it is necessary to take into account the extreme longevity of the object (100 years or more) and the high cost of keeping up protective systems that have deteriorated to such a degree that they do not serve their protective and decorative purpose. Here it seems reasonable to select the system with the highest possible lifetime with some assurance of simple and inexpensive repairs in long cycles. Protective systems for sporting goods, on the other hand, have a life expectancy of a few years and are exposed to service conditions only seasonally. Moreover, they are seldom repaired, but requirements concerning their decorative qualities are rather high.

The type and quality of the protective system has to be selected to adequately meet these demands of longevity, seasonal use, and decorative value of the system through the whole life of the equipment. This equipment is usually discarded for reasons other than deterioration because of corrosive effects.

Table 3.13 contains data about the decorative and protective longevity of the most common systems used for the protection of metallic equipments against atmospheric corrosion.

3.4.4. Mechanisms and Kinetics of Deterioration of Protective Systems

In this section we will use the classification of principles presented in Section 3.4.2.

3.4.4.1. Deterioration of Protective Systems Based upon Artificial Changes of the Surrounding Microclimate

Useful changes of the microclimate near the surface of metallic goods are derived from the knowledge of factors governing the initiation and propagation of corrosion. A protective microclimate has to prevent any formation of a liquid film of electrolyte solution on the surface. This can be achieved by:

- Keeping the temperature of the object safely above the dewpoint.
- Removing water vapor so that even at the extreme possible temperature decrease, no condensation occurs.
- Lowering the accessibility of stimulating agents from the outside so that no hygroscopic stimulative compounds may settle or be formed by the initial step of the corrosion process.
- Enclosing the object in an inert atmosphere without oxygen.
- Add an inhibitive agent to the environment, usually a vapor-phase inhibitor.

All these principles are used, as shown by the following examples, which include simple explanations of their failure.

Heating is frequently used to protect electrotechnical components operating in humid environments. If a special heating element is used and gets out of order, the protective system may fail.

Drying of the atmosphere is often used in packaging metallic equipment and in stores. The object gets enclosed in such a manner that dry and

Table 3.13. Protective (P) and Decorative (D) Longevity of Common Coatings on Steel (years)

Corrosivity of atmosphere (Table 4.3)	Zinc, cadmium electroplate 12 μm, chromated		Zinc, hot dip, 60–80 μm		Zinc, aluminum, flame-sprayed, 150–200 μm		Air-dried paint on shot-blasted surface	
	P	D	P	D	P	D	P	D
I	>100	>10	>100	>10	>100	>10	>30	>30
II	50–100	2–5	>100	2–5	>100	2–5	>20	5–10
III	5–10	<1	30–50	<1	>50	<1	5–15	<5
IV	2–4	<1	15–20	<1	30–50	<1	3–10	<3
V	1–3	<1	10	<1	10–30	<1	1–5	<2

clean atmosphere is maintained. The walls of the package should have a very low permeability for water vapor. Well-sealed metallic foils or thin sheets serve best, but plastics [polyolefines, poly (vinyl chloride), etc.] are used most often.

To prevent any condensation of the initially contained or penetrated water, a small amount of adsorbent (such as silica gel) is placed into the package. If the air-drying procedure is carried out in storage rooms, it is necessary to keep the relative humidity so low that even at the fastest possible cooling rate, no condensation takes place. When bringing a new object into the store, its temperature must not be lower than that inside, to prevent condensation on its surface.

The barrier effect of the envelope serves also to exclude access by solid and, to some extent, by gaseous stimulants. Only in special cases are packaging materials treated with compounds to lower the permeability of gaseous stimulants (e.g., chlorophyline for hindering the access of H_2S to the object).

A protective system based on conditioning the atmosphere will fail due to leaks in the walls of the package, saturation of the adsorbent so that humidity rises, or interruption of the automatic conditioning device. If an inhibitor is the main conditioning agent, the system can fail by its exhaustion. Extreme air pollution is seldom the controlling factor of failure. In infrequent cases, hygroscopic dust particles settled on the surface before applying the microclimatization protection can lead to failure.

3.4.4.2. *Deterioration of Corrosion-Resistant Metals*

Two principles may render metallic material completely resistant in most types of atmospheres: immunity or passivity. Immunity is an inherent characteristic of noble metals, which for thermodynamic reasons are not able to enter into reactions with the corrosive components of air, including the common types of industrial pollutants. Gold and the platinum group metals are fully resistant, even in the most aggressive atmospheres, because of their immunity. The less noble silver reacts with sulfur compounds contained in the atmosphere (hence the use of chlorophyline-treated packing materials for preserving silver or silver-plated parts from attack).

Complete corrosion resistance of passive metals is the result of the presence of a very thin, highly resistant oxide film. Typical passive metals are tantalum, titanium, chromium, stainless steels and other alloys with sufficiently high chromium content, and aluminum. The degree of resistance of the passive film is different for the single types of metals and alloys. If the film is destroyed (mostly locally), corrosion starts at points and may become important from the technical viewpoint. The higher the aggressivity of the atmosphere, the higher the probability of disrupting the passive layer. The most active species of stimulants able to destroy passivity are chlorine and its compounds, but sulfur compounds also may induce local depassivation. The corrosivity of atmospheres toward passive metals is preferently controlled by the pollution degree. High humidity (long periods of surface wetting) and rain definitely decrease the aggressivity by washing off accumulated particles containing soluble, depassivating anions. Sheltered exposure of aluminum leads to an increase of the corrosion with time and finally to much higher corrosion than in the open air. Local depassivation results in the formation of corrosion products and more-or-less deep pittings. As passive metals are applied basically to maintain the decorative properties of equipment, this formation of corrosion products means their deterioration. Loss of mechanical strength by pitting is less critical but in some cases may lead to stress concentrations and induce cracking.

A special type of atmospheric corrosion of (mostly) passive metals, which endangers their mechanical strength, is stress corrosion cracking. Some rare pollutants selectively attacking the single type of statically stressed passive alloy may induce this very dangerous effect.

Normal pollutants are unable to induce corrosive attack of inert noble metals such as gold, platinum, and rhodium.

3.4.4.3. *Atmospheric Deterioration of Metallic Coatings*

Metallic coatings serve protective and decorative purposes. The purpose of use determines which of these functions predominates.

Generally, it is necessary to distinguish between coatings more and less noble than the base metal. As relatively unnoble unalloyed steel is the commonest substrate, the deteriorative effects of the atmosphere on metallic coatings will be discussed mainly for coatings on this base metal. Other metals, such as copper, aluminum, and zinc alloys, are also broadly used metallic substrates. The problems of their deterioration can be derived from the principles valid for steel/metallic coating systems.

A noble metallic coating on steel necessarily must be coherent, free of voids, cracks, pores, and so on. If this is not the case, the probability that corrosion (rusting) of the underlying steel will begin is high, especially in a corrosive atmosphere. Many "noble" types of metallic coatings are of the passive metal type. Common examples are electroplated nickel, chromium, and multilayer types of coatings with chromium top coats for decoration. If these coatings are voidless and remain passive, no deterioration takes place. They may, however, be attacked in a similar way to compact passive metals, by local pitting that results in penetrating the coating to the corrodible substrate and in rusting. This is a very common mechanism of deterioration with electroplated decorative chrome coatings, which are mostly a multilayer system of electroplated copper with subsequent nickel layers of different properties and a thin Cr top. There are other reasons for a loss of coherence, including internal stresses induced by, for example, temperature changes, which may result in cracking. Rusting of the steel substrate is a very serious form of deterioration of this coating type, most frequently induced by local corrosion and hence showing a close relation to the degree of pollution of the environment (e.g., by de-icing salts). To reduce this undesirable display of corrosion, decorative coating systems have been developed showing lower pitting and rusting porbability but at the same time losing their luster sooner. Nickel, tin, and lead and some alloy coatings (Ni–Sn) are nobler than the steel substrate. These coatings are produced principally by electroplating or, as with lead and tin coatings by hot-dip processes. With the exception of nickel and Ni–Sn alloy coating, they cannot be applied for decorative purposes in more-corrosive types of atmospheres. For protective purposes they should be fully coherent and their thickness sufficient to warrant that no voids become active after the coating has partly corroded away. Very small pores in lead coatings may become sealed by the corrosion products of the coating metal.

From the "pure" thermodynamic or electrochemical viewpoint, aluminum coatings should be very unnoble. The easy passivation of aluminum and the high stability of passivity are the reason aluminum coatings behave similarly to those of other noble metals. Also, chromium would be, if not passive, very unnoble. Thin sprayed or hot-dipped aluminum coatings showing an inherent porosity usually rust slightly during the initial exposure period in corrosive atmospheres. Later, the small pores become sealed by the voluminous corrosion products formed during a short "active" corrosion period of the coating particles near the voids and the now coherent coatings become fully passive and highly protective. Corrosivity due to activating pollutants (mainly chlorides) helps to shorten this period. If local depassivation exposes the steel substrate and the corrosion products are not able to fill the pitting, rusting begins and the protective capacity of the coating gets progressively lost but may be "saved" by artificial sealing by paint or other sealer. In this stage the activating pollutants in the air are definitely detrimental.

Metals for unnoble coatings in general use are mainly zinc, cadmium, and recently Zn–Al alloys. They are produced primarily by electroplating (Zn, Cd), flame spraying (Zn, Al, Zn–Al), and hot-dip processes (Zn, Al, Zn–Al). The protective value of these coatings is based on their comparatively low corrosion rate in the prevailing types of atmospheres and their capacity to protect the nobler substrate in small voids by the electrochemical principle of cathodic protection. This effect is a sacrificial one; as the exposed surface of the steel substrate becomes larger, the corrosion rate of the neighboring coating is accelerated, and finally the void gets so large that the electrochemical protection principle ceases. The higher the corrosivity of the polluted atmosphere, the larger the distance of active electrochemical protection but also the higher the corrosion rate of the sacrificial "anodic" coating. Only in rare cases is the radius of action of this protective effect more than a few millimeters. This is well known to anyone who has observed the deterioration of galvanized (hot-dip zinc-coated) sheet (e.g., on roofs). During the initial exposure period, no rusting is observed even at the cut edges. Later, the attack starts and spreads from the edge into the remaining zinc-coated parts of the surface. This spreading is more rapid in polluted and hence corrosive atmospheres. Cadmium coatings behave similarly to zinc. Their corrosion rate is much the same. Their electrochemical effect is, however, less pronounced, but the properties of the corrosion products make cadmium more acceptable for some purposes than the cheaper zinc coatings.

Unnoble metallic coatings deteriorate by loss of mass (thickness) and by exposing the underlaying substrate, which begins to rust. Only in mild atmospheres can they be applied for protective and simultaneously unstressed decorative purposes. Electroplated zinc and cadmium coatings can be produced in a highly glossy form that is relatively stable in indoor conditioned atmospheres with no or very rare surface wetting. This decorative effect may be preserved by chromate treatment (passivation) of the coating. This treatment is a chemical one and serves mainly to protect the coating's appearance during the usually more corrosive conditions during transport and storing. The chromate film is destroyed by washing off, abrasion, and so on. Acidic atmospheric pollutants accelerate this process.

3.4.4.4. *Deterioration of Inorganic Nonmetallic Coatings*

Two sorts of coatings are in this category: conversion coating produced by artificial "corrosion" reactions leading to the formation of protective layers (insoluble, adherent compounds) and coatings formed from chemically resistant inorganic materials that can be bonded to the metal.

The most important representatives of the first group are artificial oxide layers on aluminum (and some aluminum alloys) produced preferentially by the electrochemical process of anodic oxidation, mixed oxide layers chemically formed on magnesium alloys, and phosphate coatings on steel, zinc, and aluminum. Less important are oxide and other conversion coatings on steel.

The other group is represented by the widespread vitreous enamels on ferrous substrates, aluminum and some other metals, inorganic cold-cured coatings based on hydrolyzable organic and inorganic silicates, portland cements, and other inorganic compounds able to produce insoluble compounds bonded to surface of the substrate (e.g., hydroxide salts).

The protective capacity of all these coating types in aggressive atmospheres depends on the following:

- Adhesion (strength of bonding to the substrate).
- Coherence.
- Degree of insolubility in the electrolytes formed during atmospheric exposure.

The process of deterioration of these coatings is influenced by mechanical and chemical effects. Abrasion, mechanical shock, vibration, or abrupt changes of temperature often lead to a loss of the bonding strength of the coating or to disruption of its coherence and exposure of the substrate. Conversion coatings such as oxides produced by anodic oxidation of aluminum are less prone to this type of damage than are coatings from inorganic materials having no direct chemical relation to the substrate.

If the coating has not lost its protective characteristics because of mechanical destruction, it deteriorates through chemical reactions. The mass and degree of insolubility of the coating is in direct—and the aggressivity of the surface electrolyte from the atmosphere in indirect—relation to its longevity. Relatively thick (20 μm and more), dense (after sealing) oxide coatings on aluminum are highly insoluble, even in electrolytes formed from strongly polluted industrial atmospheres. High-quality oxide coatings are able to preserve decorative properties for more than 10 to 20 years. On the other hand, thin, more-soluble oxide coatings on magnesium resist only a few months under the same conditions. Phosphate coatings exposed to polluted outdoor atmospheres are destroyed within hours. The same is true of the commercial blue, black, or brown oxide coatings on steel.

Also, the thick nonconversion inorganic coating types are more or less slowly destroyed by the chemical action of the atmosphere. Silicate coatings hydrolyze or get changed by other reactions into more soluble products that are rinsed off by rain. Typical examples are the well-known zinc-silicate coatings. They deteriorate by dissolving of the surface layer, get thinner, and finally the substrate becomes exposed and corrodes.

Air pollution contributes mainly to the chemical part of the destruction of the coatings. If their coherence has failed because of mechanical reasons, the deterioration of the coating is accelerated by the stimulating influence of the corrosion of the substrate, the products of which often cause loss of adhesion of the coating by "underrusting" and peeling off.

3.4.4.5. *Deterioration of Organic Coatings*

The mechanism of the protective action of organic coatings against atmospheric corrosion is not fully clear, so the deterioration of their protective and decorative capacity by polluted atmospheres can be explained in rough outline only. Organic coatings are assumed to protect the metallic substrate by:

- A barrier effect.
- Inhibitive action.
- Bonding forces between the coating and the metal, blocking the "available" surface for corrosion.

These mechanisms are never strictly separated and form a closed system in that one characteristic protective quality is more or less connected with the others.

a. *The Barrier Effect*

Even a physically coherent organic coating is permeable for electrically neutral molecules. Water and oxygen, especially have the ability to pass through most organic coatings. The mechanism of permeation is mostly a physical process and in many cases may be described by the laws of diffusion. The higher the polarity of the organic binder (most organic coatings contain besides that an appreciable amount of inorganic matter—pigments and fillers that contribute to its desired physical and chemical properties), the easier water in vapor or liquid form and oxygen may be expected to pass through it. Nonpolar binders such as polyolefines and their derivates (e.g., fluorinated, chlorinated) show the lowest permeability, but even these are not a fully impervious barrier. The use of nonpolar binding substances with low diffusion rates for water and oxygen for coatings brings about a very serious difficulty. Their adhesion to the substrate metal is usually very poor, and the coating has to be bonded to the metal by a special adhesive layer between the two principal components of the system.

The diffusion rate of oxygen and water vapor through most of the common organic coatings would suffice to initiate and maintain the corrosion process of a similar velocity to that of an unprotected metal. This is true as long as isolated films of organic coatings are studied (i.e., films not bonded to the metallic surface by adhesive forces). Bonding to the substrate changes the picture. The metal-coating interface is practically impenetrable, even to neutral molecules. The barrier effect is therefore closely connected with the "bonding-effect" part of the protective action of organic coatings. Most protective paints contain inorganic pigments and fillers dispersed in the organic vehicle. The volume ratio of binding vehicle to pigment strongly influences the permeability of the film. At a critical ratio, the permeation of molecules passes a minimum value. Higher or lower ratios increase the permeability.

Even though diffusion of water vapor and of oxygen through most organic films seems to be rapid enough to ensure initiation of corrosion under the coating, it is necessary to stress that in real atmospheric conditions the relatively thick coatings (>60 μm to several millimeters) take up and lose water with the changing weather conditions during the day. These changes are limited to the interface between coating and atmosphere and to the neighboring part of the coating.

A very important component of the barrier effect of organic coatings is their impermeability for electrically charged particles of the environment—the ions. This is due to electrokinetic phenomena in the coating film. In contact with water, the film attains a definite electrokinetic charge, mostly negative, that makes it impervious to negative ions. As electrical equilibrium (the Donan equilibrium) must be maintained outside and inside the film; neither negatively charged ions (anions) nor positive ones (cations) can penetrate the coating. Thus the third controlling factor of atmospheric corrosion is kept off the metal coating interface, and even though water and oxygen may be present near the metallic surface, the absence of stimulating ions (mainly chlorides and sulfates) prevents the initiation of the process. Thus the barrier effect is closely connected with the inhibitive action of organic coatings.

Stimulants of the atmosphere such as SO_2 may dissolve in the surface electrolyte with or without dissociation. Their diffusion through the organic binder should be possible as long as they remain in the undissociated state. As most paints contain pigments able to catalyze oxidation of SO_2 to SO_4^{2-} or to react directly with the dissolved molecular SO_2, the probability of diffusion of SO_2 molecules through the coating to its interface with the metal is very low.

Also, the inorganic or pigment component of the paint may affect the transport of stimulating ions through the paint to the metal surface. If the pigment reacts with the ion (e.g., ZnO or PbO with SO_4^{2-}), diffusion is low initially, but as the transformation continues, the chemical "filtering" effect ceases and the stimulating ion may pass more easily.

The barrier effect of a protective organic coating is important, but it does not suffice to warrant the necessary protective effect. Its most important component is the impermeability of the coating to stimulating ions due to electrokinetic phenomena. All this is true for fully coherent coating films. If this characteristic gets lost by aging of the organic binder or by mechanical damage under film, corrosion starts at the newly formed voids.

b.   Inhibitive Action

It is important that priming paints (i.e., those contacting the base metal) inhibit corrosion of the metal. The paint must therefore contain inhibitive agents, mostly pigments. Two principles of inhibition by paints are most common: adsorption and passivation. Some pigments act simultaneously by both mechanisms.

A typical adsorptive inhibitor is red lead applied in classical unsaturated-oil-based vehicles. This pigment reacts with the vehicle, forming lead soaps of lower fatty acids (e.g., azelaic), both very effective inhibitors of the adsorption type.

Metal chromates with controlled low solubility are widely used passivating inhibitors (e.g., zinc, lead, and strontium hydroxide chromates with balanced composition so as to achieve the necessary degree of solubility and other characteristics).

Molybdates, tungstates, and phosphates are known to act by primary sorption and secondary passivation. For the passivation step of the inhibitive process, the sorbed species' capacity to catalyze reactions of oxygen transported from the environment is utilized.

c.   The Bonding Effect

This factor was mentioned previously when discussing the barrier effect of organic coatings. High primary adsorption of the correctly formulated primer to the substrate forms a very strong bond. Hence it is of supreme importance that the primer contact a pure metallic surface freed of any impurities such as grease, corrosion products, dust, and moisture. From the thermodynamic viewpoint, corrosion may undercut the paint only if the free energy of reaction is higher than the bonding energy. This is very important for understanding the philosophy of accelerated testing of paints artificially cut by scratches, These tests are evaluated by measuring the surface on which the paint has lost its adhesion and make it possible to distinguish very precisely between well and poorly prepared surfaces, good and bad types of primers, or complete paint systems. Corrosion undercutting of the paint is a very common form of paint deterioration, starting at points where the coating has lost its coherence by mechanical or other causes.

The mechanisms and kinetics of deterioration by environmental affects are very complex with organic coating. Unlike metallic and inorganic coatings, more than time of surface wetness and type and degree of air pollution are controlling environmental factors. The organic vehicle deteriorates by aging, which is a general term describing very diverse changes of the macromolecular structure induced by chemical and physical effects. The chemical influences of water and chemically active pollutants of the air (SO_2, H_2S, O_3, H_2SO_4, NO_x, Cl_2, NH_3) may induce and help to propagate undesired changes of the bonding vehicle. Also, temperature and radiation effects of ultraviolet light are important. Aging of organic coatings changes the characteristics that are necessary for their anticorrosive effect. The coating may become thinner because of chalking, which sets pigment particles free as the binder weathers off. As mentioned previously, the most dangerous corrosion may start at pores, destroying the paint by undercutting. Cracks formed by aging lead to the same results as mechanical damage of paints that have become brittle.

Air pollution contributes significantly to the final and most serious form of deterioration of organic films, the corrosion of the metallic substrate leading to undercutting and peeling off, the process starting at voids. As organic coatings are preferred to protect steel and cast iron, this process of underrusting may be rather rapid, because corrosion stimulated by aggressive air pollutants forms rather voluminous rust, and its spreading through the activity of salt (mostly sulfate) nests may easily break the paint/metal bond. This does not mean that the other effects of stimulating air pollution are negligible; the saturation of reactive pigments and the inducing of hydrolytic reactions of the vehicle may be of some importance. Their main effect on paint deterioration, however, is corrosion of the base metal. Hence it may be postulated that the more resistant the substrate to the stimulating activity of pollutants, the less dangerous is air pollution to the deterioration of organic coatings.

High air pollution negatively affects the quality of paints even before the painting process begins. In industrial regions it is very difficult to keep the metal surface clean and free of corrosion products. Even very short exposure of a freshly cleaned (e.g., shot-blasted) steel surface leads to the formation of "latent" corrosion centers that are potential points of later paint failure. Hygroscopic dust particles closed under the paint cause the initiation of osmotic blisters (i.e., points without any adhesion with an aggressive solution present under the film). In atmospheres containing higher H_2S concentrations, classical oleoresinous primer containing inhibitive lead-based pigments cannot be used. They do not dry, as the lead compound particles become covered by a PbS film and the catalyzing activity of Pb soaps transferring oxygen necessary for the film-forming process gets lost.

SYMBOLS

A	constant
B	linear corrosion velocity
F	Faraday constant
j	current density
j_a	current density for active dissolution
j_1	density of the cathodic reaction is limited
j_p	density of passive metal
k	rate constant
K	amount of corrosion
M	constant in eq. 16
n	number of electrons transferred
n_1, n_2	constants in eq. 16
R	gas constant
t	time
T	absolute temperature
V_k	corrosion velocity
X	SO_2 pollution level
Y	log corrosion rate
Z	average temperature
α	transition coefficient
ζ	overpotential
τ	average wetting time
τ_m	monthly sum of wetting time
ϕ	potential

REFERENCES

1. Report of the Committee on Corrosion and Protection, Her Majesty's Stationery Office, London, 1971.
2. Kuliš, M., *Werkst. Korros.*, 27, 870 (1976).
3. Evans, U. R., *The Corrosion and Oxidation of Metals*, Edward Arnold, London, 1960.
4. Uhlig, H. H., *The Corrosion Handbook*, Wiley, New York, 1958.

5. Akimov, G. V., *Teoria i metody issledovania korrozii metallov* (Theory and Testing Methods of Metallic Corrosion), Mashgiz, Moscow, 1945.

6. Tomashov, N. D., *Teoria korrozii i zashishty metallov* (Theory of Corrosion and Protection), Izdatelstuo Akademii Nauk SSSR, Moscow, 1961.

7. Tödt, F., *Korrosion und Korrosionsschutz*, Walter de Gruyter, Berlin, 1961.

8. Shreir, L. L., *Corrosion, Metal/Environment Reactions, Corrosion Control*, Newnes-Butterworth, London, 1977.

9. Bartoníček, R., *Koroze a protikorozní ochrana kovu* (Corrosion and Protection of Metals), Academia, Prague, 1966.

10. Čermáková-Knotková, D., and Vlčková, J., *Werkst. Korros.*, 21, 16 (1970).

11. Vannerberg, N. G., and Sydberger, T., *Corros. Sci.*, 10, 43 (1970).

12. Schwarz, H., *Werkst. Korros.*, 16, 93 (1965); 16, 208 (1965).

13. Fyfe, D., Shanahan, C. E. A., and Shreir, L. L., *Corros. Sci.*, 8, 349 (1968).

14. Honzák, J., and Kuchyňka, D., *Werkst. Korros.*, 21, 342 (1970).

15. Bartoň, K., Kuchyňka, D., Bartoňová, Ž., and Beránek, E., *Corros. Sci.*, 11, 937 (1971).

16. Sydberger, T., and Vannerberg, N. G., *Corros. Sci.*, 12, 775 (1972).

17. Vernon, W. H. J., *Trans. Farad. Soc.*, 23, 162 (1927); 27, 264 (1931); 29, 35 (1933).

18. Patterson, N. S., and Hebbs, L., *Trans. Farad. Soc.*, 27, 277 (1931).

19. Bartoň, K., *First International Congress on Metallic Corrosion*, Butterworth's, London, 1961, p. 685.

20. Sydberger, T., and Ericson, R., *Werkst. Korros.*, 28, 154 (1977).

21. Michailovskij, Y. N., and Strekalov, P., *Zashch. Met.*, 8, 573 (1972).

22. Okada, A., and Shimada, N., *Corrosion*, 30, 97 (1974).

23. Florianovich, G. M., Sokolova, L. A., and Kolotyrkin, J. M., *Electrochim. Acta*, 12, 879 (1967).

24. Riggs, O. L., *Corrosion*, 25, 130 (1969).

25. Grauer, R., *Chimia*, 24, 269 (1970).

26. Preston, R. S., and Sanyal, B., *J. Appl. Chem.*, 6, 26 (1956).

27. Sereda, P. J., *Mater. Res. Stand.*, 719 (1961).

28. Haynie, F. H., and Upham, J. B., *Mater. Prot. Perform.*, 9(8), 35 (1970); 10(11), 18 (1971).

29. Michailovskij, Y. N., et al,. *Zashch. Met.*, 7, 534 (1971).

30. Bartoň, K., et al., *Koroze a ochrana materiálu* (Corrosion and Materials Protection), 17, 85 (1973); 20, 82 (1976).

31. Bartoň, K., Bartoňová, Ž., and Beránek, E., *Werkst. Korros.*, 25, 659 (1974).

32. ČSN 03 8211 (Czechoslovak standard: measurement of sulfur dioxide pollution in the atmosphere).

33. Tomashov, N. D., and Lokotelov, A. A., *Zavod. Lab.*, 24, 157 (1960).

34. Jirovský, I., Kokoška, I., and Průšek, J., *Werst. Korros.*, 27, 16 (1976).

35. Misawa, T., et al., *Corros. Sci.*, 14, 279 (1974); 14, 131 (1974).

36. Bartoň, K., and Beránek, E., *Werst. Korros.*, 10, 377 (1959).

37. Schikorr, G., *Aluminum*, 43, 108 (1967).

4

MIST ELIMINATION

A. Bürkholz,

Bayer AG, Germany

4.1. INTRODUCTION

In this chapter, mist is taken as consisting of droplets below 30 μm. Mists are formed mainly out of the gaseous phase when the temperature falls below the dewpoint of one or other component. These condensation mists usually have droplet diameters of ~1 μm, but they can also change and grow as a result of aging.

Mists are also formed with all atomizing processes. Whereas the majority of the droplets so formed are generally above 30 μm and are called spray, a fine component is also formed with a size that varies according to the atomizing energy. Since, however, mists wet the walls of pipes and these liquid films can be atomized by the gas current, even fine mists are nearly always accompanied by a certain proportion of spray.

In addition to dust separation, the problem of separating mists has, over recent years, grown considerably in importance. Mists are generally suspended in a gas in quantities ranging from a few milligrams to several g m^{-3} and are carried along by this gas. The carrier gas flow can be a process gas, a waste gas, or even a circulating current. Depending on the type of mist, it may either be merely a nuisance or actually be harmful. In process gases, for example, mists can corrode parts of the equipment or contaminate catalysts or products. In waste gas, they contribute, like dust, to the general pollution of the atmosphere. Because of the small size of the droplets, mists are predominantly respirable.

Because of their vapor pressure, liquid aerosols should disappear after a certain time (i.e., become converted into gaseous impurities). However, with acid aerosols, for instance, it has been proved that the lifetime is quite a long one. During this period, and particularly under solar radiation, secondary reactions can take place in the aerosol particles which can stabilize them and increase their toxicity.

Although our level of knowledge is only in the initial stages in this respect, the demand for the best possible conservation of the air is no longer disputed by anyone. From a technical point of view, it would be possible even now to satisfy this demand almost completely, regardless of how strict the requirements are. Its realization, however, would founder on the immense costs involved. Even to maintain the present level of waste air purification, industry has to invest vast sums of money every year. Efforts must therefore be made to employ the available sums to optimum effect (i.e., where there is a particular problem of separation, it is important to use the separator best suited for the particular case).

The separability of a particle depends to a very great extent on its size. Figure 4.1 shows the size distributions of some mists in the form of cumulative size distribution curves. These curves indicate the particular "oversize fraction" related to weight or volume, as a function of droplet diameter.

The curves shown with a dashed line are fractional separation efficiency curves for a few separators. The separating value η_F indicates the probability with which a droplet of given size will be retained in the separating system. It can be seen that sufficiently large droplets are always separated completely, whereas correspondingly small ones are often not separated at all. Now if a given mist is to be separated completely, the entire separation curve must run to the left of the cumulative size distribution curve. If, conversely, the separation efficiency curve runs completely to the right of the cumulative size distribution curve, no separation

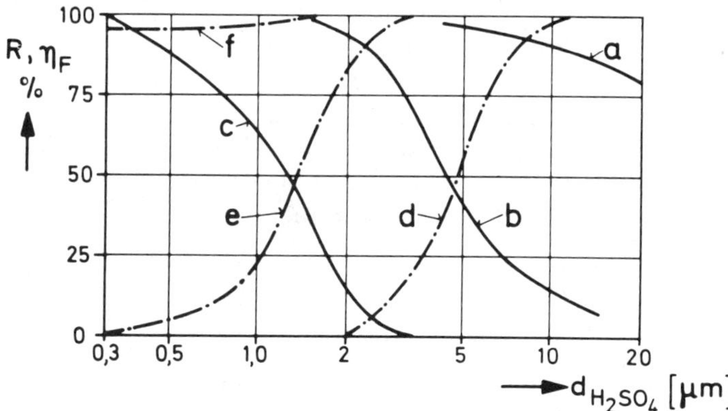

Figure 4.1. Cumulative size distribution curves of acid droplet mists and fractional separation efficiency curves: (*a*) spray after SO_2 drying tower; (*b*) fine spray after SO_3 absorber; (*c*) condensation aerosol; (*d*) fractional separation efficiency curve of packed bed; (*e*) fractional separation efficiency curve of wire mesh; (*f*) fractional separation efficiency curve of electrostatic precipitator.

is taking place and the separator is completely ineffective. It frequently happens that two curves overlap and partial separation occurs. The total separation efficiency η_0 can then be calculated from the fractional separation efficiency η_F and residue value R to give

$$\eta_0 = \sum \eta_F \frac{\Delta R}{\Delta d} \Delta d \tag{1}$$

The example in Figure 4.1 shows very clearly the enormous increase in the expenditure for separation that results from what seems to be a relatively harmless refinement in the droplet spectrum. This is because it is necessary to go over from a cheap packed column via a wire-mesh demister to an expensive electrostatic filter. We can say roughly that if the droplets are finer by a factor of 10, the costs will be higher by at least a factor of 10. This presupposes that the optimum separator is selected in each case, but to do this it is necessary to know the droplet spectrum and fractional separation efficiency curve. Both pieces of information can be obtained only by using droplet measuring methods.

4.2.   DROPLET MEASURING METHODS

A simple method, frequently used until now for detecting the presence of droplets in a gas current, was, for example, to place a stick transversely in the gas current and then to record any precipitation. However, with the gas velocities of several m s^{-1} generally found in gas ducts, it is at best possible to record the presence of a spray, whereas the fine mist droplets from the gas flow are carried past the obstacle and do not become precipitated.

Unlike dust particles, droplets cannot be collected and transported but must be measured on the spot. Suitable measuring apparatus has become available in recent years.[1] As with a dust measurement, a sampling probe generally has to be placed into the gas duct.

4.2.1.   Sampling the Gas

To obtain representative samples, the sampling should, if possible, be carried out isokinetically (i.e., at equal velocity). If not, the same errors are obtained as are found with nonisokinetic dust measurements.[101]

With a number of droplet measuring devices, the gas flow rate is either specified or at least variable only within certain limits. The aperture width of the sampling nozzle should not be below 7 mm. Isokinetic sampling is therefore only rarely possible with droplet measurement. On the other hand, it is mainly droplets below 3 to 5 μm that are of interest in separation measurements, and this fine fraction is only very slightly falsified by non-isokinetic sampling.

For the same reason, it is generally not necessary—and not possible either-to take a large number of samples distributed over the cross-section of the pipe, as is specified for the accurate determination of dust contents. Nevertheless, with droplet measurements as well, the sampling point must, if possible, be selected in such a way that it is not located after bends or obstructions.

Unlike dust measurements, droplets can also become lost in the sampling probe or in the measuring device itself. Any contact with a surface makes the droplet disappear. Surface contact may be the result of sedimentation in excessively long sampling probes or of being flung out where the gas is deflected around a bend. For this reason, there should be no throttles in the sampling flow before the measuring device. When drawing off vertically to the direction of the gas flow, the head of the sampling probe should be provided with a collecting plate on which the coarser droplets can precipitate when the gas is deflected (Figure 4.2).

Whereas it is mainly the coarse fraction of the mist droplets that becomes lost through contact with a wall surface, the fine fraction can change as a result of exchange processes with the gas phase. By cooling the sampling flow, new droplets of mist can condense, and by heating it, fine droplets can evaporate. The equilibrium time in each case depends on the type of mist, although it can take place very rapidly as a result of the large specific surface of the droplets. Thus water mists below 15 μm can virtually not be measured.[2]. To measure other types of mists may require that the sampling train has the same temperature as the gas being

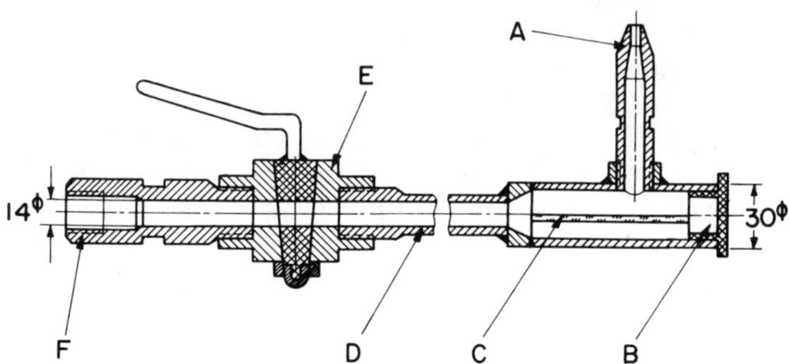

Figure 4.2. Probe for droplet sampling: (*a*) probe nozzle; (*b*) plug; (*c*) glass slide; (*d*) shaft; (*e*) valve; (*f*) fitting for cascade impactor.

measured. A good summary of setups for gas sampling can be found in Ref. 3.

The condensation of water vapor in the sampling flow is often unavoidable. By connecting up a coarse separator beforehand, it may be possible to separate the condensates without affecting the sample droplets. With mists of high vapor pressure, it may be necessary to carry out an additional analysis of the respective gaseous components in the sampling flow as a check on the measurement of the droplets.

4.2.2. Droplet Measuring Apparatus

Measuring Apparatus with Droplet Separation

4.2.2.1. Cascade Impactors

A cascade impactor consists of a series of nozzles of decreasing diameter connected one after the other, after each of which is a removable impaction plate. If a gas flow containing droplets is now sucked through the series of nozzles, the coarser droplets will, because of the increasing velocity of the stream, be precipitated on the front impaction plates and the finer droplets on the subsequent ones (Figure 4.3). The main advantage of the cascade impactor is that, every stage can be categorized by a certain average size of the precipitated droplets. This categorization is determined by calibration measurements, although it can also be calculated. Droplet size measurement is therefore reduced to determining the amount of precipitation on the impaction plates. Where the numbers of droplets are very low, this can be done by counting, although it is more rational to use chemical or physical methods of analysis. Figure 4.4 shows one version of the apparatus.[4] It is sufficiently handy and robust to be used for industrial-scale measurements as well as laboratory tests.

Since the cascade impactor permits measurements to be made precisely in the important mist range 0.3 to 15 μm, numerous measuring problems, such as separator studies, can be successfully solved. Compared with noncollecting measuring apparatus, the impactor has the advantage that, for example by chemical analysis of the droplet precipitate, it is possible

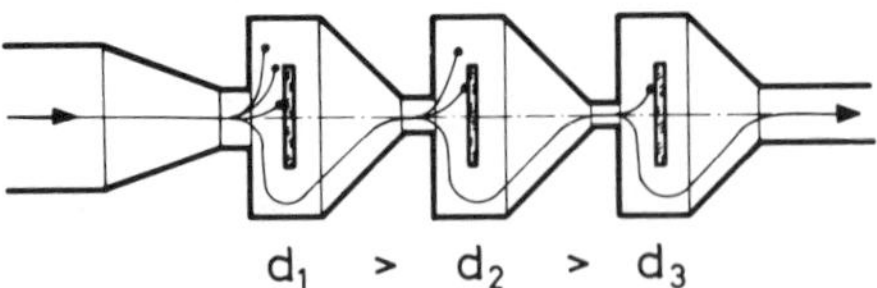

Figure 4.3. Schematic diagram of three impaction stages.

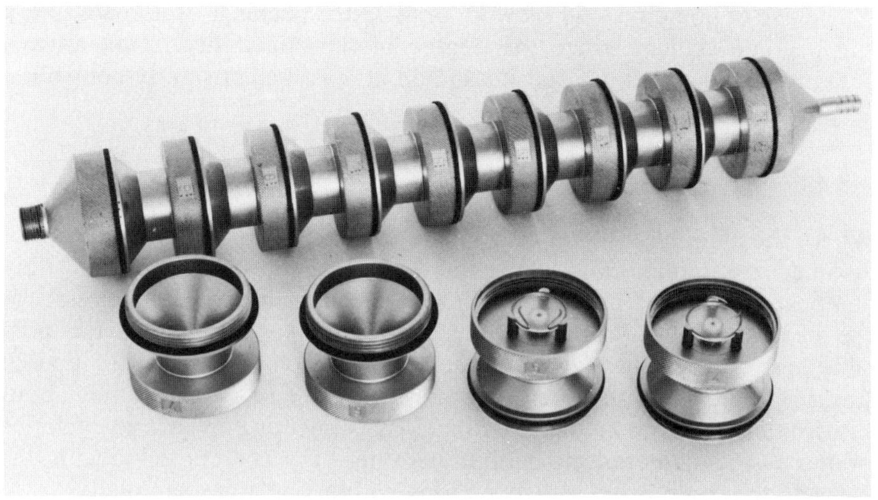

Figure 4.4. Bayer cascade impactor.

to determine not only the size distribution but also the type of aerosol concerned. This is often very important, especially with investigations of waste gases.

A cascade impactor was patented for the first time in 1945 in England by K. R. May.[5] In the following years, a few other versions were described in literature and presumably also successfully used (e.g., by W. E. Ranz and J. B. Wong,[6] R. J. Mitchell and J. M. Pilcher,[7] and J. A. Brink[8]., Nevertheless, the measuring device remained relatively unknown. Not until the last few years has interest in the cascade impactor really risen. Witness of this are numerous experimental and theoretical studies (e.g., Refs. 9–13).

4.2.2.2. *"Frit"-Type Cascade*

The cascade impactor becomes useless when the volume of droplets rises above a few g m^{-3}, since even with sampling times of less that 1 min, the impaction plates become flooded. Fritted glass or ceramic plates have a larger capacity. In principle, the amount of droplets retained in the fritted filter can be determined by weighing, but it is more usual to rinse the fritted plates and analyze the resultant liquid.

By connecting up several fritted glass or ceramic plates of decreasing pore size one after the other, it is again possible to attain fractionated droplet separation.

The use of homemade glass wool or asbestos packings as aerosol filters can become critical, since the degree of reproducibility is not always guaranteed. The packing density should at least be constantly controlled by pressure loss measurements.

4.2.2.3. *Cyclone Collectors*

Where the amount of liquid is too high, fritted glass or ceramic filters can also become flooded, and where the dust content of the gas is high, they can become clogged up. In such cases it is advisable to use cyclones for the measurement of droplets. Depending on the gas throughput, it is possible here to vary the size of the separable droplets between 0.5 and 20 μm. From the amounts precipitated, it is possible to say something about the droplet spectrum in this size range. By connecting up several cyclones of decreasing diameter one behind the other, it is again possible with this cascade to obtain droplet distributions with a single gas sampling.

4.2.2.4. *Impingement Separators*

With impinger separators, too, a nozzle ends shortly before an impactor surface. Impingers collect all particles, dust, or droplets above a certain size.[14] They cannot, therefore, really be regarded as measuring devices.

4.2.2.5. *Aerosol Measuring Apparatus for the Range below 1 μm*

Impactors separate droplets down to about 0.3 μm and fritted glass or ceramic filters down to 0.2 μm. Only a few liquids have such a low vapor pressure that stable mists having even smaller droplets can be produced with them. For these rare cases, the technician will try to use measuring apparatus that was developed for the investigation of dust aerosols. For collecting this very difficult to separate fine suspended matter, use is made in the aerosol centrifuge according to W. Stöber and H. Flachsbart[15] of high centrifugal forces, and in the thermal precipitator of thermal diffusion forces.

Measuring Apparatus without Droplet Separation

All measuring methods described so far had in common that the droplets under investigation were separated in one way or other from the carrying gas flow. In addition, methods have been developed in recent years for measuring the droplet size, as far as possible without interfering with the gas flow. Optical methods have been particularly successful. They have already been investigated in so many other places that a selection of literature is not made here.

4.2.2.6. Holography

Because of the outstanding depth of field with holographic photographs, it is possible to record all particles true to size in elongated clouds of droplets. The stream of droplets is guided in front of a specially suited photographic plate and is exposed with a short laser impulse. The droplets defract the light and produce with the undefracted light an interference pattern on the photographic plate. After developing the plate, this interference pattern is termed the hologram of the droplets. By illuminating the hologram with laser light, a real three-dimensional picture of the droplet distribution is formed. By scanning the image space (e.g., with a television camera) the individual images of the droplets appear on the screen one after another. In principle, the photograph can be made without lenses. Because the depth of the object space decreases with the particle size, the minimum detectable particle size is ~ 1 μm.

The great advantage of holography is the short exposure time, only 10^{-8} sec. This enables exposures to be made of fast-moving or rapidly changing groups of droplets. The disadvantage is in the relatively large amount of work involved in counting out the droplets.

4.2.2.7. Scattered Light Measurement

A further optical method for aerosol measurement, which has enjoyed reasonable commercial success, involves recording the scattered light of individual particles. For this, an aerosol stream reduced with carrier gas is passed through a measuring cell in such a way that there is usually only one particle in the focus of the light or laser beam. Every individual particle gives off scattered light which is focused onto a photocell and produces a voltage impulse whose size is dependent on the size of the droplet. The intensity of the scattered light is proportional to the sixth power of the particle diameter for very small particles and to the second power of the particle diameter for large particles. In between is the transition region of Mie scattering, where the intensity of the scattered light is dependent in a complicated way on the angle of scatter, the aperture angle, the refractive index, and the size of the droplet.

These scattered-light measuring devices can nevertheless be regarded as a promising development. There are devices on the market capable of measuring several thousand particles per cm^3 down to 0.3 μm. Prototypes with laser light reduce the illuminated measuring volume to 10^{-8} cm^3, enabling some 10^6 particles cm^{-3} to be counted. This gradually eliminates the largest disadvantage of the process: the high dilution of the aerosol flow.

4.2.2.8. Photometers

With the photometer, a beam of light traverses the particle-laden gas flow and a measurement is taken of the decrease in light intensity due to absorption and scattering by the particles. The light becomes increasingly weak with growing particle concentration and decreasing particle size.[16] Although photometers cannot really be regarded as apparatus for measuring droplet size, they are admirably suited to keeping a watch on, for example emissions in flues. It is not usually necessary to take a sample of the gas flow.[17]

4.2.3. Separation Efficiency Measurements

When characterizing a separator, we often give the total separation efficiency and relate it to a certain type of mist. Even if accurate data are provided on the droplet distribution, the total separation efficiency remains a specific and not a general property of the separator. The fractional separation efficiency, on the other hand, with additional information such as gas throughput and pressure loss, clearly defines the performance of the separator.

Particularly simple separation measurements can be achieved with the cascade impactor. The flow is sampled before and after the separator and passed in each case through a cascade impactor of identical design. For a given impactor stage, which should correspond to droplet size d, let us call the precipitation quantity before the separator A and after it B. For this particle size, the separation efficiency is then calculated as $\eta_F = (A - B)/A$. In this way, a measuring point is obtained for the fractional separation curve. The other impactor stages provide additional points (Figure 4.5).

Since cascade impactors are themselves inertia separators, they measure the particles according to their aerodynamic diameter. This is of particular advantage for comparisons with other inertia separators.

4.3. COLLECTION MECHANISMS

Gas flows containing droplets are two-phase flows. Separation of the entrained droplets from the gas flow is achieved when their flight paths

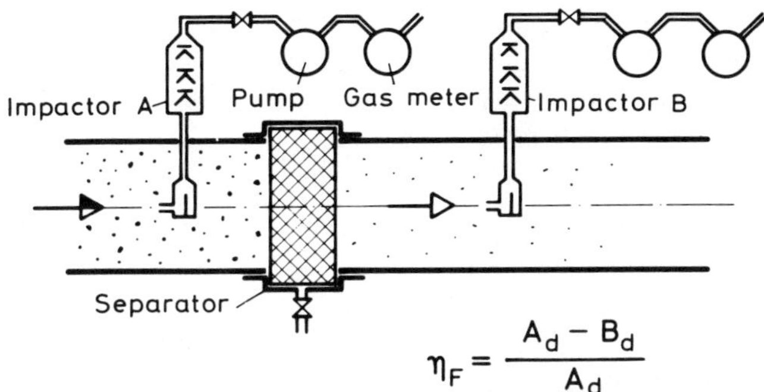

$$\eta_F = \frac{A_d - B_d}{A_d}$$

Figure 4.5. Schematic setup for separation efficiency measurements with cascade impactors.

deviate from the flow lines of the gas, due to the effects of inertia or other forces, and finally end on an impaction surface. Under certain conditions, the droplets may also get onto collecting or impaction surfaces without being influenced by external forces but as a result of their Brownian movement (diffusion separation) (Figure 4.6).

As with the collection of droplets, all the aerodynamic laws of the movement of droplets are highly analogous to the movement of dust particles. For a more fundamental insight into the nature of particle dynamics, I would recommend the excellent monograph by W. Strauss, *Industrial*

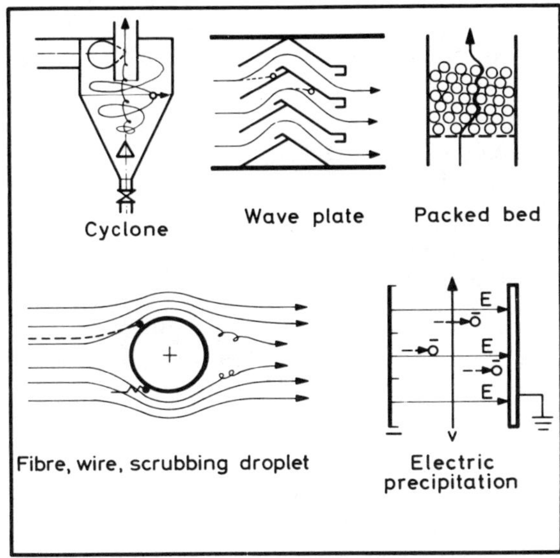

Figure 4.6. Methods for droplet separation.

Gas Cleaning.[97] In principle, droplet separators act like dust separators. Since, however, droplets generally remain adhered and coalesce once they have been separated, it is also possible to use simple impact separators such as wire mesh or wave plates.

4.3.1.   Inertial Separation

If droplets are to be flung out, the gas flow must be sufficiently accelerated and then deflected as sharply as possible. The drag force of the gas attempts to carry the droplet with it along the curved path, but the inertia of the droplet resists the change in direction. Since the frictional force is at best proportional to the surface of the droplet, whereas the so-called force of inertia is proportional to its volume, the larger the droplets are, the better they are separated by centrifugal forces. The separating range that is technically possible begins at $\sim$0.2 μm. Above 30 μm, particle separation provides virtually no difficulties at all. Because of the smallness of the mist droplets, their sedimentation velocity is so minimal that the force of gravity can be disregarded as an external field force.

A droplet entrained by the gas flow will move in such a way that centrifugal force G and frictional force F balance each other:

$$G = F$$

where

$$G = m\frac{d\mathbf{u}}{dt} \quad \text{and} \quad F = \tfrac{1}{2}\rho(\mathbf{u}\text{-}\mathbf{v})(u - v)AC_D \tag{2}$$

where m = mass of droplet
$\mathbf{u}, u$ = velocity of droplet
$\mathbf{v}, v$ = gas velocity
ρ = gas density
A = droplet cross-section
C_D = drag coefficient of droplet

If the trajectory of the droplet can be approximated to a circle of radius of curvature R and if C_D can be defined by Stokes' law, then (see also 4.4.1.3)

$$\frac{u_r}{v} = \frac{\rho_p v d^2}{18\mu R} = \psi \tag{3}$$

where u_r = relative velocity of droplet
ρ_p = droplet density
μ = gas viscosity

The expression is called the inertial impaction parameter ψ. The greater the relative velocity of a droplet to the velocity of the gas (i.e., the greater is the value ψ), the more readily the droplet will be ejected.

ψ represents the ratio of two lengths:

$$\psi = \frac{\lambda'}{R} \tag{4}$$

In Stokes' range, λ' is the length of the stopping distance of a spherical particle of diameter d and density ρ_p, when it is shot with initial velocity v into a still gas of viscosity μ.

The radius of curvature R is usually about the size of the impaction body, as for example the droplet diameter with the scrubber, wire and fiber diameter with a corresponding filter, plate distance with the wave-plate separator, or radius of gas outlet pipe with the cyclone.

If the length of the stopping distance of the droplet is greater or even several times larger than the size of the impaction body, there is reason to expect that separation of the respective particle could be possible. In this case, therefore $\psi > 1$. The dimensionless number ψ contains the most important separation parameters. For small particles, ψ must also contain the Cunningham correction $C \cong 1 + (2\lambda/d)$, where λ is the mean-free-path length of the gas molecules.

The simple derivations given above are very useful for making rough calculations and estimates of particle separation. More accurate calculations must be based on the full equations of force, which represent a system of simultaneous ordinary differential equations. Owing to the relative velocity, $\mathbf{u}$ and thus the droplet trajectory can only be calculated when the velocity field $\mathbf{v}$ of the gas is known. Even then, it cannot generally be expected that the droplet passes along a simple analytical curve. It is, however, possible to calculate the trajectory of the particle in stages using a computer.

By a pure dimensional analysis, it can be shown that the trajectory of the particle or the precipitation on simple impaction bodies can be described by three dimensionless numbers. These are generally the inertia parameter ψ, a Reynolds number Re, and the density ratio ρ/ρ_p or the ratio of particle diameter to fiber diameter.

With impaction systems such as wire and fiber filters, there is also another factor characteristic to the system. With fiber filters, the packing density α is often taken, and with model filters, the total relative projected area p. It was found that the pressure loss factor of the separator could be used for all inertial separators as the characteristic value for the system.

With a stationary velocity field of the gas flow, all particles having the same characteristic values and starting from the same point follow identical trajectories. Their probability of separation is thus equal. Different starting positions, however, give different probabilities of separation. For every set of characteristic values, there will therefore be a particular probability of separation which assumes a value between 0 and 100%. If only one parameter is varied (e.g., the inertia parameter ψ), the probability of separation can be shown as a monotonic curve $\eta_F = \eta_F(\psi)$, usually rising from 0 to 100%. By varying the parameter Re, for example, we get a group of curves from this separation efficiency curve.

4.3.2.   Separation by Diffusion

Sufficiently small particles participate in the random so-called Brownian movement in just the same way as do the molecules of the carrier gas. The square of the diffusional path x is inversely proportional to the droplet size d and directly proportional to diffusion time t:

$$x = \left(\frac{4kTt}{3\pi^2 \mu d} \right)^{1/2}$$

k is Boltzmann's constant and T the absolute temperature. If the droplet encounters in its zigzag movement a boundary surface of the flow, the surface of a fiber, for example, it can be regarded as separated. The paths of the droplet are short and for this reason, separation by diffusion only works in narrow porous systems (e.g., in fritted glass or ceramic candles or fine fiber filters). Since the probability of separation is also proportional to the residence time of the droplet in the porous system, the gas velocities should be at most a few cm s^{-1}. This has the disadvantage that diffusion separators, unlike inertia separators, are very large and necessitate correspondingly high investment costs.

The most advanced theoretical studies of separation by diffusion were made with fiber filters and single fibers in laminar fluids. Because of the collection by diffusion, the separation efficiency of a single fiber begins to rise again with decreasing inertia parameter, after passing through a minimum. With a filter, the minimum can be kept very flat (i.e., the separation efficiency very high). When calculating the filter separation efficiency, the single fiber can in this case no longer be regarded as an isolated individual fiber. A way out of this difficulty is the cellular model of Happel and Kuwabara. More details are given, for example, by Friedlander,[18] Emi et al.,[19] and Yeh and Liu.[20]

4.3.3.   Electrical Precipitation

With electrical precipitation, charges of the same sign are forced on the particles and then sucked out of the gas stream by an electric field. The collecting electrodes are either parallel plates or the inner surfaces of a parallel group of tubes. The ionizing electrodes consist of wires or pointed rods.

The ionizing electrode must have a sufficiently high voltage for charge carriers to become free in a corona discharge. These charge carriers collect on the droplets until an equilibrium potential is reached. For diffusion charging, the following equation applies for the charge on the droplet:

$$q_1 = ne = 10^6 de \tag{5}$$

Under the influence of the electric field, however, the droplet charge q is proportional to the field strength E and the square of the droplet diameter d:

$$q_2 = \frac{d^2}{4} Ef(\epsilon) \tag{5'}$$

Where e is the elementary charge and ϵ the dielectric constant. Opinion is that for particles smaller than 1 to 2 μm, charging takes place predominantly by diffusion (Refs. 21–23). The field strength between wire and plate is a function of the discharge current, the distances L wire/plate and W wire/wire and of the ionic mobility u.[24] Together with Stokes' law, the electrical force qE gives the droplet a drift velocity ω, which is of the order of cm s^{-1}. In the absence of all mixing processes, the separation efficiency η_F would be equal to $\omega B/vL$, where B is the height of the plate. However, as a result of the electric wind produced by the corona discharge, the gas is constantly thoroughly mixed. This means that here, too, the separation of a droplet becomes a statistical process and the collection efficiency η_F is given by the formula

$$\eta_F = 1 - \exp\left(\frac{-\omega B}{vL}\right) \tag{6}$$

Many efforts have been made in recent years to improve this formula from Deutsch. As yet, however, no confirmation has been obtained from the results of accurate measurements.

4.4.   DROPLET SEPARATORS

4.4.1.   Inertial Separators

4.4.1.1.   Cyclones

Probably the best-known inertial separator is the cyclone. It is a kind of pot into which the gas usually enters at the top (usually tangentially) and, after several circulations, leaves axially through a gas outlet tube. Through the centrifugal forces, the droplets are flung against the outside wall, where they can run downward.

Cyclones can be manufactured in all sizes and are used successfully for a wide variety of separation problems.[25] They have the advantage of being able to take large liquid loadings and avoid clogging. In principle, all the cyclone types known from dust separation can also be used for the separation of droplets. To prevent reentrainment of liquid, the bottom should be screened off by a cone. To avoid wall creep, the gas outlet tube should have a conical skirt on the outside. Since the outgoing gas still has a considerable angular momentum, subsequent precipitation of the mist occurs in the gas outlet tube. For this reason, a further separator, such as a baffle plate, should be positioned after the gas outlet tube.

In a normal design, the radius of the gas outlet tube r_i is one-third that of the outer radius r_a. The pressure drop coefficients are 10 to 20, related to the gas velocities in the outlet tube. Pressure drop and droplet size for 50% separation d_{50} can be calculated (see, e.g., Barth[26] or Muschelknautz[27]). Calculation and experiment generally agree very well.[92] For the 50% efficiency droplet, we get from the conditions of equilibrium (definitions of v_r and u_i are given under symbols)

$$d_{50} = \left(\frac{18\mu v_r r_i}{\rho_p \mu_i^2} \right)^{1/2} \tag{7}$$

Figure 4.7 illustrates the measured separation curves of a droplet cyclone.

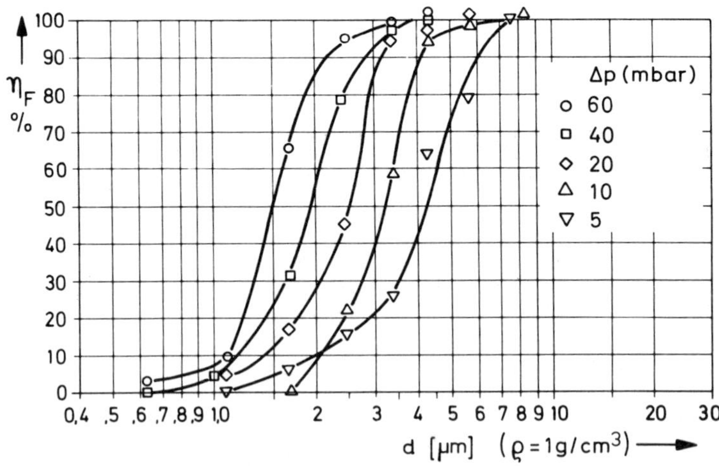

Figure 4.7. Fractional separation efficiencies of a cyclone (r_i = 50 mm) at different pressure losses.

4.4.1.2. Packed Beds

Packed beds used to be used to a large extent for the purification of process gases. Common examples are the coke boxes in sulfuric acid production, which act as dust and mist separators. Coke filters are still installed for industrial purposes today, although it is more common nowadays to use regularly shaped packing, such as Raschig rings, Pall rings, saddles, and so on.

Because of their relatively large surface area, the main application of packed beds is in processes of mass transfer. In most cases, the direction of flow through the beds is from bottom to top. Owing to the complexity of the flow field, it has not so far been possible to carry out exact calculations of the pressure drop or still less of the droplet separation. Measurements of the pressure drop are, however, available in large numbers and the model of a gas flow through a porous system combines the relevant physical quantities.[28-32]

With the repeated deflections of the gas flow in the bed, the entrained droplets are flung out and the separated liquid trickles off in a downward direction. Where the amount of liquid is too large, this may result in flooding: the liquid accumulates in the bed and is finally pushed up through it.[32,33] If the gas velocity is too high, the bed may become loosened up or packing materials and liquid be dragged along with it.

When used as mist separators, the gas can also be made to pass through the packed bed horizontally, as this brings advantages for the liquid drain.

Fractional separation efficiencies were determined for the first time by Jackson and Calvert.[34] Other investigations are reported in Ref. 93. From them, Figure 4.8 shows fractional separation curves of a Raschig ring packing for various velocities of the arriving gas. The shape of the curves are typical of impaction separation.

If these curves are plotted as a function of the inertial parameter ψ, the set of curves moves somewhat closer together. There nevertheless remains a high dependence on the Reynolds number, which here probably points mainly to a lack of thorough mixing and turbulence of the gas flow. With packings of greater pressure drop coefficient, on the other hand, the effect of the Reynolds number becomes so small that the measurement points can be represented by one common curve. Figure 4.9 shows curves of this kind for various packing heights of 8×8 mm glass Raschig rings.

As the packing height increases, the curves shift more and more into the field of lower ψ values. From the measured separation efficiency $\eta_0(\psi)$ at a packing height H_0, the separation η at a different packing height can be calculated according to the formula

$$\ln (1 - \eta)/\ln (1 - \eta_0) = H/H_0$$

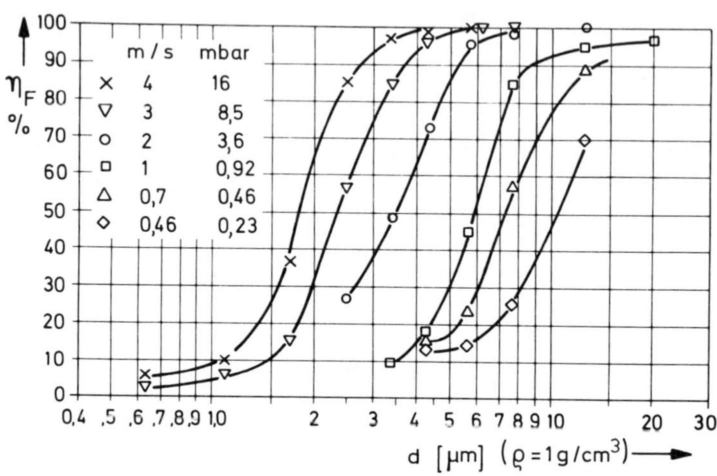

Figure 4.8. Fractional separation efficiencies of a packed bed (height 250 mm) of Raschig rings (25 × 25 mm) at different gas velocities and pressure losses.

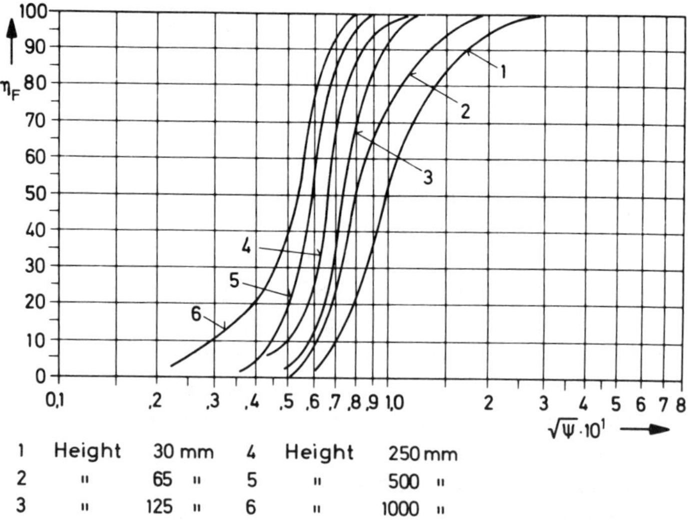

Figure 4.9. Fractional separation efficiency curves of glass Raschig rings (8 × 8 mm) at different packing heights as a function of the inertial impaction parameter.

204

Improved separation at greater packing height is, of course, obtained at the cost of a correspondingly higher pressure drop.

Pall rings and saddles have very similar separating properties to Raschig rings and can be regarded as equal when it comes to separating droplets above ~2 μm.

With mist concentrations of only a few mg m^{-3}, the large surface area of the packing materials may promote the disappearance of fine droplets as a result of evaporation. Packed beds cannot, however, be regarded as diffusional separators in the true sense of the word.

4.4.1.3. *Wave-Plate Separators*

Wave-plate separators can be described as a series of parallel impaction panels, usually placed in a vertical position. The gas passing through them is deflected one or several times through predetermined angles. The droplets that are flung out as a result are transported as a film by the gas flow into sheltered areas, where they can run off.

Wave-plate separators are both modern and economical devices and are now marketed by a number of firms. The various designs on the market differ in the profile of the wave plates and the positioning of the collecting channels. Where the gas flow arrives horizontally, flooding or reentrainment only occurs where the liquid loading is very high. Even then, it is still possible to collect this reentrainment with a second series of wave plates placed behind the first.

In special cases, the wave plates may become clogged by the precipitation of solids (e.g., due to sublimation). When this happens, it is advisable to spray the wave plates from the front—or even on both sides— with a washing solution.

The material most suited to the field of application can be used for the wave plates. When fitting them, attention must be paid to the correct direction of flow and to ensuring a good seal with the wall of the housing and a good liquid train.

Figure 4.10 shows the profiles and dimensions of a few wave plates. Their separation performance can be roughly calculated very simply by approximating the gas deflections, in accordance with Figure 4.11, to segments having an angle of deflection α and radius of curvature R.

During the residence time t of the gas, the droplet travels along the distance

$$s' = u_r t = u_r \alpha \frac{R}{v} \tag{8}$$

radially outward. Over this width, the outgoing stream of gas is free of

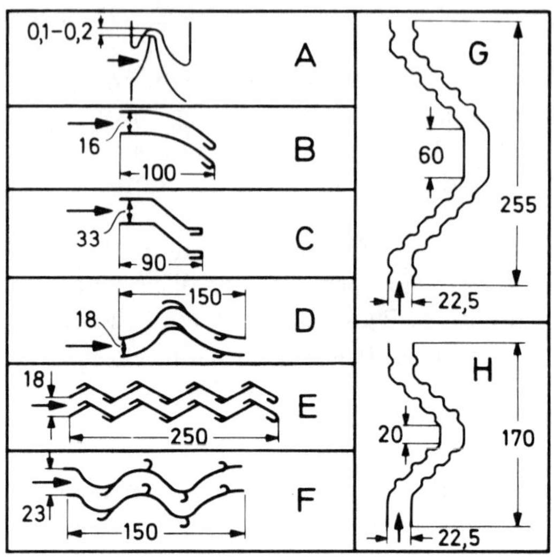

Figure 4.10. Wave-plate separator profiles. Dimensions in millimeters.

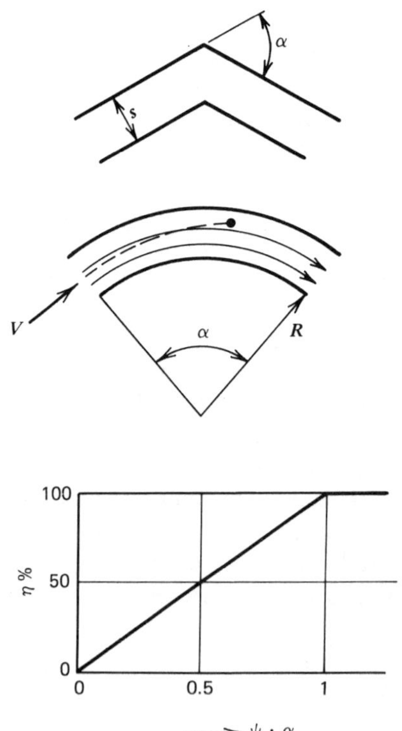

Figure 4.11. Droplet separation in a bend.

droplets. If the distance between the wave plates is s, the separation efficiency is then, from Ref. 36 with formula (3):

$$\eta_F = \frac{s'}{s} = \frac{\rho_p v d^2}{18\mu s}\,\alpha = \psi\alpha \tag{9}$$

If the wave-plate separator has n bends each with the angle α, then

$$\eta_F = 1 - (1 - \psi\alpha)^n \approx 1 - e^{-n\alpha\psi} \tag{10}$$

However, this holds only if the gas is thoroughly mixed after each bend, and this will rarely be the case. When making the calculation, s should not be taken as the distance apart of the wave plates at the point of entry but as the effective width of the stream of gas when forced through obstructions or constrictions. To the same extent as the stream is constricted, so its effective velocity increases.

As an example, Figure 4.12 shows fractional separation curves of a wave-plate separator where the arriving gas has velocities between 2.8 and 19 m s^{-1}. In the dimensionless representation of Figure 4.13 it can be seen that the conformity between measured and calculated values is fairly good. Since the wave plates in the real separator are never exactly equidistant, however, measured values fall below calculated values in the upper region.

The best results for separation are obtained where the distances between the wave plates are very small, type A. However, these close-spaced wave plates have a strong tendency to become clogged by entrained dust, even when sprayed with a washing liquid.

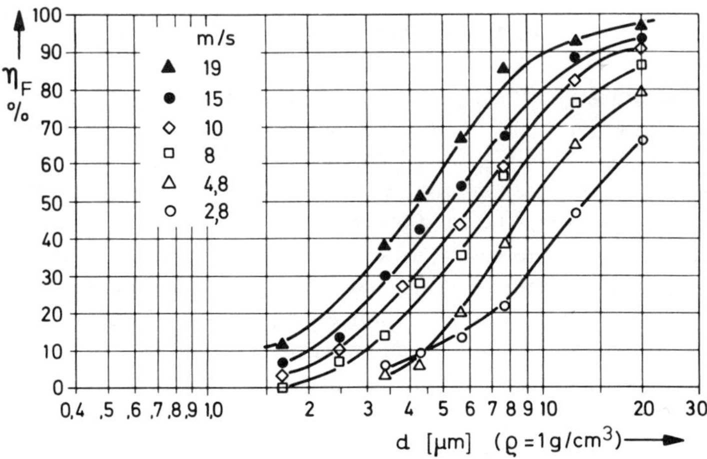

Figure 4.12. Fractional separation efficiencies of a wave-plate separator (Type H) at different gas velocities.

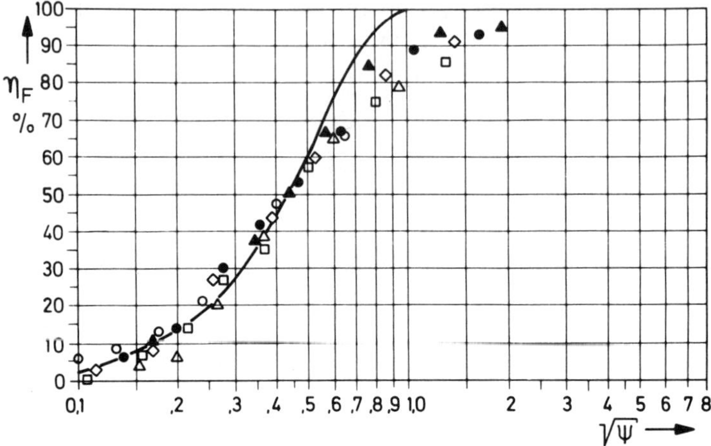

Figure 4.13. Measured values of Figure 4.12 as a function of the inertial impaction parameter. The curve represents the calculated separation efficiency values according to formula 10.

Where the distance between the wave plates is too large and deflection of the gas too low, there is a risk that droplets on the innermost flow lines do not become separated, which means that separation will never be complete. Inadequate pressure drop of the wave-plate separator can therefore have a very deleterious effect on the separation efficiency.

Detailed investigations of wave-plate separators are described in Refs. 36 and 94. A special design is also described in Ref. 37.

4.4.1.4.   *Wire-Mesh Mist Eliminators*

Wire-mesh mist eliminators have been widely used for many years as droplet filters. They are made up of several layers of wire mesh (Figure 4.14) which are mostly profiled and follow each other at a distance of ~2 mm. The thickness of the wire is usually around 250 μm and the thickness of the filter 100 to 300 mm.

The filter thus forms a system of high porosity and large surface area. Wire-mesh mist eliminators are therefore used to a large extent—like packing beds—for gas/liquid exchange processes. Depending on the field of application, the different companies offer filters of a wide variety of metal compounds and plastics. The filters can almost be tailor-made and thus be used almost everywhere; all necessary advice can be obtained from the manufacturing firms.

Usually, the filter is placed horizontally (e.g., above exchanger columns). The droplet-laden gas then flows through the filter from bottom to top, generally at 2 to 5 m s^{-1}. The collected liquid forms a film on the wires, which runs together at the gussets and drips downward against the gas flow.

Higher gas velocities and large quantities of liquid may lead to flooding. In such cases, the liquid accumulates in the filter and is finally forced upward through it. When this occurs, the pressure drop of the filter rises considerably. Flooding can also happen in packed beds; this was described by Souders and Brown.[32] The maximum velocity is often approximated (e.g., York,[38] York and Poppele[39]) by the formula

$$v = K \left(\frac{\rho_P - \rho}{\rho} \right)^{1/2}$$

where K is ~0.35 and when ft, lb, and s are used. The actual correlation can be taken from the experimental values in Figure 4.15.

The flooding and separating behavior of wire-mesh mist eliminators, packing materials, and wave plate separators has been described by, for example, Calvert et al.,[40] Germerdonk and Günther,[47] and Bell and Strauss,[42] Here the liquid to be separated consisted of a coarse spray. A flooded wire-mesh mist eliminator changes its separating properties and actually functions more as a scrubber.

With true mists, it is rarely possible that the liquid loading of the gas flow is high enough to lead to flooding. Excessive gas velocity does, however, lead to the collected liquid being atomized and to its reentrainment by the gas flow. These sprays form droplets generally larger than 100 μm.[43] The wire-mesh mist eliminator then acts as an agglomerator and requires the subsequent connection of a coarse or oversize separator. Because of the better liquid drain, it is often advisable to design the wire-

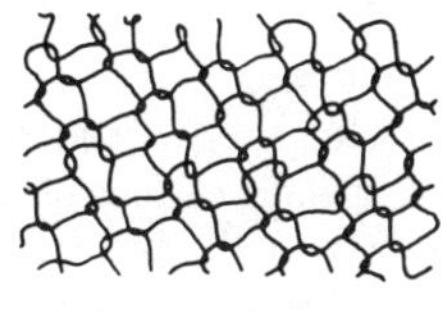

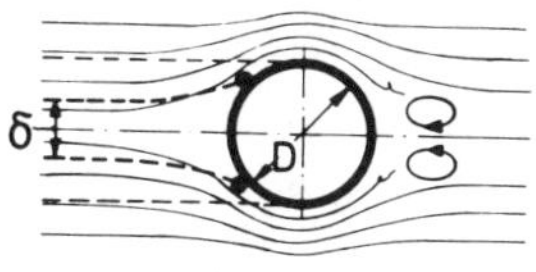

$$\varepsilon = \delta/D = f(\psi)$$

Figure 4.14. Projected area of one wire-mesh layer. Collection mechanism.

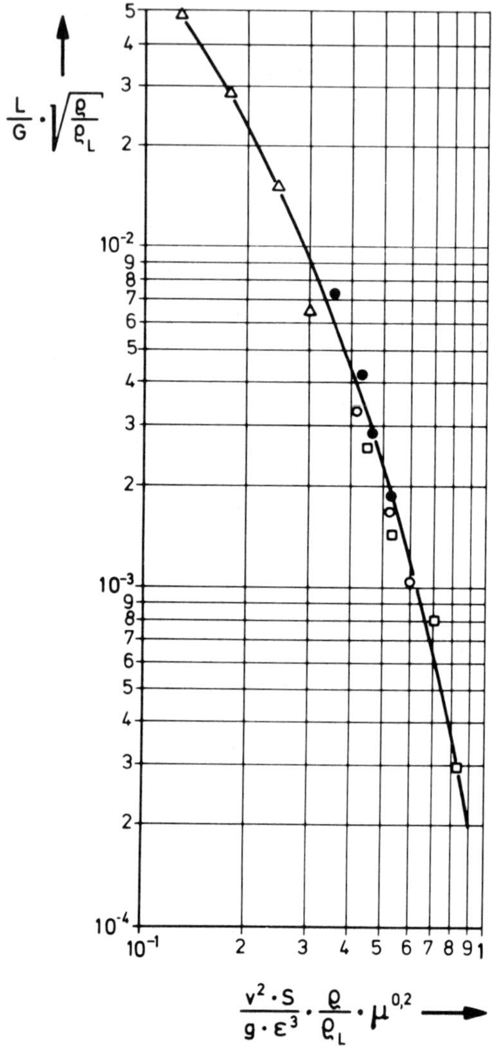

Figure 4.15. Experimental flooding curve on wire meshes: Δ: $\epsilon = 0.96$, $S = 640$ m^{-1}; ●○□: $\epsilon = 0.975$, $S = 360$ m^{-1}. ϵ, porosity of wire mesh; S(m²/m³), specific surface of wire mesh; L(kg hr^{-1}), quantity of liquid; G(kg hr^{-1}), quantity of gas; ρ(kg m^{-3}), gas density; ρ_L(kg m^{-3}), density of liquid; g(m s^{-2}), acceleration of gravity); μ(cP), viscosity of liquid.

210

mesh mist eliminator vertically and have the gas flow horizontally through it.

As far as actual mist separation is concerned, wire-mesh and fiber filters work according to a common principle, whereby the gas flows around a number of cylindrical elements. Airborne droplets have the tendency to continue on their forward path and to impinge on the wire or fiber. Depending on size, all particles, which were originally to be found within certain central flow lines are collected, while all particles outside are carried by the gas past the cylinder. The separation efficiency ε is defined as the ratio of the distance δ between the undisturbed limiting flow lines to the diameter of the cylinder D (see Figure 4.14).

The separation efficiency on a cylinder has been calculated by several authors (e.g., Langmuir and Blodgett,[44] Ranz and Wong,[45] Davies and Peetz,[46] Griffin and Meisen,[47] George and Poehlein,[48] Löffler and Muhr,[49] Löffler.[98] The latter derived an approximate formula containing ψ and Re, also allowing for interception. The interception effect is based on the glancing contact of the impaction body by the droplet and is equal to the ratio of droplet diameter to impaction body diameter. In general, measured values (e.g., by Wong et al.[50] or May and Clifford[51] show adequate agreement with the calculated values.

If we regard the separation in a wire-mesh or fiber filter as the sum of separation at the individual wires, it is possible to approximately calculate the separation in the filter.[43]

If the filter consists of n layers and if p_1 is the relative projected area of the wires of one layer, the separation efficiency η_F of the filter is

$$\eta_F = 1 - (1 - \epsilon p_1)^n \tag{11}$$

This can be approximated for multilayer filters using the formula which also applies for the general case:

$$\eta_F = 1 - e^{-p\varepsilon} \tag{12}$$

where $p = np_1$, the total relative projected area of the filter, and ε is the separation efficiency of an isolated cylinder.

The factor p can be calculated from the specific weight G of the demister and its thickness h. Per cross section of flow, we then get for the total length L and surface area S of the wire of density ρ_f,

$$L = \frac{Gh}{(D^2/4)\pi\rho_f}$$

$$S = D\pi L = \frac{4Gh}{D\rho_f} \tag{13}$$

However, only the projected length of the wires on the cross section of flow actually has a separating effect:

$$p = Sf_1 f_2 \qquad (14)$$

where f_1 is the ratio of projected surface to surface of a wire: $f_1 = 1/\pi$. f_2 must take into account the profiling of the wire layers; f_2 usually lies between 0.7 and 0.85.

The factor p can also be calculated from the pressure drop or the pressure drop coefficient ζ of the filter. For $p = 1$, the value ζ should be equal to the drag coefficient C_D of the individual wire. In Figure 4.16, ζ values for $p = 1$ are compared with C_D values from Ref. 52. It can be seen that the values above 2 to 3 m s^{-1} become almost constant and lie between 1.5 and 2 for $p = 1$. With wire-mesh mist eliminators, p is usually between 5 and 10.

With p as parameter and the ε values obtained by experiment from (51), we obtain, according to formula 12, the separation curves shown in Figure 4.17. It can be seen that, because of the statistical nature of the separating process, complete separation is not reached on thin filters with $p = 1$, even with large ψ values. References in the literature to wire-mesh mist eliminators are scarce and mainly concern investigations on pressure drops, flooding speeds, and total separation efficiencies.[53–55]

Detailed investigations on separation were first carried out by Carpenter

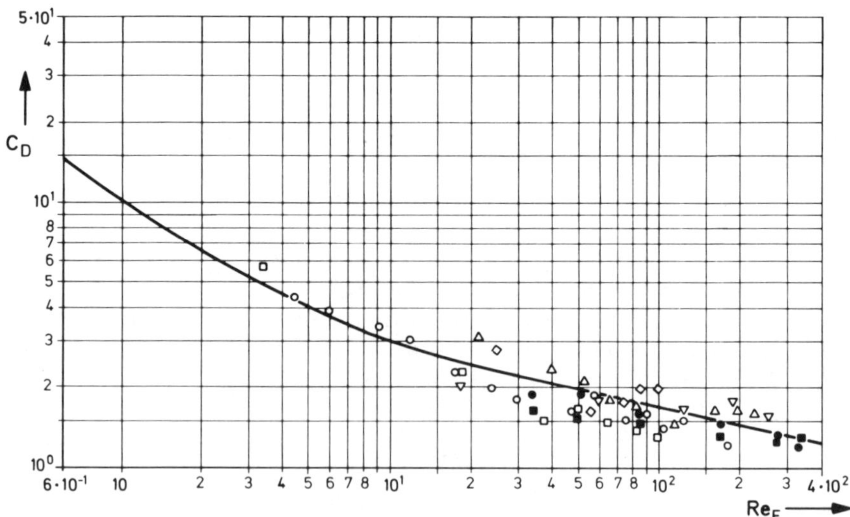

Figure 4.16. Drag coefficients of a cylinder as a function of Reynolds number. Points are calculated from pressure loss measurements on wire filters; curve is from H. Schlichting.[52]

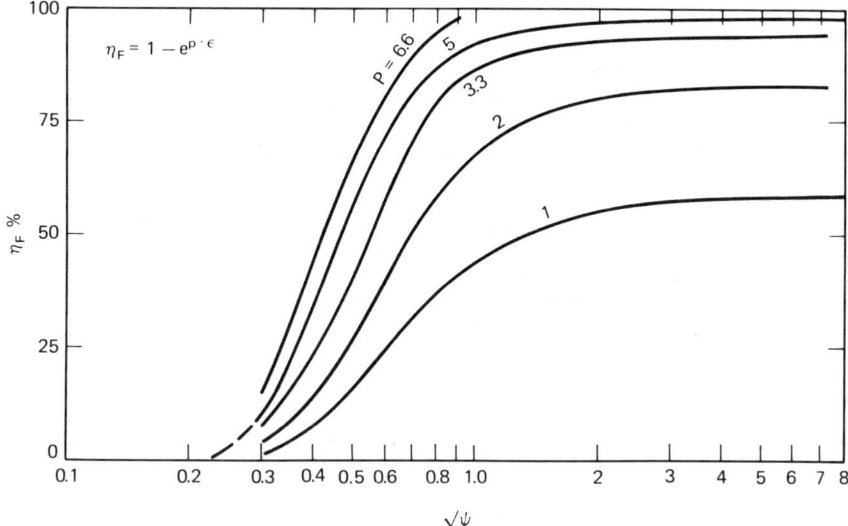

Figure 4.17. Fractional separation efficiency curves of wire mesh filters, calculated according to formula 12 with experimental ϵ-values from May and Clifford.[51]

and Othmer.[56] More recent results can be found in Refs. 43, 95, and 96. From these results, Figure 4.18 shows separation curves of a commercial wire-mesh mist eliminator at differing gas velocities. In Figure 4.19, the measured values are compared with calculated separation curves obtained with the simple approximation formula 12, but using the ϵ values according to the data from Löffler and Muhr.[49]

The agreement is good, which means that it is possible to calculate sufficiently accurately in advance the separation performance of a wire-mesh mist eliminator.

When wire-mesh mist eliminators are pressed too strongly or become partly clogged, the undisturbed gas flow around the individual wires slowly changes to a flow pattern through a porous system. This generally impairs the separating properties.

4.4.1.5. Fiber Filters

Over recent years, a large number of fiber filters have come onto the market which work by the principle of inertial separation. The thickness of the fiber is usually below 100 μm; the fiber material is metal or plastic and can be adapted to a large extent to the particular conditions. Asbestos fibers also have excellent separating properties.

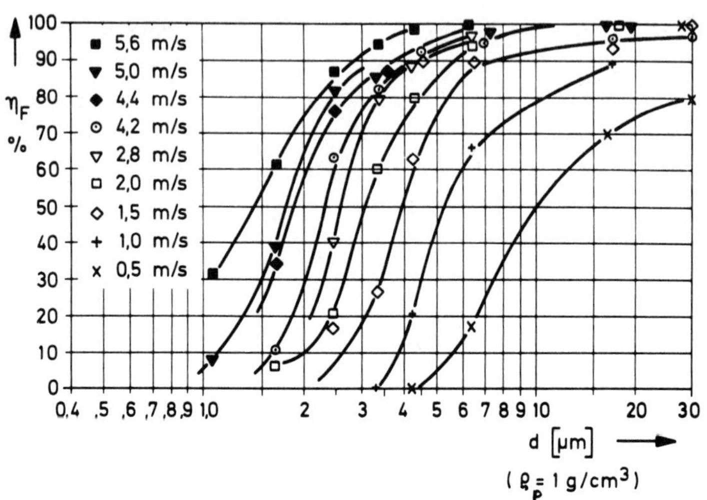

Figure 4.18. Fractional separation efficiencies of a wire mesh at different gas velocities.

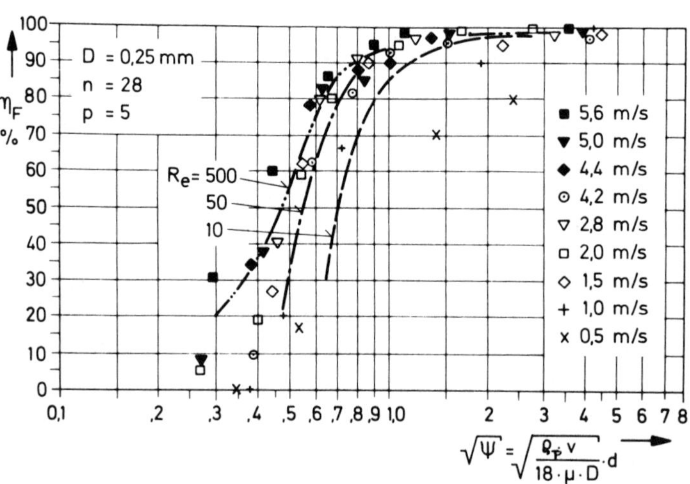

$$\sqrt{\Psi} = \sqrt{\frac{\rho_T \cdot v}{18 \cdot \mu \cdot D}} \cdot d$$

Figure 4.19. Measured values of Figure 4.18 as a function of the inertial impaction parameter. Curves represent calculated efficiencies according to formula 12 with ε-values from Löffler and Muhr.[49]

214

Firms usually supply fiber packings in a flat frame with a supporting screen on both sides. The velocity of the arriving gas should be a few m s^{-1}, as with wire-mesh mist eliminators. Figure 4.20 shows fractional separation curves for a fiber filter. Here, too, separation increases as the velocity of the arriving gas rises and decreases as the droplet size becomes smaller.

Approximate calculations of the separating properties of fiber filters can also be obtained by formula 12 (Figure 4.21). The starting point is again the collection on the individual fiber, although this can no longer be regarded as isolated on account of the lower filter porosity. Stenhouse and colleagues[57-59] therefore base their calculations on the cellular model of Happel and Kuwabara. Because of the smallness of the fibers, the diffusion effect can also have an influence on separation. More recent calculations of separation efficiency are therefore generally based on combined formulae. Reports on experimental results on fiber filters are given in Refs. 36, 60, and 61.

4.4.1.6. Scrubbers

With the scrubber, the gas to be cleaned has to pass through a dense curtain of droplets of a scrubbing liquid. The scrubbing liquid is atomized either by a jet, by a spinning disk, or by the gas flow itself. The gas flow must then flow around the droplets, which act as impaction bodies. The droplets do not, however, remain stationary, but are carried along by the stream of gas. This reduces their relative velocity and effectiveness as

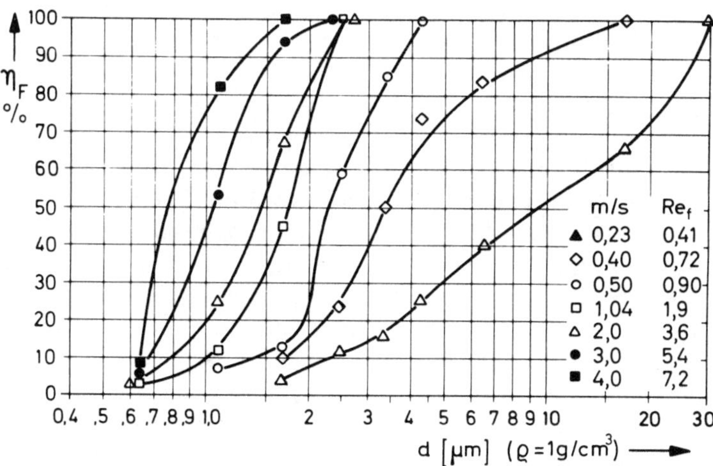

Figure 4.20. Fractional separation efficiencies of a high-velocity fiber filter.

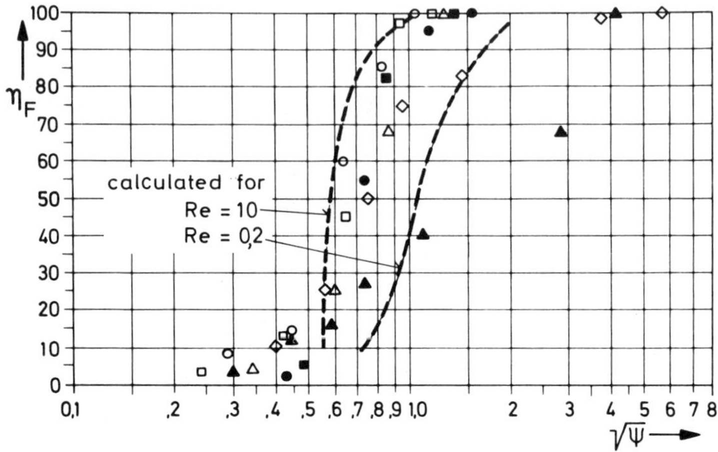

Figure 4.21. Measured values from Figure 4.20 as a function of the inertial impaction parameter. Curves represent calculated efficiencies according to formula 12 with the factor $p = 30$ and ϵ-values from Davies and Peetz.[46]

impaction bodies. Smaller scrubbing droplets have a better collection effect than larger ones, but also come to rest more rapidly in the gas flow. Optimum droplet size would be between 10 and 50 μm.

Particle separation on the isolated individual droplet is known through calculation and measurement to be a function of the inertia parameter ψ. ψ is formed with the relative velocity and the diameter of the scrubbing droplet. Nevertheless, there are no real separation calculations available for scrubbers, since nothing is known about the droplet spectrum (droplet sizes, numbers of droplets, trajectories of droplets). A few semiempirical considerations were made (e.g., Barth,[62] Calvert and colleagues,[63–65] Güntheroth,[66] and Boll.[67]

Scrubbers are used principally as dust separators. Here, there is a large amount of good empirical material available (Wicke and Krebs,[68] Holzer[69]), which can be transferred to mist separation. The role of wettability is still not entirely clarified, but it ought not to have too great an influence. The big advantage of these separators is that they can be used as dust and mist separators and as gas scrubbers at the same time. Their effectiveness ranges far into the submicrometer range. Figure 4.22 shows fractional separation efficiencies of a few types.

The outstanding efficiency of wet scrubbers, even for submicrometer particles, was the reason why separating mechanisms other than inertia were considered. Of particular interest seem to be investigations on the part played by condensing water vapor, which could lead to an increase in the size of the particles and thus to improved separation (Lancaster and Strauss,[99] Davis and Truitt,[70] Holzer[71]).

When choosing the most suitable type of scrubber, the numerous publications which have appeared recently (Refs. 73–76) can be of considerable help. According to Wicke and Krebs,[68] scrubbers can be divided into five categories (Figure 4.23). Scrubbing towers are, in principle, similar to packing beds with a high volume of liquid. Jet scrubbers work like water jet pumps and produce the scrubbing droplets out of nozzles. Their separating performance is independent of the gas throughput. Swirl scrubbers and venturis, on the other hand, are highly dependent on the gas throughput since the gas flow has to atomize the scrubbing liquid. With rotary atomizing scrubbers, a dense cluster of droplets is produced via a rotating disk and shot transversely into the gas flow. With venturis, this happens through nozzles. As a result of the very high gas velocities here, the scrubbing droplets are atomized even further.

In addition to these types, cyclones, wave-plate separators, wire-mesh mist eliminators, and packed columns can, of course, also be made to work as scrubbers by injecting a liquid. All scrubbers have to be followed by a coarse separator to remove the scrubbing droplets (Bell and Strauss[42]).

Problems arise with foaming. It often also happens that a wastewater

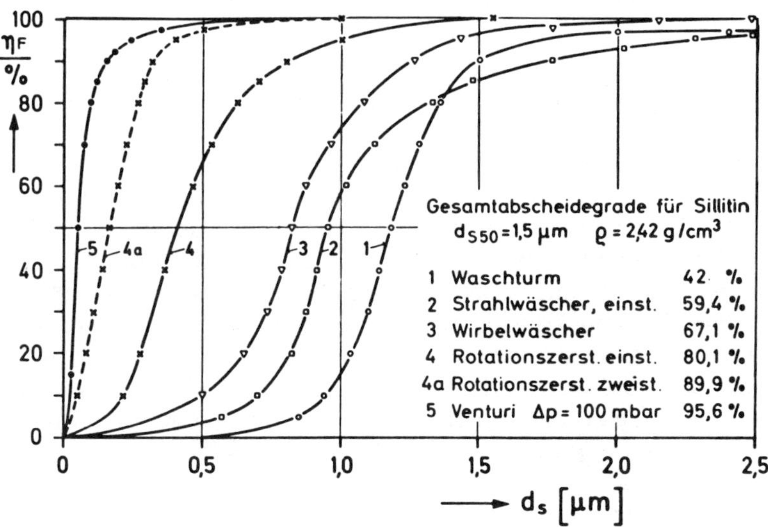

Figure 4.22. Fractional separation efficiency curves of scrubbers: 1, washing tower; 2, jet scrubber; 3, swirl scrubber; 4a + 4, rotational scrubbers; 5, venturi. Particles sizes for ρ_p = 2.42 g cm^{-3} (Holzer[69]).

	1	2	3	4	5
Typ	Waschturm	Strahlwäscher	Wirbelwäscher	Rotations-zerstäuber	Venturi

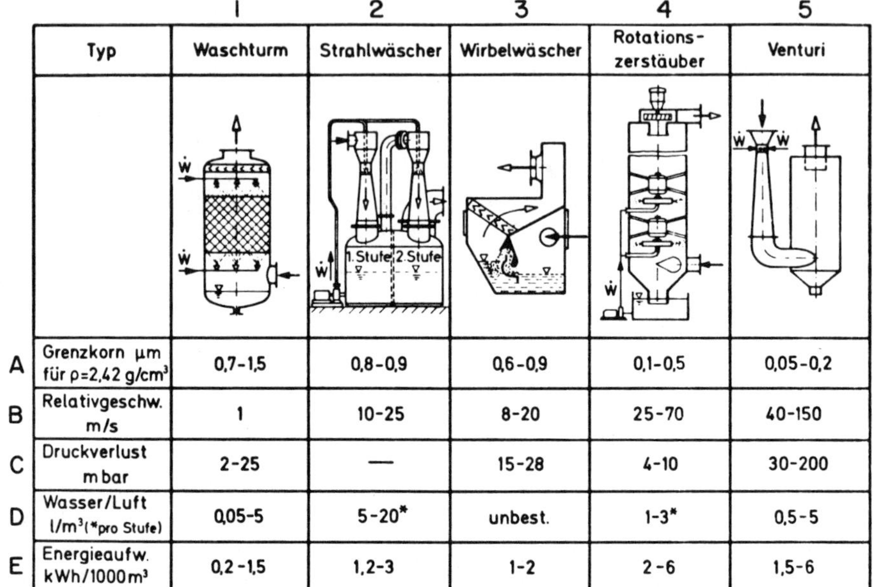

		1	2	3	4	5
A	Grenzkorn μm für ρ=2,42 g/cm³	0,7–1,5	0,8–0,9	0,6–0,9	0,1–0,5	0,05–0,2
B	Relativgeschw. m/s	1	10–25	8–20	25–70	40–150
C	Druckverlust m bar	2–25	—	15–28	4–10	30–200
D	Wasser/Luft l/m³ (*pro Stufe)	0,05–5	5–20*	unbest.	1–3*	0,5–5
E	Energieaufw. kWh/1000 m³	0,2–1,5	1,2–3	1–2	2–6	1,5–6

Figure 4.23. Comparison of typical operational values of wet scrubbers. Numbers 1–5 as in Figure 4.22. From Holzer (69). Row A: size limit in μm for ρ = 2.42 g cm^{-3}. Row B: relative velocity, m s^{-1}. Row C: pressure drop in mbar. Row D: water/air ratio in l m^{-3} per stage. Row E: energy usage in kWhr/1000 m³.

problem occurs in addition to the waste air problem, and with venturis, a noise problem, too.

The effectiveness of a scrubber depends to a very large extent on the operating conditions, such as pressure drop and volume of scrubbing liquid. Here the individual types can also differ considerably in their economy. After numerous investigations by Wicke and Holzer,[68,69] there is a certain minimum energy requirement for every separating efficiency, below which it is not possible to go. The smaller the required droplet size for 50% separation, the greater the energy requirement (Figure 4.24). To obtain an accurate view of the economic aspect, it is important to include, in addition to the straight operating costs, the different purchasing prices of the numerous scrubbers available on the market.

4.4.1.7. "Frit"-Type Filters

These are usually available on the market in candle form and consist of a sintered, porous ceramic body. The gas flowing through it is diverted several times in the narrow passageways of the pores, whereby droplets of mist are flung out. Figure 4.25 shows fractional separation curves for several gas throughputs. The occurrence of a separation minimum indicates that, in addition to inertial separation, separation by diffusion has also taken effect.

Fritted glass or ceramic candles are good mist separators. Nevertheless, they have a number of disadvantages which restrict their use. The gas throughput per candle is small, although the pressure drop is relatively high. Therefore, a filter often needs hundreds of candles, necessitating a considerable amount of laborious work in installing and sealing them all. Besides this, the ceramic bodies are fragile and can easily become clogged up by dust.

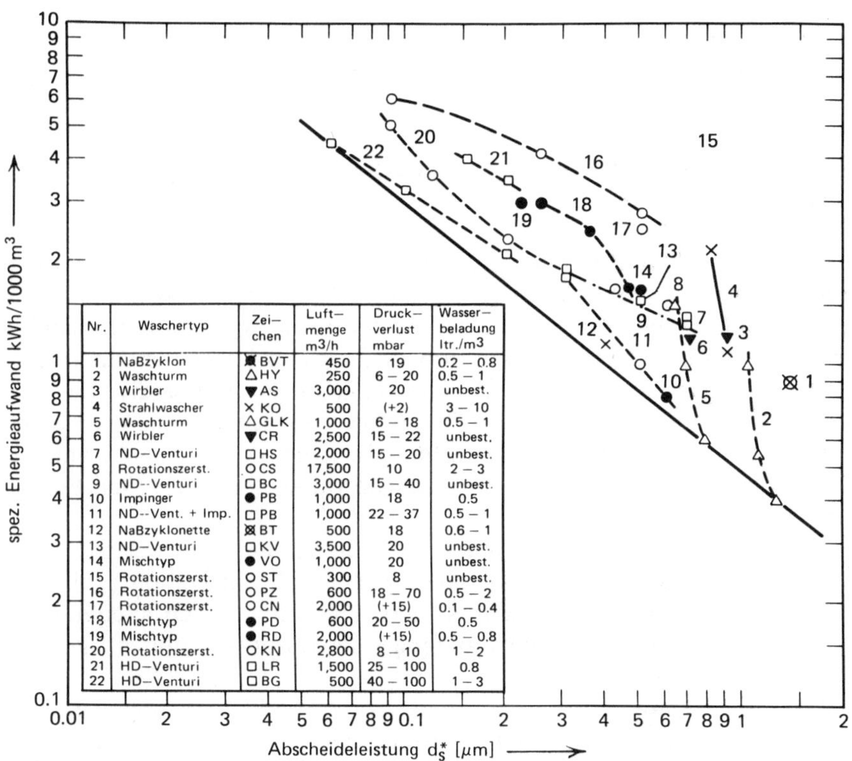

Nr.	Waschertyp	Zei‑chen	Luft‑menge m3/h	Druck‑verlust mbar	Wasser‑beladung ltr./m3
1	NaBzyklon	▨ BVT	450	19	0.2 − 0.8
2	Waschturm	△ HY	250	6 − 20	0.5 − 1
3	Wirbler	▼ AS	3,000	20	unbest.
4	Strahlwascher	× KO	500	(+2)	3 − 10
5	Waschturm	△ GLK	1,000	6 − 18	0.5 − 1
6	Wirbler	▼ CR	2,500	15 − 22	unbest.
7	ND−Venturi	□ HS	2,000	15 − 20	unbest.
8	Rotationszerst.	○ CS	17,500	10	2 − 3
9	ND−Venturi	□ BC	3,000	15 − 40	unbest.
10	Impinger	● PB	1,000	18	0.5
11	ND−Vent. + Imp.	□ PB	1,000	22 − 37	0.5 − 1
12	NaBzyklonette	▨ BT	500	18	0.6 − 1
13	ND−Venturi	□ KV	3,500	20	unbest.
14	Mischtyp	● VO	1,000	20	unbest.
15	Rotationszerst.	○ ST	300	8	unbest.
16	Rotationszerst.	○ PZ	600	18 − 70	0.5 − 2
17	Rotationszerst.	○ CN	2,000	(+15)	0.1 − 0.4
18	Mischtyp	● PD	600	20 − 50	0.5
19	Mischtyp	● RD	2,000	(+15)	0.5 − 0.8
20	Rotationszerst.	○ KN	2,800	8 − 10	1 − 2
21	HD−Venturi	□ LR	1,500	25 − 100	0.8
22	HD−Venturi	□ BG	500	40 − 100	1 − 3

Figure 4.24. Power consumption and separation efficiencies of wet scrubbers. Numbers 2 and 5 are of type 1; 4, type 2; 3 and 6, type 3; 8, 15, 16, 17, and 20, type 4; 7, 9, 13, 21, and 22, type 5. ρ_p is 2.42 g cm^{-3}. Vertical axis: specific energy use. Horizontal axis: cut size. Column titles are number, scrubber type, symbol, air flow rate, pressure drop, and water usage (Holzer[69]).

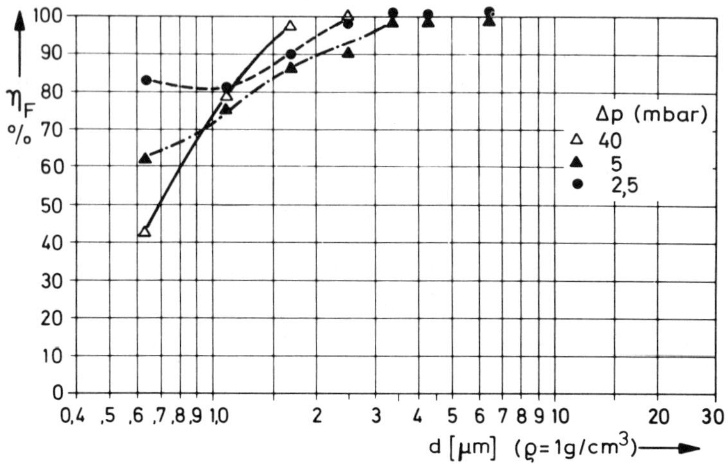

Figure 4.25. Fractional separation efficiencies of a ceramic "candle."

4.4.1.8. *Characterization of Inertial Separators by Characteristic Values*

It was mentioned in Section 4.3.1 that the separation of droplets with all inertia systems depends on three "dynamic" dimensionless numbers (ψ, Re, ρ_P/ρ) and a characteristic value specific to the system (e.g., packing density α). As long as the field of flow of the gas is known or can at least be assumed, it is possible to calculate, at least approximately, the separation efficiency of the system. This is the case for cyclones, wire and fiber filters, and wave-plate separators. In general, good conformity is obtained between calculated and measured values (see e.g., Figures 4.13 and 4.19).

These approximate calculations are a valuable aid, particularly where there are no measurements available. For simple separation problems, an approximate calculation of this kind will generally be adequate for designing or correctly utilizing wave-plate separators, cyclones, and wire and fiber filters.

With the other inertia systems, such as packed beds or scrubbers, we are dependent on values obtained in past experience or on measurements. There are, in fact, a large number of measurements now available. To summarize this individual information and set up an interrelationship, use is also made in these cases of empirical approximation formulas.

After numerous tests by the author,[77,78] the introduction of a new dimensionless value, the inertia separation parameter ψ_A, proved to be very suitable for representing the results of separation. Given

$$\psi_A = \psi \, \mathrm{Re}^{1/3} \, \zeta^{2/3} \cdot 3 \tag{15}$$

where ψ represents the inertia impaction parameter, Re the Reynolds number, and ζ the pressure loss factor:

$$\psi = \frac{\rho_P v d^2}{18\mu D}$$

$$\mathrm{Re} = \frac{\rho v D}{\mu}$$

$$\zeta = \frac{\Delta p}{\frac{1}{2}\rho v^2}$$

where ρ_P = droplet density
  ρ = gas density
  μ = gas viscosity
  v = gas velocity
  d = droplet diameter
  D = characteristic size of impaction body
  Δp = pressure drop

   Explicitly, we get from formula 15,

$$(\psi_A)^{1/2} = \rho_P^{-1/2} \, \rho^{-1/6} \, \mu^{-2/3} \, \Delta p^{1/3} \, D^{-1/3} \, d \cdot \tfrac{1}{2} \tag{16}$$

If

$$\eta_F = \eta_F[(\psi_A)^{1/2}] \tag{17}$$

then it must follow that

$$\eta_F = \eta_F[(\Delta p)^{1/3}d] \tag{18}$$

Figures 4.26 through 4.29 are intended to show that practically common curves are obtained for separating systems when the measuring points are plotted against the product $(\Delta p)^{1/3}d$ instead of against d or $\psi^{1/2}$. In place of the group of curves as shown in Figures 4.7, 4.8, 4.12, 4.18, and 4.20, we obtain in this representation one single separation curve characteristic of the separating system.

   From the curves in Figures 4.27 through 4.30 it can be seen that the separation efficiency values $\eta_F = 50\%$ and $\eta_F = 95\%$ correspond to certain values of the product $(\Delta p)^{1/3}d$:

$$(\Delta p)^{1/3}d_{50} = K_{50} \tag{19}$$
$$(\Delta p)^{1/3}d_{95} = K_{95}$$

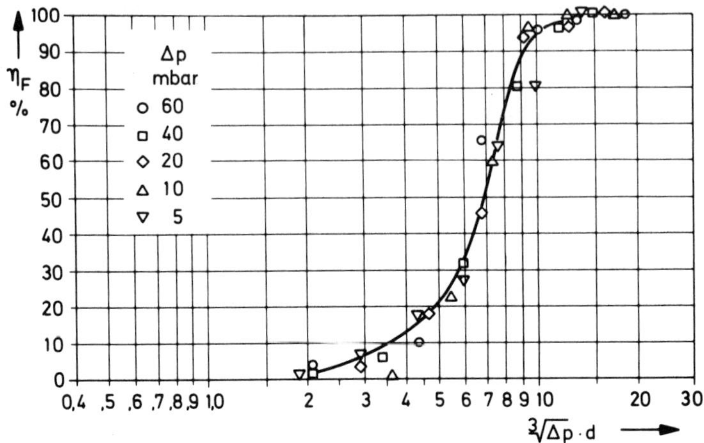

Figure 4.26. Characteristic fractional separation efficiency curve for a cyclone, Figure 4.7.

Elevated to the 3. power, K_{50} and K_{95} have the dimensions of an energy, and thus can be regarded as characteristic separation energies. Figure 4.30 illustrates experimentally for some packed beds the connection between d_{50} and Δp suggested by formula 19. The pressure losses are plotted against the droplet sizes for which a 50% and a 95% separation are obtained. All measuring points lie on straight lines.

K_{50} and K_{95} characterize the separators not only according to their

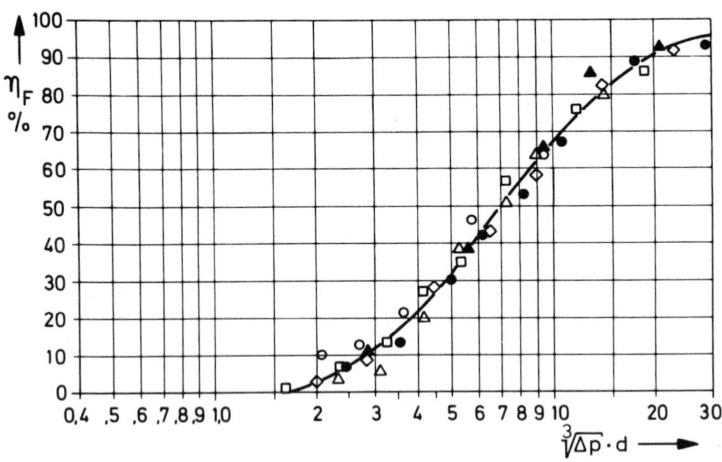

Figure 4.27. Characteristic fractional separation efficiency curve for a wave-plate separator, Figure 4.12.

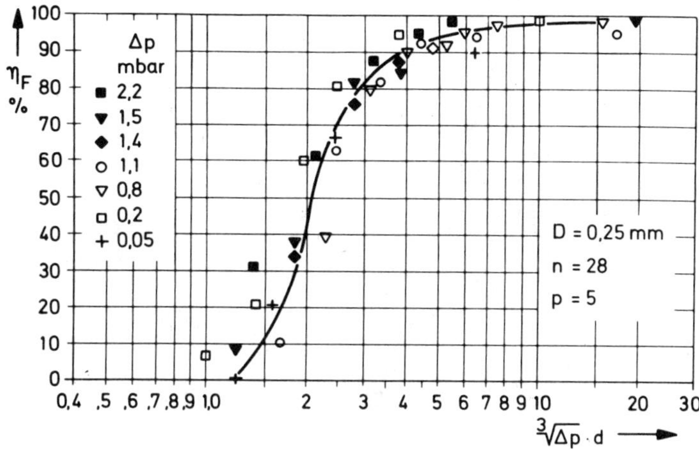

Figure 4.28. Characteristic fractional separation efficiency curve for a wire mesh, Figure 4.18.

efficiency but also classify them according to the particular energy required. The lower these characteristic values, the smaller will be the droplets that can be separated and the lower is the pressure loss required.

Characteristic values derived from the author's measurements are shown in Tables 4.1 through 4.6 for cyclones, packed beds, wave-plate separators, wire and fiber filters, and venturis. For the pressure drop Δp,

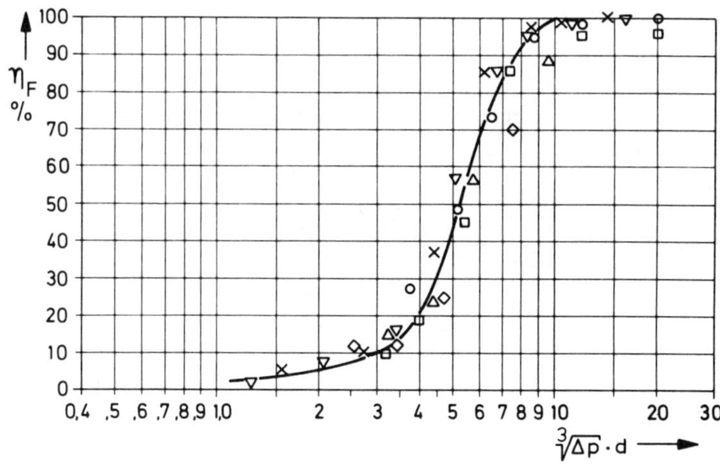

Figure 4.29. Characteristic fractional separation efficiency curve for a packed bed, Figure 4.8.

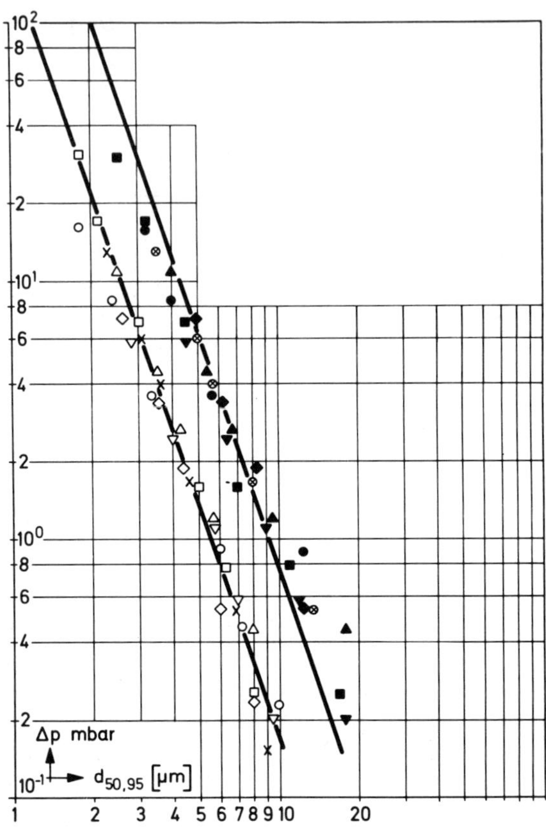

Figure 4.30. Pressure losses of packed beds as a function of drop size to be separated with 50% and 95% efficiency.

Table 4.1. Characteristic Separation Values of Cyclones

Cyclone	ζ	r_i (mm)	K_{50}	K_{95}	$(\psi_A)_{50}^{1/2}$
I	8	100	7.2	12	1.55
II	11	50	6.8	10	1.85
III	12	25	4.7	8	1.60
IV	15	17	4.2	7	1.64
V	6.5	12	3.0	5	1.32
VI	14	10	3.7	6	1.72
VII	6	5	3.0	5	1.75

Table 4.2.   Characteristic Separation Values of Packed Beds

	H (mm)	ζ	D (mm)	K_{50}	K_{95}	$(\psi_A)_{50}^{1/2}$
Saddles	260	90	25	5.5	9.0	1.9
25 × 25 mm	490	175	25	6.1	9.5	2.1
Pall rings	260	45	25	5.2	10.0	1.8
25 × 25 mm	520	90	25	5.6	9.0	1.9
Raschig rings	250	150	25	5.4	8.5	1.8
25 × 25 mm	480	300	25	5.6	9.0	1.9
Raschig rings	500	1000	10	5.5	8.0	2.5
10 × 10 mm						
Raschig rings	250	400	8	4.5	6.5	2.2
8 × 8 mm						

the unit mbar was used and for the droplet size d_{50} and d_{95}, the unit μm. K_{50} and K_{95} then have the same numerical values as d_{50} and d_{95}, respectively, when Δp becomes $= 1$ mbar.

With cyclones, the characteristic values become systematically smaller as the radius of the exit duct decreases. The high values for large cyclones point to the fact that cyclones gradually become uneconomical for large gas throughputs. The characteristic values for packed beds and wave-plate separators do not differ very much. Here, too, however, it can be seen that better separation is obtained where the distances between the wave plates are smaller. Wire and fiber filters follow the same trend.

The figures for the pressure loss factor ζ given in Tables 4.1 through 4.6 are valid only for the range of the square law of resistance. From the characteristic values K_{50} and K_{95} it is very easy to calculate the droplet sizes d_{50} and d_{95} according to eq. 19. Somewhat more universal characteristic values can be obtained by plotting the fractional separation ef-

Table 4.3.   Characteristic Separation Values of Raschig Rings

Height (mm)	ζ	K_{50}	K_{95}
1000	1600	6.3	8
500	900	6.0	8
250	500	5.2	7
125	350	4.5	6.5
62	140	4.3	6.5
30	65	3.8	8.5

Table 4.4. Characteristic Separation Values of Wave-Plate Separators

Type	ζ	D (mm)	K_{50}	K_{95}	$(\psi_A)_{50}^{1/2}$
H	2.5	22	7.0	30	2.5
G	4.5	20	5.5	25	2.0
F	8.5	23	6.0	12	2.1
D1	4.5	18	5.5	15	2.1
D2	8.5	18	5.5	15	2.1
E	17	18	5.5	12	2.1
A	—	0.15	1.7	6	3.1
C	1.6	33	7.0	—	2.2
B	4.5	17	15	70	5.8

ficiency against the square root of the inertia separation parameter ψ_A, as shown in Figure 4.31. The position of the separation curve is fixed by the numerical value $(\psi_A)_{50}^{1/2}$ for 50% separation. This characteristic value $(\psi_A)_{50}^{1/2}$ is entered in each of the Tables 4.1 through 4.5 in the last column. It is evident that geometrically similar separators such as cyclones have

Table 4.5. Characteristic Separation Values of Wire and Fiber Filters

	ζ	D (mm)	K_{50}	K_{95}	$(\psi_A)_{50}^{1/2}$
Regular wire filters					
I	2	0.5	4.2	50	5.2
II	4	0.5	4.0	10	5.0
III	5.5	0.5	3.4	9	4.3
IV	11.0	0.5	2.5	4.5	3.3
Wire meshes					
I	10	0.25	2.2	5.0	3.5
II		0.25	2.0	4.5	3.2
III	10	0.25	2.5	4.0	4.0
Fiber filters					
I	80	0.04	2.0	4.0	5.8
II	1000	0.026	2.5	3.5	8.5
III	15	0.05	1.5	4.0	4.0
IV	30	0.05	1.5	4.0	4.0

Table 4.6.   Characteristic Separation Values of Venturis

Type	Δp (mbar)	d_{50} (μm)	$d'_{50}{}^{a}$ (μm)	K'_{50}	$K' = \Delta p\, d'_{50}$
I	100	0.19	0.25	1.2	25
	80	0.22	0.29	1.2	23
	60	0.24	0.31	1.2	19
	35	0.54	0.60	2.0	21
II	100	0.09	0.16	0.7	16
	75	0.14	0.22	0.9	17
	50	0.30	0.39	1.4	19
III	47	0.46	0.60	2.2	29
	32	1.10	1.10	3.5	35
IV	38	0.46	0.54	1.8	20
	20	0.77	0.85	2.3	17
V	20	0.80	0.90	2.4	18
	14	1.15	1.20	2.9	17

^{a} d'_{50} with Cunningham correction factor.

virtually the same characteristic values. The corresponding curve in Figure 4.31 then characterizes not only the individual separator but also the type of separator. Using formula 16 the relevant droplet sizes d_{50} can be calculated from $(\psi_A)_{50}^{1/2}$. If d is inserted in μm, D in mm, and Δp in mbar, at a droplet density of 1000 kg m^{-3} and normal air conditions, we get

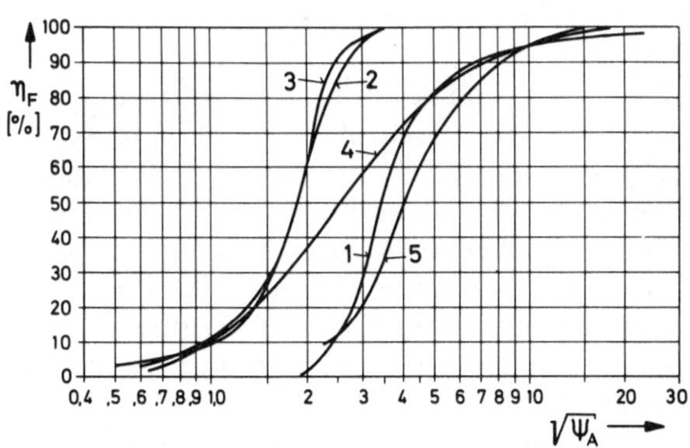

Figure 4.31. Characteristic fractional separation efficiency curves: 1, woven wire mesh; 2, Raschig ring bed; 3, cyclone; 4, wave-plate separator; 5, fiber filter.

from formula 16 the specially adapted numerical equation

$$d_{50}\,(\mu m) = (\psi_A)_{50}^{1/2}\left[\frac{D(mm)}{\Delta p(mbar)}\right]^{1/3} \tag{20}$$

For different gas conditions, these droplet sizes change according to the relationship

$$\frac{d_{50_1}}{d_{50_2}} = \left(\frac{\rho_1}{\rho_2}\right)^{1/6}\left(\frac{\mu_1}{\mu_2}\right)^{2/3} \tag{21}$$

In special cases and for venturis generally, instead of formula 19, the relationship

$$(\Delta p)^{1/2}d_{50} = \text{const.} \tag{22}$$

was found to hold more true for the pressure loss dependency. Retaining a constant characteristic value $(\psi_A)_{50}^{1/2}$, we would get for this from eq. 20 a size D of the scrubbing droplets, which decreases as the pressure loss rises:

$$D \sim \frac{1}{(\Delta p)^{1/2}}$$

For the working point of venturis of optimum design, we even find that

$$\Delta p d'_{50} = \text{const.} \approx 16\ (\text{mbar}\ \mu m)$$

holds true. d'_{50} is here the aerodynamic droplet size corrected with the Cunningham factor.

4.4.2. Diffusion Separators, Fiber Filters

Filters of this kind are available on the market from a large number of firms. The best-known are the Brink filters. For reasons of space, the filters are usually constructed in candle form. Depending on the gas throughput, dozens of candles can be placed in one housing. The fiber packing is positioned between two supporting fabrics or between perforated panels. Depending on the application, glass or plastics fibers can be used as filter material. The thickness of the fibers is usually less than 50 μm. The liquid is forced by the gas flow through the packing and can then flow downward on the gas outlet side.

With all diffusion filters, there is the danger of becoming clogged by dust or sediment. Where the sediment is water-soluble, the filters can be freed again—at least partially—by spraying with water.

In normal operation, the pressure drops are at 20 mbar and are proportional to the velocity of the arriving gas (a few cm s^{-1}). Blockages become evident as a cause of increased pressure. Diffusion filters have separation efficiencies of mostly 90 to 100%, even for submicrometer particles. Unlike inertia separators, they retain their separation efficiency also with lower gas throughputs.

Fiber filters seem to be used in large numbers for the separation of acid mists. As far as their separation performance is concerned, they have to compete with scrubbers and electrical filters. Reports on experience in operation are given by Fairs[19] and Brink and colleagues.[80–82]

4.4.3.   Electrostatic Separators

Electrical filters are used mainly for dust separation and have been in use now for many years. Earlier, they consisted exclusively of parallel plates with wires stretched between them. Nowadays, parallel bundles of tubes are also used.

Mist separation is generally less of a problem than is dust separation. Droplets normally have more favorable electrical properties than dust particles do, and the precipitation electrodes do not have to be cleaned off. It may nevertheless be advantageous to sprinkle the collecting electrodes with a liquid (wet electrical filters).

In most cases, the gas flows through the filter from the bottom to top. The path of the gas in the filter is usually a few meters, with large units above 10,000 m^{3}hr^{-1}. The distance of the ionizing electrode from the collecting electrode is then normally 15 to 20 cm. The voltage is selected high enough for the spark-overs to remain just below a certain limited value. The gas current is equalized by a distributor plate with a pressure drop of up to 10 mbar.

Compared with scrubbers, electrical filters have the advantage of low water and energy consumption and also there is no risk of clogging up as there is with fiber filters. The only disadvantage of electrical filters is most probably their high purchase price.

The gas velocity should be only a few m s^{-1}. At higher velocities, the separation performance for droplets below 3 μm decreases rapidly. The lower the gas velocity, the better the separation. Figure 4.32 shows as an example both measured and calculated fractional separation efficiencies of an older plate-type electrical filter from a sulfuric acid plant.[36] Figure 4.33 shows fractional separation efficiencies of a wet-tube-type electrical filter in the case of lactam mists.[35] Surprisingly, very few studies

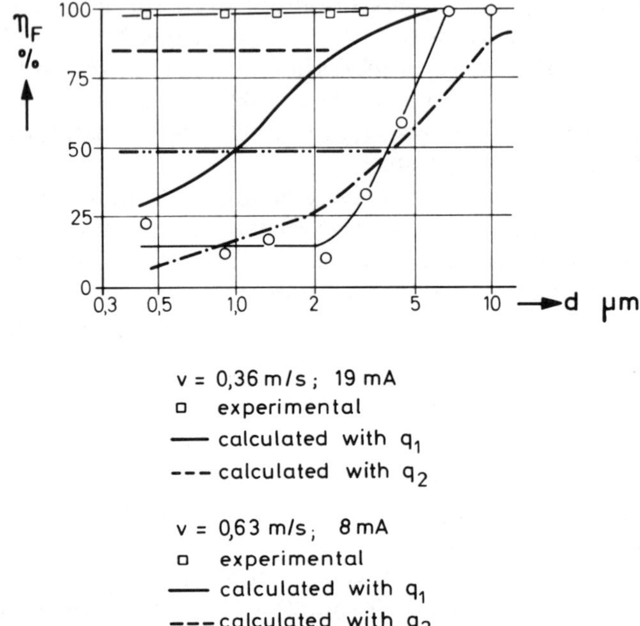

Figure 4.32. Fractional separation efficiencies of an electrostatic precipitator (plate type).

have been made of the fractional separation performance of electrical filters.[83,84,100]

Electrical filters are used everywhere where large quantities of gas are involved, where the mists have a large submicrometer fraction and where high demands are made on the purity of the waste gas. Generally speaking, above 60,000 m^3 hr^{-1} will be selected for electrical filters and lower for scrubbers or fiber filters. The firms who supply such apparatus generally have enough experience to master the problems that arise in this connection.

4.4.4.   Other Separators

In principle, it is also possible for droplet separation to use the tubular cloth filters employed for dust separation. This has the advantage that, because of the liquid drain, cleaning of the fabric is not necessary as long as dust is not present as well. On the other hand, the effect of the "filter cake" is missing, so that the collection efficiency remains low. For this

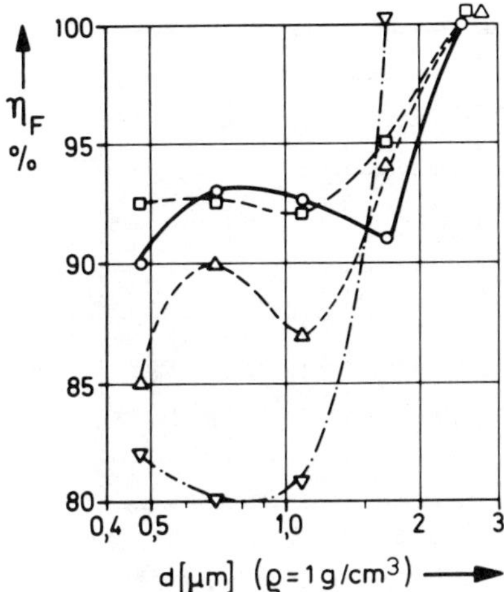

Figure 4.33. Fractional separation effiencies of an electrostatic precipitator (tube type).

reason, various companies have sprinkled single- or multilayer filter cloths on the market.[85,86] With these, the water loads are similar to the true wet scrubbers at 0.5 to 4 1 m^{-3}. The gas velocities are just a few cm s^{-1} and the pressure drop is between 20 and 40 mbar.

Despite the use of double cloths, the filters seem unsuitable for the separation of very fine aerosols. Figure 4.34 shows separation curves for a dry and a sprinkled filter, for which the data are far below those given by the manufacturer. The filter was designed for 500 m^3 hr^{-1} gas throughput and comprised seven double tubes of polypropylene fabric, each 1 m^2 and 2.2 m long. The aerosol to be separated consisted of condensation mist from solvent vapors.[87]

Recently, descriptions have also appeared in literature of so-called foam scrubbers as aerosol separators.[88] Here the gas to be scrubbed is conveyed via a distributor plate vertically upward through a bed of foam. At minimal pressure drops of maximum 3 mbar, even submicrometer particles are supposed to be collected. The gas velocities are supposed to be a few m s^{-2} and the water load 0.02. A throughput of up to 10,000 m^3 hr^{-1} can be attained per unit. No details are given regarding separation of the entrained foam. When viewed against the energy requirement, the indi-

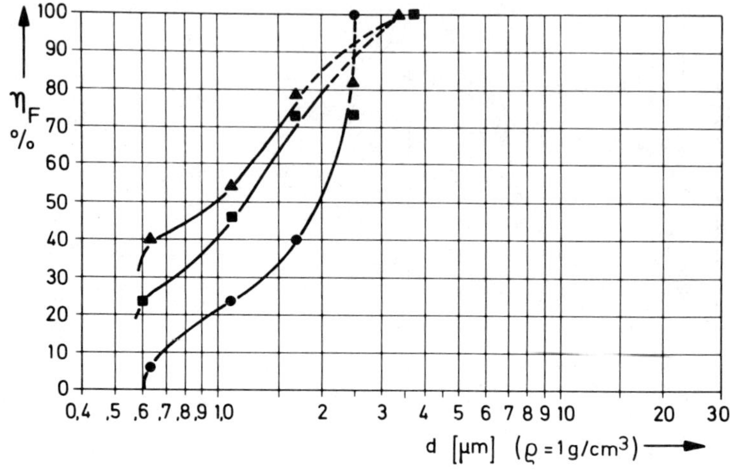

	Gas [m³/h]	Water [m³/h]	Pressure loss Δp [mbar]	
●	500	2	11	Single layer
▲	500	2	29	Double layer
■	700	2,8	30	Double layer

Figure 4.34. Fractional separation efficiencies of an irrigated bag filter.

cated collection efficiency would seem somewhat improbable, although accurate measurements are not yet available.

The rotary impaction filter from Soole[89] is on a more comprehensive physcial bases. Here the gas to be scrubbed has to pass through a rapidly turning wheel rim in which a large number of fine wires are arranged radially. As with the wire-mesh mist eliminator, the mist droplets are collected out of the gas as a result of the relative velocity between gas and wire system. A description is given of prototypes of the filter with gas throughputs up to 4.72 m³ s⁻¹ and 1000 rpm. With it, it should be possible to separate all droplets above 10 μm from the gas, although this would mean that the filter does not work any better than a normal wire-mesh mist eliminator or a wave-plate separator. Because the rotating parts necessarily increase the cost of such a system, they are only likely to be worthwhile for special applications.

Another method that is not yet commercial uses fluidized beds as aerosol separators. Since in this case the upward velocity of the gas has to be smaller the smaller the particles of the filter medium, inertial separation can no longer be considered. For this reason, in addition to the interception mechanism, electrical and diffusion forces are given in order to explain the observed particle separation in the submicrometer range.[90] Bubbles, which are frequently found in the fluidized bed, may interfere with the separation. As a rule, only 90% is reached with this method. As a corrective measure, it has been proposed to use perforated plates to destroy the bubbles or multistage fluidized beds.[91]

Instead of the fluidized bed, it is also possible to use a fixed bed as an aerosol separator and thereby to achieve separation efficiency levels up to 99%. Correctly designed, all these arrangements have the advantage of a low pressure drop of only a few mbar. On the other hand, the low gas velocities of at most a few cm s^{-1} mean that only relatively small quantities of gas can be scrubbed.

Finally, a very radical and very costly method: thermal combustion of the gas. It is restricted to waste gases and is generally used for the elimination of odors.

In addition to the separators described here, new types of separators have occasionally appeared on the market with ostensibly quite unheard-of separating efficiencies. Unfortunately, none of these "wonder separators" have yet been able to establish themselves in practice. This means that we must still depend on finding the most suitable separator for the particular separating problem. The smaller the particles to be separated, the larger are the costs for investment and energy. There are no tricks for avoiding this rule.

4.4.5. Separators for Spray

Separators for spray are coarse separators, their separating range generally lying above 30 μm. They often have to be used as a second separator after the proper mist or dust separators. A brief outline will be given here.

Long horizontal pipes, in which the droplets can sediment out of the gas flow, can act as coarse separators. Sometimes, the liquid can then be collected in a baffle. At projections, edges, and bends, however, new spray is produced from the liquid film on the walls of the pipes and blown into the gas flow.

Complete spray separation is attained in cyclones, wave-plate separators, and packed beds. In line with the size of the spray droplets, these separators do not need to be "keenly" designed. It is therefore often enough to have simple deflection separators, such as the so-called "knock-out drums."

With wet scrubbers, it is usual to collect the scrubbing droplets entrained by the gas in a subsequent cyclone. Wave-plate separators are, however, just as effective.

A detailed description of spray separators by W. Strauss can be found in Refs. 97 and 102.

4.5. COST COMPARISONS AND OPERATIVE RANGES

As with dust separators, the costs incurred in using a mist separator can be divided into three groups. First, there are the straight purchasing or fabrication costs for the separator itself. Added to this are the installation or connecting-up costs, such as the laying of gas ducts. With an inertial separator, it may also be necessary to purchase a suitable blower to produce the pressure drop.

These additional costs are often overlooked, yet they may amount to considerably more than the straight purchasing costs. Should it be necessary to stop the plant while the separator is installed, the loss of production involved may far exceed all other costs. It will therefore not always be possible to decide on one or other type of separator from the point of view of straight purchasing cost. Questions of reliability, cleanliness, space requirement, and noise also play a part.

Above all, the separator must conform completely to the operating conditions. Particular requirements made on the material, such as high-grade steel or special coatings when subjected to corrosive gases as well as special-purpose designs, can increase the buying cost considerably. The size of the apparatus is also important for the price. The larger the unit, the less expensive is generally the cost per 1000 m^3 hr^{-1} gas flow. However, it is not possible to extrapolate all units upward in this way.

Very roughly, it can be said that, correctly designed, cyclones, wave-plate separators, and packed beds can be constructed fairly cheaply. Somewhat more expensive are wire-mesh mist eliminators and wet scrubbers. The wet scrubber will usually be selected where dust separation and gas scrubbing are to be carried out in addition to mist separation. Diffusion fiber filters and electrical separators normally require high investment costs.

The actual price is nevertheless subject to fluctuations through the market situation and the existence of several firms offering the equipment. In this respect, it must be taken into account that guarantees as far as separating efficiencies are concerned are not always adhered to by the selling companies. It is up to the buyer to make a thorough study of the particular separating problem involved and, by making an optimum choice of separator and considering in good time the cost factor when planning new installations, to keep the costs as low as possible.

A further point when calculating the economic aspect are the maintenance costs for the unit. Wave-plate separators, packed beds, and wire-mesh mist eliminators can gradually be clogged up by dust, rust, or reaction products and may have to be cleaned from time to time. The risk of becoming blocked is particularly great with fiber filters and ceramic fritted candles. The latter may even break or become leaky. Cyclones and wet scrubbers normally work without any problem. Wet scrubbers nevertheless require a scrubbing liquid—usually water—which is often circulated and then has to be reprocessed, either continuously or at regular intervals. Reprocessing of the scrubbing liquid can lead to the recovery of valuable raw materials but, with waste gases, may turn the waste air problem into a wastewater problem. Electrostatic separators for droplet mists generally work reliably and without problem. Because of the high cost of the apparatus and the sensitivity of such units, errors in control, such as the intrusion of hot gases following the breakdown of cooling equipment, can have a serious effect on costs. Cost accounts can look particularly bad in cases where the costs of unscheduled downtime have to be added to straight repair or maintenance costs. It would therefore be true to say that servicing costs cannot generally be given as a fixed item.

In contrast, energy costs are normally well known and can actually be specified in terms of figures. Everywhere where the flowing gas in the separating system undergoes a certain pressure drop, energy must be raised to maintain the gas flow. This energy comsumption is directly proportional to the quantity of gas throughput and the pressure drop in the separator. Based on the same quantity of gas in each case, the pressure drop is thus a direct measure of the energy consumption.

There is no point whatsoever in comparing separators that work in completely different particle-size ranges. For practical reasons, mists are therefore divided into the submicrometer range and the range above it.

Droplet separation that is almost complete even in the submicrometer range can be achieved only with electrical filters, fiber filters, and venturis.

Electrical separators need only a very small pressure drop. However, to make the gas flow uniform, a pressure drop of some mbar is frequently produced intentionally be means of a distributor plate connected up before the separator. Since, however, the amount of energy to be expended is small, electrical filters are the most favorable of these separators as far as energy is concerned.

Diffusion fiber filters also require little energy for operation since they generally work with a maximum pressure drop of 40 and generally only 20 mbar. The greatest pressure drop, which varies between 50 and 200 mbar depending on droplet size, is required by venturis. Thus, whereas with the electrical filter, the low operating costs can be set off against the high initial investment, the low purchasing price for the venturi is balanced out again by the high energy costs. It has therefore not been possible to say that the advantages quite definitely lie with one particular type of separator. In the case of venturis, however, the high noise level may well be regarded as a negative factor in view of noise pollution.

In the range just above 1 μm, complete separation of mists is best carried out with wet scrubbers or with fiber filters—possibly even wire filters—under high gas velocities. All these separators act as impaction separators and the amount of energy that has to be produced is greater the smaller the size of the droplets. This is shown in Figure 4.24 for the wet scrubbers. In this case, the energy is either calculated directly from the electrically installed energy or from the energy of the liquid pumps (presumed efficiency 50%), or from the blower output (with 70% efficiency). A pressure drop of $\Delta p = 1$ mbar thus corresponds to a power consumption of 0.04 kwhr/1000 m^3.

With wet scrubbers of medium energy, wire-mesh mist eliminators, wave-plate separators, cyclones, and packed beds, it is generally possible in the range below 5 μm to obtain partial separation and in the range above 5 μm complete separation. All these separators make use of the inertia effect of the droplets; the collection efficiency attained is a function of the energy applied. This is shown by Figure 4.35, where the pressure drop is plotted for a few types against the droplet sizes for 50% and 95% separation, d_{50} and d_{95}. If these sizes are kept constant, the required pressure drop then depends merely on the characteristic values K_{50} and K_{95} of the particular separator ($\Delta p \sim K_{50}^3$, K_{95}^3).

As far as energy is concerned, fiber filters are the most favorable, followed by wire-mesh mist eliminators. Wave-plate separators and packed beds are virtually equal from the point of view of energy. For cyclones, the energy requirement is higher the greater the size of the cyclone.

These separators, with the exception of cyclones, are conventionally run in a pressure range 1 to 10 mbar. As shown in Figure 4.35, this would automatically also give the operating range of the separators with regard to droplet sizes. Where Δp(max) = 10 mbar, it follows from

$$(\Delta p)^{1/3} d_{95} = K_{95}$$

$$d_{95}(\text{min}) = \frac{K_{95}}{2.15}$$

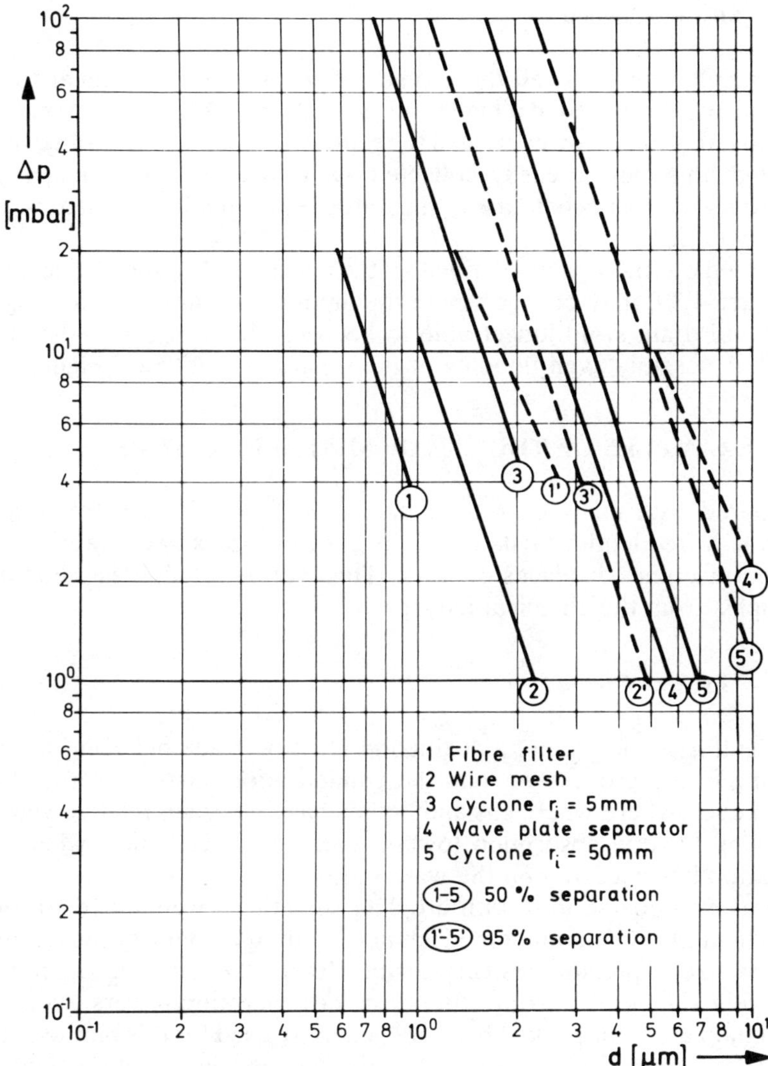

Figure 4.35. Pressure losses of inertia mist separators as a function of drop size to be separated with 50% and 95% efficiency.

237

In principle, however, all these types of separators can, similar to venturis, also be operated at a higher pressure drop. With one or other of the systems, this may, however, lead to reentrainment of the separated liquid, although this can be easily collected again at low cost by means of a subsequently connected coarse separator. The separator then acts as an agglomerator.

The separating capacity defined by the constants K_{50} and K_{95}, as shown in Figure 4.30, may change where the amounts of liquid are too high. If the liquid drain is no longer able to keep up, there will be a holdup of liquid or possibly even flooding of the separator (see, e.g., Figure 4.15).

4.6. EXAMPLES OF THE USE OF MIST SEPARATORS

Compared with dust separation, the problem of mist separation arises much less frequently. Mists chiefly occur in process gases, waste gases, and on machines at places of work. This will be briefly described using examples from the chemical industry.

4.6.1. Process Gases

With sulfuric acid plants, a distinction must be made between the spray forming in the process gas and the condensation mists. A spray is produced everywhere where gas and liquids have a certain relative velocity: in drying towers, absorption towers, and also in gas ducts where condensation has occurred on the walls.

Spray is a coarse mist with droplets sometimes well above 10 μm. It is particularly dangerous for machinery and equipment because, in line with its size, it precipitates very easily. Yet for the same reason, it is also relatively simple to separate. In drying or absorption towers, wire-mesh mist eliminators, or packed beds are generally used for this purpose. Such separators nevertheless become largely useless when acid condensation on the subsequent pipe walls can provide a source of renewed spray. Sensitive equipment such as blowers should therefore be protected by having easy-to-clean coarse separators (e.g., wave-plate separators, knock-out drums, etc.) connected up directly in front of them.

When the temperature falls below the dewpoint, condensation mists are generated as very fine droplets below 2 μm out of the SO_3 and the residual moisture left in the process gas. Because of their lower tendency to precipitation, they are often less harmful for the equipment, but are all the more difficult to separate. Where the dust content of the gas is low, diffusion filters of glass fibers can be used. There are a large number of references in the literature to practical experience obtained with these filters.[102,103] Excellent separation efficiencies can also be obtained with packings of asbestos wool.[35] For process gases, electrical filters are used to a small extent, whereas wet scrubbers are ruled out.

With all separators in sulfuric acid plants, attention must be paid to the fact that sulfate slurry can accumulate and that the gases are highly corrosive.

The presence of acid spray can be proved by the "stick test," which is well known to the people in the sulfuric acid industry. The presence of a fine condensation mist in the gas, on the other hand, is best seen by illuminating with a torch, whereby the path of light becomes clearly visible.

In addition to sulfuric acid plants, acid mists are also formed to a large extent in phosphoric acid works. Reports on experiences obtained with suitable separators are given, for example, by J. A. Brink.[103]

The hydrogen gas that is formed in large quantities in chlorine production contains not only dusts and mercury residues but also very small quantities, only a few milligrams, of very fine alkali droplets below 2 μm. These alkali traces can form very hard sediments on compressors and lead to the necessity for frequent repairs. For separation in such cases, the injection of water into the gas flow and a subsequent combination of dense wire filter and wave-plate separator have been shown to be successful.[35]

Mist separation becomes very difficult when caking can occur on dry surfaces as a result of sublimation out of the gas phase. Phenomena of this kind can occur in the production of maleic or phthalic acid. It may be possible to prevent the separator from clogging by spraying it on both sides with a suitable liquid.

Large numbers of droplet separators also have to be used in petrochemical plants in order to prevent, for example, the entrainment of droplets out of reaction or distillation towers. With the continuing trend toward larger and larger production units, such as crackers, the breakdown of just a simple installation like a separator can lead to undreamed-of costs. This can happen, for example, due to flooding and droplet entrainment on a wire-mesh mist eliminator when its porosity is considerably reduced as a result of caking or sagging in the middle (see Figure 4.15).

4.6.2.　Waste Gases

Waste gases from acid production can be successfully freed of mists with the aid of wet scrubbers. Harmful, gaseous components are also often washed out during this scrubbing process. In titanium pigment factories, it is usual to install electrical filters. They enable the acid concentration of the mist to be lowered to less than 30 mg m^{-3}. The plume of waste gas then contains mainly water vapor and becomes invisible after only a few yards.

Waste gases from the production of pigments and dyestuffs contain gaseous acid and organic components in addition to mists of sulfuric acid, hydrochloric acid, and pigment and dyestuff dusts. For scrubbing such mists, suitable combinations of wet scrubbers and jet scrubbers can be set up, using not only water but also other scrubbing liquids (e.g., alkaline).

It is precisely wet scrubbers that have proved to be highly successful over the last few years for the scrubbing of waste gases. In the chemical industry, they represent well over half of all the units installed. Equally large is the number of manufacturers and types of equipment moving on to the market.

In the production of plastics and synthetic fibers, solvents and auxiliaries are released, which also produce fine condensation mists below 3 μm as the waste gases cool. In some cases, the mist particles are neither solid nor liquid, but on precipitation (e.g., on filters) give a greasy paste. For separating in such cases, use is made chiefly of high-performance wet scrubbers or wet electrical filters.

Oil-lubricated vacuum pumps, which are used in large numbers for the evacuation of drying ovens, emit dense oil mists with the waste air. The majority of the droplets are below 1 μm and the amounts can be up to a few g m^{-3}. Entrained dust particles or coking products from the oil clog up the conventional ultrafilters almost instantaneously. Wire filters with high gas flow[35] have, depending on the amount of dust involved, a longer service life, extending into weeks or months. For such applications, wet scrubbers also require hardly any maintenance, although they often cause problems in reprocessing the scrubbing liquid. Provided that the waste gases cannot form explosive mixtures, the use of electrical filters is also possible.

A special problem is created by the often-acid films on the walls of waste gas stacks. These films form as a result of condensation or the precipitation of spray and mist droplets. The liquid film is pushed upward with the gas flow and ejected at the end of the stack in the form of large droplets, which subsequently fall down in the immediate vicinity of the stack. Special separators have been developed and introduced on the market for sucking off these films from the end of the stack. An alternative solution is to spray the wall of the stack with water and thereby prevent the liquid film from rising upward.

The examples given here are from the author's experience and can therefore not be representative of the whole field of mist separation. Further examples from practice can be found above all in the reviews by W. Strauss[102] and J. A. Brink.[103]

Research and development in the field of mist separation has risen considerably in recent years. The majority of separation problems can be solved with conventional means, yet experience has shown that, in practice, a separator is still very often used at the wrong place. It can only be hoped that, as the level of knowledge increases, unnecessary costs can be avoided and the means available for preventing pollution of the air can be used in the most rational manner and with maximum efficiency.

SYMBOLS

A	droplet cross section
B	path length of gas in electrostatic precipitator
C	Cunningham correction factor
C_D	drag coefficient
d	droplet diameter
d_{50}	droplet diameter for 50% separation efficiency
d_{95}	droplet diameter for 95% separation efficiency
D	diameter of impaction body (wire, fiber, scrubbing droplet, etc.)
e	electronic charge
E	electric field strength
f_1	correction factor
f_2	correction factor
F	fluid resistance force on particle
g	gravity acceleration
G	inertial force on droplet
	specific weight of wire mesh
h	height of wire mesh
H	height of packed bed
H_0	height of packed bed
k	Boltzmann's constant
$K, K',$ $K_{50},$ K_{95}	characteristic values for droplet separation
L	distance between wire and plate in electrostatic precipitator
m	mass of droplet

n	number of charges on droplet; number of layers in wire mesh
p	ratio of total projected area of all wires or fibers to cross section of gas flow
p_1	as p for only one layer
Δp	pressure drop
q_1, q_2	electric charges on droplet
r_i	radius of exit tube of cyclone
r_a	radius of outer wall of cyclone
R	radius of curvature of particle trajectory radius Raschig rings
Re	Reynolds number
s	distance between wave plates
s'	radial path of droplet
S	specific surface of wire mesh
t	time
T	absolute temperature
u	droplet velocity
u_r	relative droplet velocity
u_i	tangential gas velocity beneath exit tube of cyclone
v	gas velocity
v_r	radial gas velocity in cyclone
W	distance between successive wires in electrostatic precipitator
α	angle of curvature in wave plate; packing density in fiber filter
δ	distance between limiting flow lines
ε	dielectric constant of droplet
ε	porosity of wire mesh; fractional separation efficiency of cylinder
ζ	dimensionless pressure loss factor
η_F	fractional separation efficiency
η_0	overall efficiency
λ	mean free path of gas molecules
λ'	stopping distance of droplet
μ	viscosity of gas
ρ	gas density
ρ_P	density of droplet
ρ_L	liquid density
ρ_f	density of material of wire mesh
ψ	inertial impaction parameter

ψ_A separation parameter
ω drift velocity

REFERENCES

1. Bürkholz, A., *Chem.-Ing.-Tech.*, 45, 1 (1973).

2. Gebhart, J., and Porstendörfer, J., Jahreskongr. 1974 Ges. Aerosolforsch.

3. Morrow, N. L., Brief, R. S., and Bertrand, R. R., *Chemical Engineering.*, 79, 85 (1972).

4. Bürkholz, A., *Staub—Reinhalt. Luft*, 33, 397 (1973).

5. May, K. R., *J. Sci. Instrum.*, 22, 187 (1945).

6. Ranz, W. E., and Wong, J. B., *Arch, Ind. Hyg.*, 5, 464 (1952).

7. Mitchell, R. I. and Pilcher, J. M., *Ind. Eng. Chem.*, 51, 1039 (1959).

8. Brink, J. A., Jr., *Ind. Eng. Chem.*, 50, 645 (1958).

9. Mercer, T. T., Tillery, M. I., and Newton, G. J., *J. Aerosol Sci.*, 1, 19 (1970).

10. Bürkholz, A., *Chem.-Ing.-Tech.* 42, 299 (1970).

11. Brink, J. A., Jr., Kennedy, E. D., and Yu, H. S., *AIChE Symp. Ser. 70*, 333 (1974).

12. Marple, V. A., Thesis, University of Minnesota, 1970.

13. Andersen, A. A., *J. Bacteriol.*, 76, 471 (1958).

14. Jackson, M. L., and Patterson, R. G., *AIChE Symp. Ser. 71*, 47 (1975).

15. Stöber, W. and Flachsbart, H., *Environ, Sci. Technol.*, 3, 1280 (1960).

16. Bürkholz, A., First Eur. Symp. Particle Meas., Nuremberg, 1975.

17. Hermann, J. and Eiberweiser, H. J., *Staub—Reinhalt. Luft*, 34, 159 (1974).

18. Friedlander. S. K., *Ind. Eng. Chem.*, 50, 1161 (1958).

19. Emi, H., Okuyama, K., and Yoshioka, N., *J. Chem. Eng. Jap.*, 6, 349 (1973).

20. Yeh, H. C., and Liu, B. Y. H., *J. Aerosol Sci.*, 5, 191 (1974).

21. Deutsch, W., *Ann. Physik.*, 68, 335 (1922).

22. Pauthenier, M., *Rev. Gen. Electr.*, 71, 265 (1962).

23. Hanson, D. N., and Wilke, C. R., *Ind. Eng. Chem. Proc. Des. Dev.*, 8, 357 (1969).

24. Heinrich, D. O., *Brennst.-Waerme-Kraft*, 7, 346 (1955).

25. Muschelknautz, E., and Herold, H., *Chem.-Ing.-Tech.*, 47, 267 (1975).

26. Barth, W., *Allg. Waermetech.*, 9, 252 (1960).

27. Muschelknautz, E., *Staub—Reinhalt. Luft*, 30, 1 (1970).

28. Barth, W., *Chem.-Ing.-Tech.*, 23, 289 (1951).

29. Brauer, H., *Chem.-Ing.-Tech.*, 29, 785 (1957).

30. Reichelt, W., and Blaß, E., *Chem.-Ing.-Tech.*, 43, 949 (1971).

31. Rumpf, H., and Gupte, A. R., *Chem.-Ing.-Tech.*, 43, 367 (1971).

32. Souders, M., Jr., and Brown, G., *Ind. Eng. Chem.*, 26, 98 (1934).

33. Sherwood, T. K., Shipley, G. H., and Holloway, F. A. L., *Ind. Eng. Chem.*, 30, 765 (1938).

34. Jackson, S. and Calvert, S., *AIChE J.*, 12, 1075 (1966).

35. Bürkholz, A., to be published.

36. Bürkholz, A., and Muschelknautz, E., *Chem.-Ing.-Tech.*, 44, 503 (1972).

37. Regehr, U., *Chem.-Ing.-Tech.*, 39, 1107 (1967).

38. York, O. H., *Chem. Eng. Prog.*, 50, 421 (1954).

39. York, O. H., and Poppele, E. W., *Chem. Eng. Prog.*, 59, 45 (1963).

40. Calvert, C., Jashnani, I. L., and Yung, S., EPA–APT Fine Particle Scrubber Symp., San Diego.

41. Germerdonk, R., and Günther, H., *Chem.-Ing.-Tech.*, 41, 649 (1969).

42. Bell, C. G., and Strauss, W., *J. Air Pollut. Control Assoc.*, 23, 967 (1973).

43. Bürkholz, A., *Chem.-Ing.-Tech.*, 42, 1314 (1970).

44. Langmuir, J., and Blodgett, K. B., U.S. Army Air Forces Tech. Rep. 5418, 1946.

45. Ranz, W. E., and Wong, J. B., *Ind. Eng. Chem.*, 44, 1371 (1952).

46. Davies, C. N., and Peetz, C. V., *Proc. R. Soc. Lond.*, A234, 269 (1956).

47. Griffin, F. O., and Meisen, A., *Chem. Eng. Sci.*, 28, 2155 (1973).

48. George, H. F., and Poehlein, G. W., *Environ. Sci. Tech.*, 8, 46 (1974).

49. Löffler, F., and Muhr, W., *Chem.-Ing.-Tech.*, 44, 510 (1972).

50. Wong, J. B., Ranz, W. E., and Johnstone, H. F., *J. Appl. Phys.*, 26, 244 (1955).

51. May, K. R., and Clifford, R., *Ann. Occup. Hyg.*, 10, 83 (1967).

52. Schlichting, H., *Grenzschichttheorie*, 3rd ed., G. Braun, Karlsruhe, 1958.

53. Massey, O. D., *Chem. Eng. July 13*, 143 (1959).

54. Saemundsson, H. B., *Verfahrenstechnik*, 2, 480 (1968).

55. Ramstetter, T. K., *Verfahrenstechnik*, 3, 534 (1969).

56. Carpenter, C. Leroy, and Othmer, D. F., *AIChE J.*, 1, 549 (1955).

57. Stenhouse, J. I. T., Harrop, J. A., and Freshwater, D. C., *J. Aerosol Sci.*, 1, 41 (1970).

58. Harrop, J. A., and Stenhouse, J. I. T., *Chem. Eng. Sci.*, 24, 1475 (1969).

59. Stenhouse, J. I. T., and Harrop, J. A., *Filtr. Sep. March, April*, 169 (1971).

60. Brink, J. A., Jr., Burggrabe, W. F., and Rauscher, J. A., *Chem. Eng. Prog.* 60, 68 (1964).

61. Nguyen, X., and Beeckmans, J. M., *J. Aerosol Sci.*, 6, 205 (1975).

62. Barth, W., *Staub*, 19, 175 (1959).

63. Goldshmid, Y., and Calvert, S., *AIChE J.*, 9, 352 (1963).

64. Calvert, S., *AIChE J.*, 16, 392 (1970).

65. Calvert, S., *J. Air Pollut. Control Assoc.*, 24, 929 (1974).

66. Güntheroth, W., *Fortschrittsber. VDI-Z.*, 3. (13) (1966).

67. Boll, R. H., *Ind. Eng. Chem. Fund.*, 12, 40 (1973).

68. Wicke, M., and Krebs, F. E., *Chem.-Ing.-Tech.*, 43, 386 (1971).

69. Holzer, K., *Staub—Reinhalt. Luft*, 34, 360 (1974).

70. Davis, R. J., and Truitt, J., *Instrum. Control Syst.*, 45, 68 (1972).

71. Holzer, K., private communication.

72. Fattinger, V., *Chem.-Ing.-Tech.*, 43, 426 (1971).

73. Batel, W., *Staub—Reinhalt. Luft*, 33, 491 (1973); 34, 52 (1974).

74. Eckert, J. S., and Strigle, R. F., *J. Air Pollut. Control Assoc.*, 24, 961 (1974).

75. Hesketh, H. E., *J. Air Pollut. Control Assoc.*, 24, 939 (1974).

76. Hanf, E. W., and MacDonald, J. W., *Chem Eng. Prog.* 71, 48 (1975).

77. Bürkholz, A., *Verfahrenstechnik*, 9, 449 (1975).

78. Bürkholz, A., *Verfahrenstechnik*, 70, 29 (1976).

79. Fairs, G. L., *Trans, Inst. Chem. Eng. Lond.*, 36, 475 (1958).

80. Brink, J. A. in R. Perry and C. Chilton, *Chemical Engineers Handbook*, 5th ed., McGraw-Hill, New York, 1973.

81. Brink, J. A., *Can. J. Chem. Eng.* 41, 134 (1963).

82. Brink, J. A., Jr., Burggrabe, W. F., and Greenwell, L. E., *Chem. Eng. Prog.*, 64, 82 (1968).

83. Maartmann, St., *Staub—Reinhalt. Luft*, 34, 353 (1974).

84. Reynolds, J. P., Theodore, L., and Marino, J., *J. Air Pollut. Control Assoc.*, 25, 610 (1975).

85. Muhlrad, W., and Remillieux, J., *Chim. Ind. Genie Chim.*, 103, 2755 (1970).

86. Yoshida, T., Kovsaka, Y., Inake, S., and Nakai, S., *Ind. Eng. Chem. Process Des. Dev.*, 14, 101 (1975).

87. Bürkholz, A., and Holzer, K., unpublished.

88. Javorsky, B. S., *Filtr. Sep.*, Mar.–Apr., 173 (1972).

89. Soole, B. W., *J. Roy Naval Scientific Service.*, 28, 5, 251.

90. Black, C. H., and Boubel, R. W., *Ind. Eng. Chem. Process Des. Dev.*, 8, 573 (1969).

91. Jackson, M., *AIChE Symp. Ser. 70*, 82 (1974).

92. Bürkholz, A., *Staub—Reinhalt. Luft*, 38, 211 (1978).

93. Bürkholz, A., *Chem.-Ing.-Tech.*, 48, 795 (1976).

94. Bürkholz, A., *Chem.-Ing.-Tech.*, 51, 1255 (1979).

95. Bürkholz, A., *J. Aerosol Sci.*, 9, 199 (1978).

96. Bürkholz, A., *Chem.-Ing.-Tech.*, 48, 887 (1976).

97. Strauss, W., *Industrial Gas Cleaning*, 2nd ed., Pergamon, Oxford, 1974.

98. Löffler, F., in W. Strauss (Ed.), *Air Pollution Control*. Pt. 1, Wiley, New York, 1971, pp. 337–375.

99. Lancaster, B. W., and Strauss, W., in Strauss, ibid., pp. 377–427.

100. Robinson, M., in Strauss, ibid., pp. 227–335.

101. Bohnet, M., in W. Strauss (Ed.), *Air Pollution Control*, Pt. 3, Wiley, New York, 1978, pp. 79–120.

102. Strauss, W., in A. C. Stern (Ed.), *Air Pollution*, Vol. 4, Academic Press, New York, 1977.

103. Brink, J. A., in P. Perry and C. Chilton, *Chemical Engineer's Handbook*, 5th ed., McGraw-Hill, New York, 1973.

5

FILTRATION OF PARTICULATES

Gordon M. Bragg

Department of Mechanical Engineering, University of Waterloo, Waterloo, Canada

The chapter deals with practical aspects of industrial filtration of particulates. The types of dusts met in practice are described and the types of equipment to remove these dusts are illustrated and their operation described. Design and selection information is given on filter cloths and their appropriate use. The formation of cakes on this cloth and the resulting pressure drops are fully explained and design information is given for complete filter systems. Some special cases such as demisting and high-temperature applications are briefly described.

5.1. INTRODUCTION

5.1.1. Mechanisms of Filtration

The modern industrial filter, in different configurations, can give both the most efficient particulate collection of any possible system[1] or one of the cheapest and most effective dust cleaning systems for normal industrial use.[2] The great breadth of application of filters is matched by a corresponding breadth of geometries, installation types, and filtration materials. Flow rates in industrial applications can run from less than $1/10\text{m}^3$ s^{-1} to over 1000 m^3 s^{-1},[3] Specialized materials are now available for applications up to 400°C.[4] In dollar terms the fabric filter represents approximately 30% of the capital expenditure on pollution control equipment.

The industrial filter has a long history of application. As early as 1852, zinc oxide was being collected in bags where the air was being drawn in by a fan.[5] In 1881, Beth patented a filter that was cleaned by shaking.[6] A baghouse capable of handling 120 m^3 s^{-1} was developed by 1892, and a reverse air flow device was patented in 1893. In the years since, an enormous range and variety of devices have been patented and developed which fall into the field of industrial dust filters.

Two quite different types of filter are in existence at the present time. The first, characterized by the standard air conditioning filter, is an open, low-pressure drop and relatively inefficient device which is normally replaced on a regular basis during a maintenance program. The other type, which is based on a relatively dense filter medium, collects the majority of dust in a cake on the surface of the filter. This second type is intended for nearly continuous operation and the cake is removed periodically during this operation. This second type of filter is the more common industrial type.

A basic bag filter of the most common industrial type is shown in Figure 5.1. It consists of a series of bags usually around 10 to 15 cm in diameter and somewhere between 4 and 10 m in length. Dirty air enters a chamber beneath the bags and is drawn through the inner surface of the bags. As a result, clean air enters the chamber exterior to the bags and is drawn

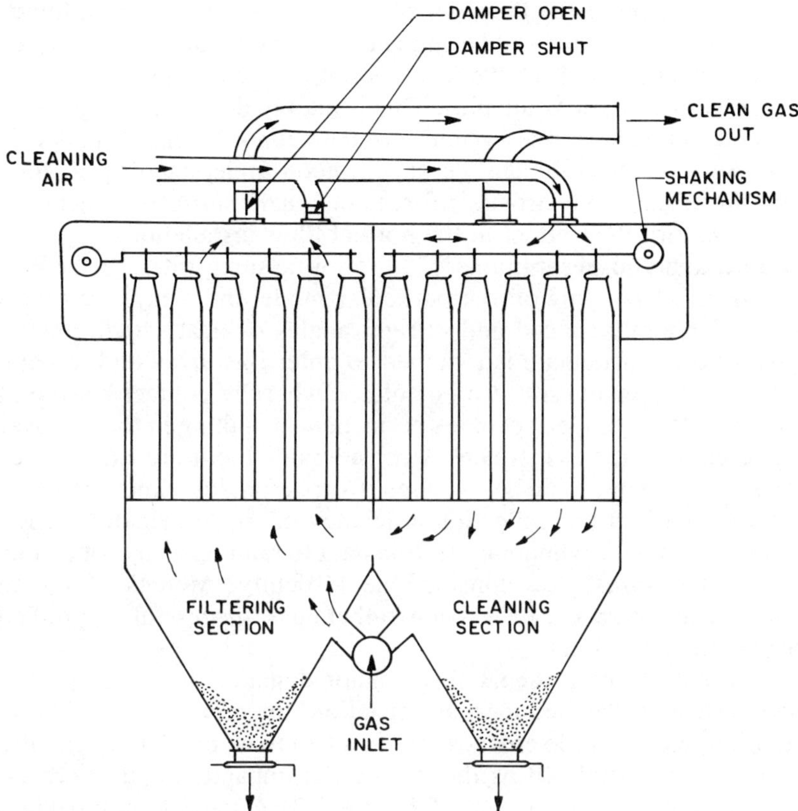

Figure 5.1. Schematic diagram of basic bag filter. This filter employs both reverse flow and shaking to clean the bags.

upward. The dust collects on the inner surface of the bag. Periodically, the bag is shaken or a reverse pressure is applied. As a result, the cake formed on the inner surface of the bag breaks up and is deposited in the hopper at the bottom of the chamber. Many variations of geometry flow direction and bag arrangement are possible. Cleaning procedures such as shaking, reverse pressuring, and air jet cleaning are employed in different applications.

5.1.2.   Filtration Theory

The basic theory of filtration has been outlined by Löffler[7] and others. The main conclusions of this theory are reviewed here to develop a physical understanding of the behavior of industrial filters. Modern filtration theory is highly developed and a large number of data on idealized systems have been documented. However, it is not possible at present to design filter media from basic principles. This is mainly due to the large number of variables present in real systems, which cannot be easily modeled together. As a result, although we have a good understanding of, for example, the effect of electrostatic forces on single particles, it is not possible to quantify their effect in the normal filter installation.

Modern analytic descriptions of filtration assume filters to be arrays of cylinders. Depending on the particular model the arrays may be regularly or randomly spaced and at right angles or nearly right angles to the main flow. Particulate is assumed to enter this array and to contact the cylinders. Mechanisms are postulated whereby contact between the particle and the cylinder becomes permanent, resulting in the removal of the spheres from the gas stream. A considerable literature exists describing the flow around a single cylinder. Corresponding experiments have been performed on the collection efficiency of single cylinders. The extrapolation of single cylinder collection data to random arrays of cylinders has, however, presented considerable difficulty. Models of the flow through cylinder arrays have been rather less successful in predicting industrial filter behavior.

In classical filtration theory three major dynamic mechanisms are assumed. These are labeled inertial impaction, interception, and diffusion. In addition, electrostatic charges are assumed to be present in real filters. The collection postulated as due to inertial impaction and interception can be described with the aid of Figure 5.2. A particle approaching a cylindrical fiber of diameter D has a diameter d and a mass m.

If we consider an isolated fiber, a streamline similar to that labeled 1–2 will be typical. If we calculate the trajectory of the center of mass of the particle, a path similar to 1–3 will result. If the path 1–3 intersects the cylinder cross section, collection by inertial impaction is said to occur. In addition, if we consider the finite diameter of the particle, an additional

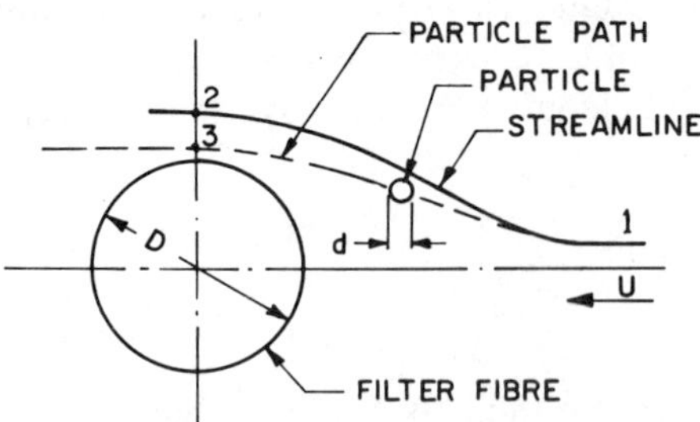

Figure 5.2. Basic collection mechanisms for particles on a cylinder.

probability of collection occurs when path 1–3 passes within distance $d/2$ of the cylinder. This addition affect has been labeled interception. The characteristic Reynolds number of the flow is

$$Re = \frac{UD}{\nu}$$

where U is the face velocity and ν the gas kinematic viscosity. In real filters this Reynolds number varies from very much less than 1 (Stokes' flow) to values of the order of 1000. As a result, the flow field with which to begin the trajectory calculation is in question. Several models of this flow have been proposed[8,9] and a discussion of the matter is given in Löffler[7] and Strauss.[10] The most commonly used model is that of Kuwabara[11] and Happel.[12] This model, which assumes a system of cylinders, is independent of Reynolds number. The model gives good correlation with experimental results for very small Reynolds numbers (0.01 to 0.05), but the derivation assumes that the filter is highly porous. This is not generally so. As a result, Löffler[7] concludes: "When the differences between the models and real filters are considered, it is clear that the theories of deposition cannot provide exact results, but at best are able to show trends and provide approximate answers."

The diffusion mechanism of collection is assumed to apply for relatively small particles (<1 μm) and presumes that the particles are diffused about their ideal trajectory and hence have an increased probability of contacting the fiber. Since the diffusivity of the particles is inversely proportional to the diameter of the particle, we may expect that in appropriately designed filters the efficiency of collection will increase with decreasing particle size. This effect normally occurs for particles less than 1 μm. As a result, filter systems can be expected to have a minimum collection efficiency at a specific particle size. Evidence for this is reviewed by Strauss.[10]

Electrostatic charging of the aerosol and of the filter can increase the collection efficiency of filters. Mathematical modeling of the forces and the resulting collection due to electrostatic effects has been attempted. The models predict that normally charged aerosols would be collected by electrostatic effects in the submicrometer range. Experiments with known charges conform to the postulated models. However, in real situations it is not possible to predict accurately the charge carried by dusts and aerosols. As a result, the theory is not widely applicable to design of filtration systems. The subject of collection mechanisms has been reviewed by Pich[13] and others.

The usual modeling procedure assumes some summation of collection mechanisms for the single fiber. This summation must be performed in such a way that "double collection" is not allowed. First, the effects of collection efficiency due to several mechanisms for a single fiber must be calculated. In addition, the collection efficiency of a filter bed must be predicted using the combined efficiencies of the fibers from which the bed is constructed. Only the Kuwabara and Happel model mentioned previously avoids this problem. An example of a typical calculation using the Kuwabara model for the flow is given in Figure 5.3. The calculation is for particles passing 2-μm-diameter fibers in air at 0.1 m s^{-1} and a solidity of 0.1. Examples of complete calculations for filters using models built up from cylinders have been compared by Bragg and Hanna.[15] Typical results using the models of Kirsh et al.[16] and Chen[17] are shown in Figure 5.4. In each case a collection efficiency minimum is predicted by the models. It is to be expected that future research in this area will enable considerably more sophisticated procedures which do not artificially separate various aspects of the particle dynamics.

Particle adhesion in most analytic models is assumed to be 100%. That this is not true experimentally has been extensively documented. This is particularly true for particles greater than 5μm. Löffler[7] has reviewed the subject of adhesion.

In addition to modeling a filter as an array of fibers, it is possible to predict the consequences of assuming that the filter is an array of channels. This approach has been considered by Clarenburg[39] and Bragg and Pearson.[40] Bragg and Pearson consider the filter as an array of helical tubes. For these tubes the probability of impaction and diffusion can be calculated (the effect of interception is small in this type of model) and predictions similar to those of cylinder arrays can be made. The advantages of this type of model are that the confining of cylinders into arrays

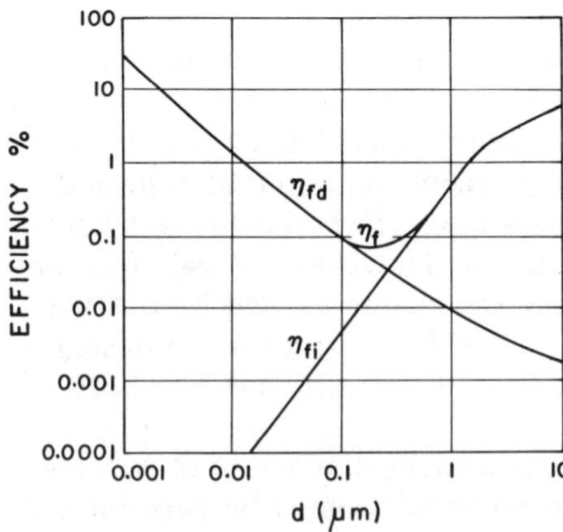

Figure 5.3. Typical results for Kuwabara flow calculation of filter efficiency. Fibers in the filter are 2 μm, the approach velocity is 0.1 m s^{-1}, the porosity ϵ is 0.9, and 100% collection after contact is assumed. Particle density is 1500 kg m^{-3}, η_{fd} is the efficiency for collection by diffusion, and η_{fi} is the efficiency for collection by impaction and interception. $\eta_f = \eta_{fi} + \eta_{fd} - \eta_{fi} - \eta_{fi}\eta_{fd}$ (after Crawford[14]).

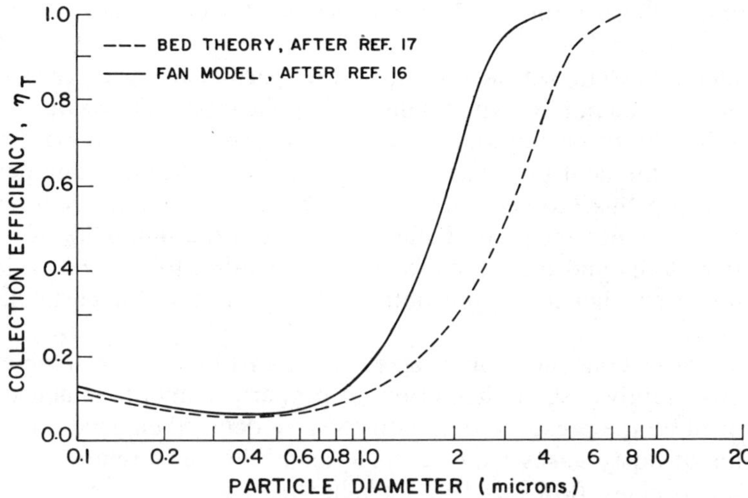

Figure 5.4. Typical results of flow models built from arrays of cylinders. Models are from Kirsh et al.[16] and Chen.[17] Approach velocity is 0.21 m s^{-1}, fiber diameter is 10 μm, porosity is 0.97; 100% collection after contact is assumed; particle density is 1000 kg^{-3} (after Bragg and Hanna[15]).

is unnecessary and that the model should be applicable to the filtration of cakes of particulate.

Experimental justification of the various collection models is quite extensive. Considerable work has been done in the last 10 years on the collection capacity of fibers and arrays of fibers. Unfortunately, this depth of experimentation has not been followed in the ranges appropriate to industrial filtration and in the caking that occurs on the normal industrial filter. Summaries of available data are given by Löffler[7] and Pich.[13]

5.2. DUSTS

5.2.1. The Character of Dusts

Dusts are solid particle aerosols suspended in a gas and ranging in size from 10^{-3} μm up to approximately 10^3 μm. The terminology used to describe dusts is not specific, and terms such as smoke, fume, dust, mist, and spray are only arbitrarily defined. The quantity of dust in the air ranges from approximately 10 μg m^{-3} to 10^3 g m^{-3}. The latter appears in cases of pneumatic conveying. Particles may also consist of more than one structure (agglomerates) held in contact by electrostatic or other interparticle adhesive forces. The properties of some typical dusts are shown in Table 5.1.

Detailed knowledge of the dust structure, particularly particle-size distribution, is often not known in industrial applications. This is due in part to the difficulty of making reproducible particle-size measurements. The modern development of impactors and automatic particle-sizing devices has decreased this problem. As a result, modern design procedures are moving from experience-based statements about the suitability of particular filter cloths and processes for particular dusts to a more basic understanding and hence a more optimum configuration for particular installations.

In the lower concentration ranges (less than 10 gm^{-3} the same calculations frequently assume that air or gas properties other than density are unchanged by the presence of the dust. More detailed calculations of the behavior of dusty gases may be made using the basic relationships for general two-phase flow (see, e.g., Wallis[20]).

5.2.2.   Particle Size

Particle size is the main variable in filtration design and knowledge about the particle size may be obtained by a variety of instrumentation. As may be seen from Table 5.1, the great majority of aerosols have particle size ranges of at least two orders of magnitude. For sizing the larger dusts, common and micromesh sieves may be used. For the smaller, respirable (less than approximately 5 μm) dusts, cyclones, centrifuges, and impactors may be used.[21] For smaller dusts (less than about 2 μm) electron microscopes and scanning electron microscopes may be used. The sample

Table 5.1.   Properties of Some Typical Dusts and Aerosols (Adapted from First and Drinker[18])

Aerosol	Size range (μm)	Concentration range (g m^{-3})
Rain	200–10,000	0.5–5
Fog	2–80	0.005–0.05
Tobacco smoke	0.01–1	Variable up to 10^{-3}
Carbon black	0.01–0.5	—
Fly ash effluent	1–200	0.5–5
Foundry air	0.001–100	0.005–10
Mine air	0.001–100	0.02–0.5
Cement dust	0.5–100	0.01–1.0

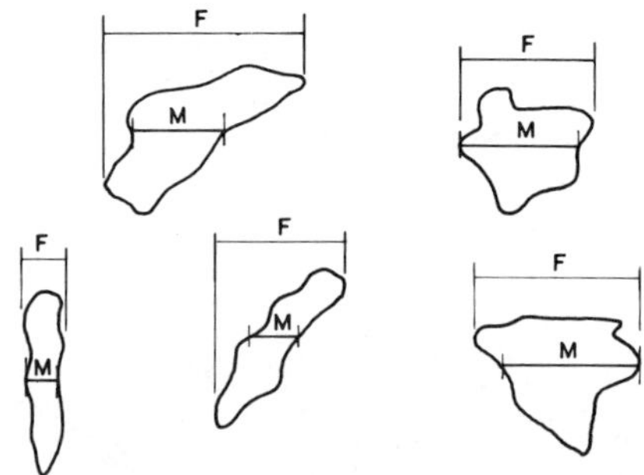

M — MARTIN'S DIAMETER
F — FERET'S DIAMETER

Figure 5.5. Martin's and Feret's diameters for characterizing nonspherical dusts.

is commonly collected on a filter and the filter is inspected under the microscope. Increasingly, automatic treatment of the visual scan is obtained. Electron microscopes can provide resolution down to approximately 0.001 μm. Particle size is usually referenced to some characteristic dimension of the particle. The two most common particle measurements are Martin's diameter and Feret's diameter. Martin's diameter is the horizontal line bisecting the area of the particle as viewed under a microscope. Feret's diameter is the extreme dimension of the particle as viewed under the microscope. These are shown in Figure 5.5. This type of reproducible characteristic dimension enables unusual shapes to be treated. In spite of this consistent treatment, difficulties often arise. In particular, the question often arises as to whether agglomerates observed under the microscope are present in the actual aerosol. A decision regarding this requires some knowledge of the process. For example carbon black particles less than 0.10 μm are frequently present in combustion. In practice, however, they almost always occur as agglomerates, even in high temperatures. As a result, filtration of carbon black is directed toward filtration of the agglomerates rather than of the much smaller individual carbon particles.

The method of particle formation determines the details of the individual particle shape. Particles formed by condensation or from vapor-phase reactions such as metallurgical fumes or from high-temperature combustion processes often assume spherical or cubic shapes. Furthermore, oxides and other metallurgical process fumes often form large flocs. The relative effect of this on particle density may be seen in Table 5.2.

Those dust particles formed by disintegration processes such as grinding are usually determined by the nature of the orginating material. They often appear as rather crystalline irregular shapes. Frequently, filtration

Table 5.2. Particle Densities of Flocs (After Whytlaw-Gray and Patterson[19])

Material	Floc density (g cm^{-3})	Normal density (g cm^{-3})
Silver	0.94	10.5
Mercury	1.70	13.6
Magnesium oxide	0.35	3.65
Lead monoxide	0.62	9.36
Antimony trioxide	0.63	5.57
Aluminum oxide	0.18	3.70
Stannic oxide	0.25	6.71

occurs at a point in the process where the gas is cooling. In addition, the filtration process itself can cool the gas. Under these conditions, particle shapes and properties can alter during the filtration process. In general, determination of filtration system properties to meet these varying conditions depends on previous operating experience.

The results obtained from measurement of the dust may be characterized by several numbers. These include the *range,* which is the size of the largest and smallest particle, the *arithmetic mean size* of the particle, the *geometric mean* size of the particle, and various median sizes. The quantitative description of particle size is extensively treated by Cadle.[21]   In addition to estimate particle-size distribution, several other variables are important in determining the collectability of various dusts. These include electrostatic charge and humidity. At present it is seldom possible to use measurements of these quantities to determine, from basic considerations, the optimum filter cloth. Previous operating experience with these parameters, however, allows appropriate designs to be selected.

5.3. FILTER EQUIPMENT

5.3.1. Filter Types

The simplest type of industrial filter is illustrated in Figure 5.6. Filter elements in this filter consist of long bags 10 to 20 cm in diameter and 0.5 to 2 m in length. Air enters the bags from a chamber at the top of the unit. The clean air exits through the bag walls and the dust falls to a hopper at the bottom of the unit. The bags are cleaned by manual shaking or by motorized shaking devices. The cleaned air immediately enters the region around the cleaning unit. Units of this type are suitable for low-temperature emissions and large dry dusts. Typical applications are in sanding and grain-handling operations.

In Figure 5.7 a typical reverse jet collector is illustrated. This device, which operates rather similarly to the open bag house, has a unit surrounding each bag and containing a compressed air supply. This unit blows jets of air against the exterior of the bag and so performs a local cleaning operation. Equipment of this type can achieve high efficiencies and high capacities for given volumes. The cleaning mechanism can require considerable maintenance. As might be expected, cloth wear in this type of unit may be relatively high. Typically felted fabrics are used in reverse jet collectors. Wool felt is often used for lower temperatures and Orlon or Dacron felt is used for higher temperatures and chemical resistance. Felted fabrics allow high-efficiency collection with this filter type, even though the reverse jet removes most of the cake on the surface because of its highly efficient cleaning process. This type of filter can use higher filtration velocities than can many other filter types.

Figure 5.8 illustrates a pulsing jet bag house. In this unit, which requires an external housing capable of operating at pressures up to 250 Pa, the dirty gas flows from the outside of the bag to the inside. The inlet duct enters the bottom of the housing, where the larger particles will fall out

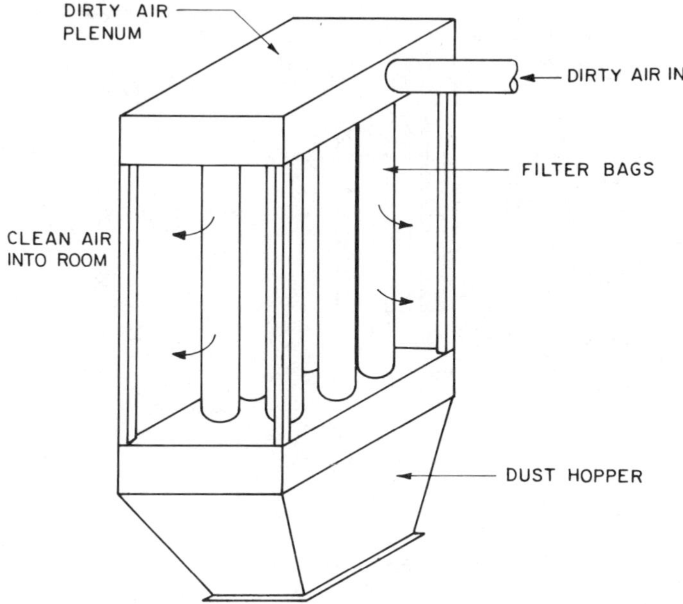

Figure 5.6. Simple open return bag house with manual shaking.

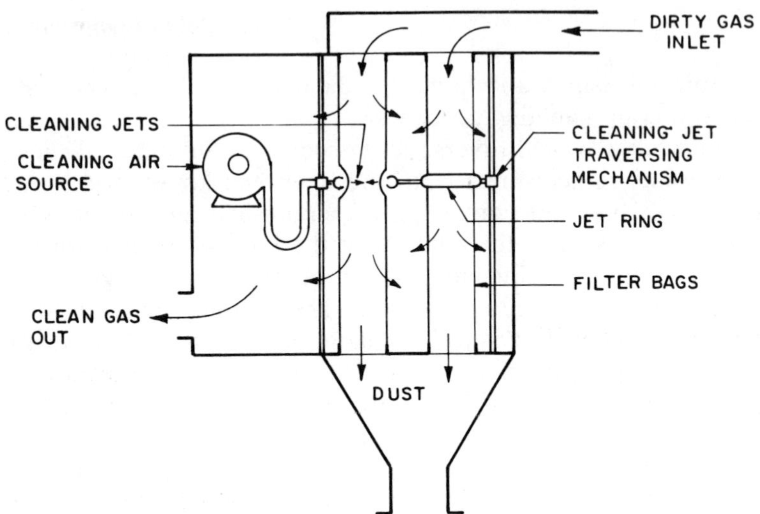

Figure 5.7. Reverse jet collector.

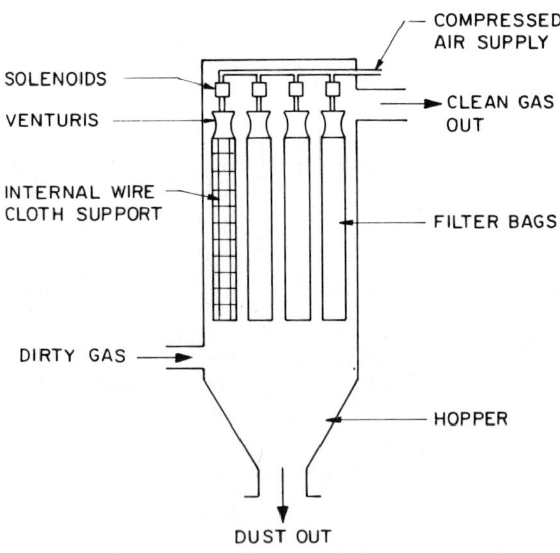

Figure 5.8. Pulsing jet filter.

in the hopper. The remaining dust will move up around the outside of the tube and the dust will be collected on the external surface of the bag. Clean air from the inside of the tubes enters the outlet duct. Cleaning in this unit is achieved by periodic compressed air pulses injected into the clean side of the filter system. This produces a reversed flow through the bag and a resulting flexing of the bag, which removes the cake. The magnitude, quantity, and timing of these pulses may be altered to achieve optimum cleaning. A compressed air supply is required to operate this system and felted fabrics are commonly used.

The reverse flow filter with mechanical shaking has already been described (Figure 5.1). In addition to these types, many variations upon the basic designs are possible. For example, Figure 5.9 illustrates a system where the bags are not cylindrical. Collection occurs on the outside of the bag and the compressed air supply for cleaning is mounted on a rotating air supply device.

5.3.2. Filter Element Geometry

Many variations of filter element geometry are possible. The most common arrangements are tubes or bags. An overall objective of the arrangements is to optimize the amount of filter cloth per cubic foot of bag house.

Approximately one-fifth of the collectors available use flat panels. The other 80% use tubes. Manufacturers[22] claim that a flat filter can obtain approximately 30% more cloth for a given volume of collector. Panels can be inspected more easily since both sides of the cloth can be seen with appropriate location of inspection panels and single panels are more easily changed than single tubes. This does not imply that complete rebagging is easier in the panel unit. Leaks are also more easily identified.

Conversely, tube equipment can have the following advantages. The extremely long length of bags that can be installed allows more cloth per square foot of floor space for the bag house. Repairs to leaking bags may be easier. Single tubes are easily changed and leaks are easily located by dust around the base of the tube. Inlet and outlet locations near the top and bottom of the bag house are the most logical arrangement. Panel units are more susceptible to clogging.

In early designs the advantages of upward flow in bag houses was obvious. With an inlet at the bottom of the bag house the heaviest dust settled immediately to the hopper. Cloth cleaning took place during shutdown of the entire bag house. As a result, the dust did not have to fall against an upward gas flow. Furthermore, downflow required the use of an extra tube sheet and an extra bottom plenum. This may be seen by

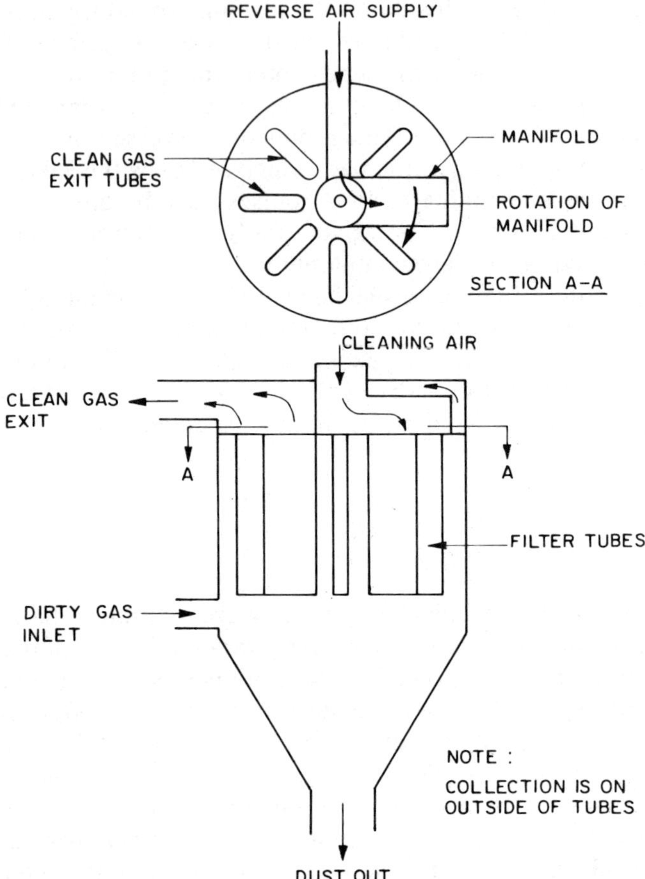

Figure 5.9. Reverse flow cylindrical collector with mechanical cycling of reverse flow.

comparing Figure 5.1 to Figure 5.6. Modern reverse jet cleaning with continuous on-line operation introduced the extra tube sheet and downward flow to encourage dust fall during filtration. Pulse-type cleaning uses almost exclusively upward flow. The seeming contradiction between this procedure and that of reversed jet cleaning may be explained by assuming that in pulse cleaning large pieces of cake fall, and hence the disadvantage of upward flow is overcome. There is some evidence[21] that downward flow distributes particulates more evenly over the length of the tube. It is obvious that upward-flow collectors having one less tube sheet are generally less expensive and that with two tube sheets, tube tension is more difficult to adjust.

With tube-type filters the filtering may occur on the inside or the outside of the tube. Reverse jet equipment provides inside filtering, and almost all downward-flow equipment is inside-filtering. Approximately 60% of all upward-flow equipment is inside filtering. The choise is not determined by the fabric. The major advantage to inside filtering is the ability to enter the filter compartment for inspection and maintenance during operation. In addition, inside filtering does not ordinarily require the use of supporting mesh. The disadvantage of inside filtering is that typically fabric flexure is larger during cleaning.

Most filter bags are 10 to 20 cm in diameter and have length/diameter ratios from 5 to 40. Normal choices of diameter and length do not affect filter performance. Dimensions are more likely to be determined by standardized sizes available and initial cost considerations. Many bag shapes are determined by the available width of filter material. One hundred centimeters is a standard available width. Some bags are woven in circular shape appropriate to bag house use. Using smaller-diameter tubes increases the filter capacity per unit of floor area occupied. The disadvantages of small bags in these installations are the possibility of bridging and the possibility of plugging at the lower end of smaller bags. Smaller bags also require more tension adjustment and inspection for the same cloth area.

The relation between the entrance velocity, V_e, and the filtration velocity, V_f, is obtained by equating entry flow to the tube and flow through the cloth. Then

$$V_e = \frac{4V_fL}{D} \tag{1}$$

Typical values are $V_e = 0.5$ m s^{-1}, $L/D = 25$, and $V_f = 0.02$ m s^{-1}. Varying the L/D ratio affects the trade-off between exit velocity and filtration velocity. Losses due to entrance velocity head losses are negligible in most design calculations. As a result, minimization of this quantity is not important. Also, filtration velocity is controlled by variables other than bag geometry, and as a result L/D and hence V_e are controlled by the requirement for stability of the vertical tube and the necessity of eliminating high scouring velocities at the entrance to the tube. High L/D ratios require greater spacing between bags to offset swaying of the bag. Tube length by itself is also controlled by the tension along the bag. This tension is obviously highest at the top of the bag, and this is where wear may be expected due to high tension. Longer bags are also more expensive per square foot. Air pulse and shaking cleaning mechanisms are also length-limited. Reverse jet cleaners require that the removed dust fall to the bottom of the hopper and not be recollected. As a result, these types are also length-limited.

Bag spacing must be optimized to prevent rubbing during mechanical shaking and access for examination and maintenance. In larger pieces of equipment, walkways between banks of bags must also be provided. For ease of maintenance banks should be no more than three or four bags deep. Larger-diameter bags should not be more than two bags deep. Minimum spacings between bags should be ~5 cm.

5.3.3.   Cleaning Mechanisms

The largest amount of research on bag house design has been done on various aspects of cleaning mechanisms. The cleaning process is designed to remove a specific amount of the deposit or cake quickly and uniformly without removing too much of the residual deposit and thereby decreasing the collection efficiency. In addition, it is desirable to minimize damage to the cloth and excessive dispersing of the removed dust particles in order to prevent refiltration. A summary of cleaning methods and their properties is given in Table 5.3.

Shaking is usually in banks of bags and the shaking occurs from the top. The amplitude of the shake may be anywhere from a fraction of a centimeter to several centimeters. The frequency is usually several cycles per second and is essentially simple harmonic motion in most cases. Accelerations range from 10 to 100 m s^{-2}. During shaking, filtration should be stopped, otherwise, the dust will work through the cloth, reducing the efficiency and possibly damaging the cloth by internal abrasion. The pressure–time cycle that results from a process of shutting off a compartment and shaking a bag is shown in Figure 5.10. The entire cleaning cycle

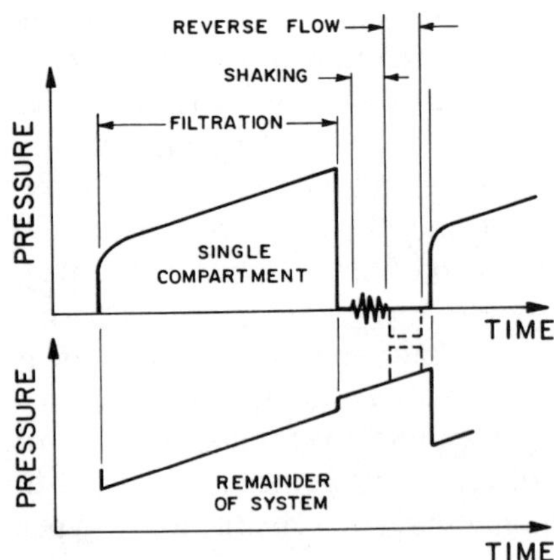

Figure 5.10. Pressure variation across a compartment during one cleaning cycle.

Table 5.3. Comparisons of Cleaning Methods (After Billings and Wilder[22])[2]

Cleaning method	Uniformity of cleaning	Bag attrition	Equipment ruggedness	Type of fabric	Filter velocity	Apparatus cost	Power cost	Dust loading	Maximum temperature[b]	Sub-micrometer efficiency
Shake	Average	Average	Average	Woven	Average	Average	Low	Average	High	Good
Reverse flow, no flex	Good	Low	Good	Woven	Average	Average	Medium–low	Average	High	Good
Reverse flow, collapse	Average	High	Good	Woven	Average	Average	Medium–low	Average	High	Good
Pulse—compartment	Good	Low	Good	Felt, woven	High	High	Medium	High	Medium	High
Pulse—bags	Average	Average	Good	Felt, woven	High	High	High	Very high	Medium	High
Reverse jet	Very good	Average–high	Low	Felt, woven	Very high	High	High	High	Medium	Very high
Vibration, rapping	Good	Average	Low	Woven	Average	Average	Medium–low	Average	Medium	Good
Sonic assist	Average	Low	Low	Woven	Average	Average	Medium	—	High	Good
Manual flexing	Good	High	—	Felt, woven	Average	Low	—	Low	Medium	Good

[a] These value judgments do not permit comparison of performance aspects, only of methods.
[b] Fabric-limited.

typically takes from 30 s to several minutes and the shaking consists of 10 to 100 cycles. Problems with shaker mechanisms include less effective shaking because of wear in the shaker mechanism and wear at the fixed end of the flexed bag.

Reverse-flow cleaning with no flexing is suitable if the dust releases fairly easily from the fabric. With this type of cleaning, the bag is usually supported by a metal grid or ring. Support is usually on the clean inside of the tube. This type of cleaning is also used with other methods such as shaking. Flow reversal may be achieved by appropriate damping or separate reversing fans. Some models (as in Figure 5.9) have a traveling apparatus that blocks off the primary flow and introduces reverse flow. The pressure history across a particular compartment is indicated in Figure 5.10. Disadvantages of reverse-flow equipment include the costs of any additional fans, ducts, or dampers and the cost of a support grid. The support grid can also make maintenance difficult.

Reverse flow with collapse allows extreme cloth flexure, which can cause wear. It can, however, result in very efficient cleaning and can be combined with other methods. Cloth bags having the dust cake on the inside are frequently cleaned this way. With this type of cleaning there is an optimum tightness of the cloth. Similar fan and dampering arrangements as in the method of reverse flow without collapse are necessary.

In the method of pulse cleaning a sharp pulse of compressed air is released in the vicinity of the fabric, giving rise to some combination of shock, fabric deformation, and flow reversal. The result is the removal of the dust deposit, with only a brief interruption in the filtering flow. The fabric flexural wear is minimized, and the installation is smaller because the fabric is in use almost all the time. The components of a pulse cleaning filter include an air compressor, a storage or surge tank, piping, solenoids, and nozzles. Woven cloths tend to be overcleaned by pulsing, resulting in excessive leakage. Felts and needled materials are more appropriate for pulsed cleaners. Bag inlet velocities are high in pulse equipment, and felts, having softer surfaces, may withstand abrasion more than cloths. Felts form a more porous dust cake having lower specific resistance, so that they may be cleaned less than cloths for the same pressure buildup. In addition, felts may be used even though they are initially more expensive, because there is little mechanical wear of the fabric in pulse equipment. The very low ratio of cleaning time to filtering time in pulse equipment makes this equipment useful at very high dust loadings, up to several hundred grams per cubic meter on large particles. As a result, this equipment is widely used for pneumatic conveying systems.

Reverse jet cleaning has the particular advantages of high filtration efficiency and high filtration velocities. It achieves this objective by using a small volume jet of moderately pressurized air spaced at moderate time intervals. Felt is commonly used in these applications. This type of filter is commonly called a Hersey filter since many designs are based on the original Hersey patents. In all models to date flow is downward and outward. For this reason the collector can be open-sided, but since it is often used on toxic materials because of its high efficiency, industrial health codes usually require that the filter gases be contained until exhausted outdoors. This type of filter is particularly useful at very low dust concentrations. This can occur if the dusts are valuable or toxic. The high efficiency and high air capacities, of the order of 50 to 150 mm s^{-1} make this device useful in these applications. For very low loadings, cake cleaning may be begun and finished manually. The air required to clean the fabric is usually supplied by a high-speed blower at pressures of 1 to 5 kPa of water. The volume of air required is of the order of 5% of the primary filtering flow. As a result, the filtration velocity need increase only slightly.

Attempts have been made to increase the frequency of agitation by cleaning using sonically assisted cake removal methods.[23] The process is effective on glass cloth but remains a very uncommon procedure.

On very small units manual cleaning is recommended. Manual cleaning however wears the cloth rapidly. Manual cleaning is seldom used for units filtering 1 m^3 s^{-1} or greater.

5.3.4.   Hopper and Disposal Equipment

Most hoppers are under a small pressure or vacuum. As a result, they require valving at the bottom to let the collected dust out without admitting air to reentrain and redeposit dust or causing air flows through the discharge.

Hoppers must be designed to prevent blockage of the flow of the dust through the exit. Most hoppers have tapering sides to facilitate this flow. The flow of solids has been extensively studied because of its importance in several fields. A review of the field is given by Orr.[24] In order to design a hopper appropriate to a given dust, a considerable amount of information must be available on the material. In particular, the angle of internal friction (β) must be known. This is the angle to the horizontal at which the dust flows in a container with a hole at the bottom. Ordinarily, a hopper should be at least as steep as this value. The angle of internal friction, however, is a function of particle shape and size and is enormously dependent on the moisture content of the dust. As a result, most hoppers with sloping sides are built at 60° to the horizontal, which is greater than typical values for most dry dusts. Hoppers can however, be blocked if dusts are damp or the hoppers allowed to overfill. Under these circumstances vibrators and mechanical rappers may be necessary. Typical values of β for dry dusts are 25° for glass beads, 37° for sand, and 39° for limestone and coal dust. If detailed data are available on the dust, it is possible to calculate the maximum mass flow through a hole of given size and the conditions under which arching will occur in a hopper.

Valving equipment at the hopper exit can be of several types. Automated equipment usually uses a rotary valve as, shown in Figure 5.11a, or for larger units rotary screw conveyors are used downstream from rotary valves (Figure 5.11b). For noncontinuous operation, manual sliding gates, trip gates, or discharge socks are used in Figure 5.11c.

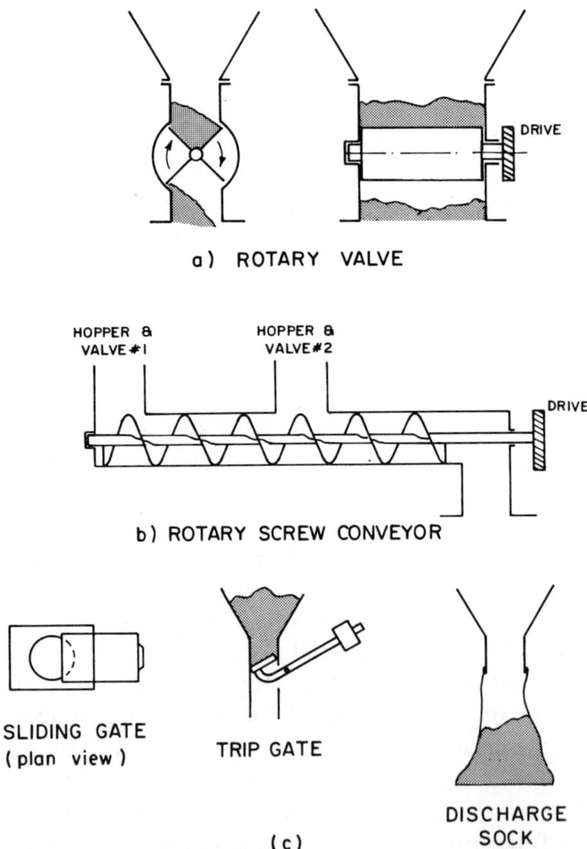

Figure 5.11. Types of hopper discharge valves: (a) rotary valve; (b) rotary screw conveyor; (c) sliding gate, trip gate, discharge sock.

5.4. FILTRATION MATERIALS

5.4.1. Filter Cloths

An important parameter in the appropriate design of a filtration system is the filter material. A choice must be made concerning the type and quantity of material. The type of material may be further broken down to fiber and fabric type. Each of these subdivisions produces different properties in the final filtration material.

The fiber type used depends upon cost, operating temperature, dust being collected, and so on. The basic division between types of fibers is that between natural and man-made fibers.

Cotton and wool fibers are commonly used in fabric filter collectors. Cotton fibers are single cells varying in length from 1 to 4 cm and in width from 12 to 25 μm. Wool fiber consists of flat shingle like scales. They are from 3 to 30 cm in length and essentially cylindrical in shape. Cotton is a standard fiber for common dusts. It is inexpensive, readily available, and an effective filter media. It has relatively poor resistance to acids and only fair resistance to alkalies. Its recommended maximum operating temperature is of the order of 105°C. Its resistance to acids is fair and to alkalies relatively poor. The use of wool is less common than in the past. In felted form, wool has been the standard fabric for use in Hersey-type reverse jet bag houses.

At present, man-made fibers are of increasing popularity in the fabric filter industry. There are a wide range of different materials from which man-made fibers can be made. Some of these are listed in Table 5.4 together with their common trade names. We will not deal with each of these fabrics in detail but will consider only some of the more popular types.

Nylon or polyamide fiber is a cylindrical fiber with smooth surfaces. The fibers are uniform in diameter and of the order of 10 μm. Fibers are from 2 to 10 cm in length. Normal nylon may be used up to approximately 105°C. It has good acid resistance and poor alkali resistance. Nylon is available in many different forms and complete range of deniers and staples. Nylon is particularly useful in the filtration of abrasive dusts and wet abrasive dusts at low temperatures. Good elastic properties make it suitable for conditions of continuous flexing.

Nomex nylon is a proprietary aromatic polyamide developed by DuPont for applications requiring good dimensional stability and heat resistance. It was developed expressly to be used at higher temperatures. Unlike glass, it is resistant to fluorides and has good abrasion and flex resistance. It is used in the cement industry, carbon black industry, and with non-

Table 5.4. Properties of Filter Materials

| Fiber | Physical characteristics | | | Relative resistance to attack by: | | | | |
	Relative strength	Specific gravity	Maximum usable temperature (°C)	Acid	Alkali	Organic solvent	Abrasion	Other attribute
Cotton	Strong	1.6	105	Poor	Fair	Good	Good	Low cost
Wool	Fair	1.3	95	Fair	Poor	Good	Good	—
Polyamide (nylon)	Strong	1.1	105	Fair	Good	Good	Excellent	Easy to clean
Polyester (Dacron)	Strong	1.4	150	Excellent	Good	Good	Excellent	—
Acrylonitrile (Orlon)	Fair	1.2	130	Good	Fair	Good	Good	—
Vinylidene chloride (Saran)	Fair	1.7	95	Good	Fair	Good	Good	—
Polyethylene	Strong	1.0	65	Fair	Fair	Fair	Good	—
Tetrafluoroethylene (Teflon)	Fair	2.3	260	Excellent	Excellent	Good	Fair	Expensive
Glass	Strong	2.5	290	Good	Fair	Good	Poor	—
Graphitized fiberglass	Weak	2.0	260	Fair	Good	Good	Fair	Expensive
Asbestos	Weak	3.0	260	Fair	Fair	Good	Fair	—
Nomex nylon	Strong	1.4	230	Good	Fair	Excellent	Excellent	Poor resistance to moisture

ferrous metals and steels. In the absence of acids, Nomex may be used in heats up to 230°C. In the combined presence of moisture and high temperatures, a loss in strength occurs. The material is available in multifilament and stapled forms. Nomex is degraded by oxidizing agents but is highly resistant to many organic solvents.

Polyester fibers are also very popular. Trade names include Dacron, Fortrel, and Kodel. Polyesters have relatively better high-temperature properties than many other synthetics but are not as good as Teflon and Nomex. Polyester is particularly suitable for filtration and dust collection under conditions of wet heat. These materials may be used up to approximately 150°C. They have excellent acid resistance properties and good alkali resistance properties. This material is available in a wide range of woven forms.

Teflon fluorocarbon is available in two forms. One is suitable for continuous service up to 260°C and the other for continuous service to 230°C. Moist-heat applications will allow similar temperatures. Teflon is inert to acids and alkalies. This is the most chemically resistant fiber produced, although alkali metals and fluorine gas may cause degradation. The material is available in a wide range of woven forms. Cake discharge is excellent and its major weakness is abrasion susceptibility. Teflon is an expensive filter material.

Glass fiber filter materials are increasingly popular. Glass fibers are 10 to 20 cm in length and 5 to 20 μm in diameter. Fabrics made from glass yarns actually gain in strength as the temperature rises up to approximately 200°C. The recommended top operating temperature is 290°C. Glass is resistant to acids of normal strength and under ordinary conditions. Overall resistance to acids is slightly above average. Hot solutions of weak alkalies will attack glass. Glass is dimensionally stable and essentially incombustible. It has, however, particularly poor flex abrasion resistance. Various chemical treatments to the glass fabric can improve the abrasion characteristics of glass. A common addition is to impregnate carbon black into the structure of the filter bag.

The materials described in the preceding section may be wound into a large variety of yarns and then woven into an enormous variety of cloths. The size and method of manufacture of the yarns determines the basic strength of the material and the weave determines the character of the surface manufactured from this yarn. In addition to normal woven fabrics, filter fabrics are frequently felted. The felts used in fabric filtration in their early stages of production are also woven. Subsequent steps, however, completely change the character of the material from that of a woven fabric. Although felts can be made simply by matting fibers together and by other non-woven methods, the use of woven base fabric called a scrim greatly increases the strength and stability of the fabric. Production of a felt occurs when a scrim is worked in such a way as to obscure the identity of the separate yarns. Several operations can cause this. In the case of wool, working in the presence of warm water and lubricants can achieve a feltlike appearance. Napping consists of pulling at the surface of the material until a considerable amount has been worked loose. Needle punching is a method of combining two or more layers of fiber into a feltlike fabric. Usually, one layer is a scrim for strength, while the other may consist of fibers to be added to the basic scrim. As a result considerable control is possible over separate properties of the finished material. For example, the scrim may contribute the dimensional stability while the top layer contributes the ideal properties for dust control. Needled fabrics are finding increasing use in fabric filtrations because they can have desirable cake-release properties.

In addition to standard manufacturing techniques, glass fiber fabrics are treated with silicones, graphite, and other finishing agents to reduce fiber-to-fiber abrasion resistance during filter cleaning.

The resulting fabric from these manufacturing processes must be resistant to abrasion wear during cycling, internal chafing, and tension. It is, unfortunately, not possible to determine from basic principles the ability of a fabric to withstand these forces in a particular application. In addition, the material must be permeable to flow. However, the permeability of filtration fabrics is so decreased by the residual dust deposit that the permeability of the clean fabric generally has little to do with the pressure drop across an operating filter. The objective in fabric design is therefore to maintain a highly permeable residual dust and fabric combination while passing a minimum amount of dust. Because of this, the pores in a fabric must be closely controlled and not exceed a certain diameter. In addition, the ability of the fabric to release the deposited dust will depend on the mode and intensity of cleaning and on the adhesive character of the fabric. The way in which fabric construction relates to deposit release has not been determined, but it is presumed to depend partly on the electrical resistances of the fibers.

5.4.2. Gas/Cloth Ratios

All basic selections of industrial filters begin with some attempt to assess what is called the gas/cloth or air/cloth ratio. This is defined as the volume flow rate of gas to be treated divided by the total collection surface to be provided. As a result, the number is the recommended filtration velocity in millimeters per second. Many handbooks contain summary data of this type. An extensive list of recommendations for shaker bag houses has been provided by Clement.[25] These data are given in Table 5.5 together with corresponding recommended duct velocities to convey the dust. Conveying velocities are discussed further in Section 5.6. The use of gas/cloth ratios in a table such as table 5.5 should be a general guide only. Final selection should be modified according to several other variables besides dust type. Since too high a face velocity results in high pressure loss, increased wear, plugging, and reduced efficiency, the choice is an important one. Choice of an unnecessarily low face velocity results in an overly large filter and increased capital cost. Alterations to the recommended values of face velocity should be made if dust loadings on particle sizes are outside the normal range.

Typical factors to be applied to the recommended values are given in Tables 5.6 and 5.7 according to one manufacturer.[26] In addition, the following factors indicate that the air/cloth ratio should be *lowered* below the recommended:

1. Longer than normal intervals between bag cleaning.
2. Fine particles without a mix of larger particles.
3. Oily or sticky particles.
4. The necessity for filter to run with some compartments occasionally closed. In this situation the recommendations are for the worst operating case.

Reverse jet or Hersey-type bag house achieve extremely efficient cleaning of filter cloths. As a result, it is possible to use rather higher face velocities than with mechanical cleaning. Some typical values based on manufacturers' recommendations[27] are given in Table 5.8. Values for pulsed jet filters tend to be similar to those for reverse jet systems.[28] This applies for felted materials only.

There is no precise analytic method for determining the best air/cloth ratio. Instead, as indicated, it is customary to select ratios based on similar previous experience. Each dust collector manufacturer has guidelines on the selection of air/cloth ratio. These are based on experience with a variety of applications and vary from manufacturer to manufacturer. Normally, these guidelines should enable estimates to be within at least 25% of the optimum design ratio. Frequently, in new applications a pilot study has been used to determine the best air/cloth ratio. Such studies can be misleading, however, unless they accurately model the proposed equipment and use a suitable aerosol.

The natural tendency to choose as high a filtration velocity as possible

Table 5.5. Peak Filtering Velocities and Minimum Dust-Conveying Velocities for Various Dusts and Fumes (Clement[25])

Dust or fume	Maximum filtering velocity[a] (mm s⁻¹)	Branch pipe velocity (m s⁻¹)	Dust or fume	Maximum filtering velocity[a] (mm s⁻¹)	Branch pipe velocity (m s⁻¹)
Abrasives	15	22.5	Fertilizer (bagging)	12	20
Alumina	11	22.5[b,c]			
Aluminum oxide	10	22.5	Fertilizer (cooler, dryer)	10	22.5
Asbestos	14	17.5–20			
Baking powder	12	20–22.5	Flint	12.5	22.5
Bauxite	12.5	22.5	Flour	12.5	17.5[h,i]
Bronze powder	10	25	Glass	12.5	20–22.5
Brunswick clay	11	20–22.5	Granite	12.5	22.5
Buffing wheels	16	2.5–20[b or d,e]	Graphite	10	22.5
Carbon	10	20–22.5	Gypsum	12.5	20
Ceramics	12.5	20–22.5[c,e,i]	Iron ore	10	22.5–25
Charcoal	11	22.5[g,h,i]	Iron oxide	10	22.5
			Lampblack	10	22.5
Chocolate	11	20[g,h,i,j]	Leather	17.5	17.5[b,c]
			Cement	7.5	22.5[c,f]
Chrome ore	12.5	25	crushing		
Clay	11	20–22.5	grinding	11	20
Cleanser	11	20[e,i,g]	(separators,		
Cocoa	11	20[g,h,i,j]	cooling, etc.)		
Coke	11	20–22.5[g,h,i]	conveying	12.5	20
Cork	15	15–17.5[c,e,i]	packers	14	20
Cosmetics	10	20	batch spouts	15	20
Cotton	17.5	17.5[b,c,e,i]	Lead oxide	11	22.5
Feeds and grain	16	17.5[e,h]	Lime	10	20
			Limestone	14	22.5
Feldspar	12.5	20–22.5	Manganese	11	25

to lower system costs should be considered in light of the observation that lower velocities are very likely to result in lower operational costs. This will be seen in lower pressure drops across the system and less wear on the bags. Filter and fabric manufacturers are continually working to increase optimum air/cloth ratios.

Dust or fume	Maximum filtering velocity[a] (mm s^{-1})	Branch pipe velocity (m s^{-1})	Dust or fume	Maximum filtering velocity[a] (mm s^{-1})	Branch pipe velocity (m s^{-1})
Marble	15	22.5	Silica	14	22.5
Mica	11	20	Soap	11	17.5[e,i]
Oyster shell	15	22.5	Soapstone	11	20
Paint pigments	10	20	Starch	11	17.5[e,i]
Paper	17.5	17.5[c]	Sugar	11	20[i]
Plastics	12.5	22.5[i]	Talc	11	20
Quartz	14	22.5	Tobacco	17.5	17.5[c,e,i]
Rock	16	22.5	Wood	17.5	17.5[c,i]
Sanders	16	22.5[d,e]			

[a] Maximum filter velocities are for shaker-type filters only.
[b] Cyclone-type precleaner.
[c] Special hoppers, gates, and valves.
[d] Spark arrester.
[e] Flame-retardant cloth.
[f] Insulate casing.
[g] Grounded bags.
[h] Special electricals.
[i] Pressure relief.
[j] Sprinklers.

Table 5.6. Particle-Size Factor for Air/Cloth Ratio

Mean particle size (μm)	Factor to be applied to values of Table 5.5
100	1.2
50–100	1.1
10–50	1.0
3–10	0.9
1–3	0.8
<1	0.7

**Table 5.7.   Dust Load Factor
for Air/Cloth Ratio**

Loading (g m^{-3})	Factor to be applied to values of Table 5.5
2–6	1.2
4–8	1.0
9–17	0.95
18–40	0.90
>40	0.85

5.5.   CAKE FORMATION AND PRESSURE DROP

In simple models of fibrous filters such as those described in Section 5.1, the pressure drop–velocity relationship is linear. This is true for a clean filter cloth. The same statement holds for a cake of collected dust of constant thickness. However, for a filter cloth with cake constantly being added, the pressure drop is a function of time and we can write the pres-

**Table 5.8.   Peak Filtering Velocities for
Reverse Jet Bag Houses**

Material	Filtering velocity (mm s^{-1})
Aluminum oxide	45
Carbon	25–35
Cement	35–45
Flour	50
Grain	60
Graphite	25
Gypsum	40
Lead oxide	30
Lime	40
Metallurgical fumes	36 (not recommended)
Paint pigment	40
Soap	45
Starch	40
Talc	45
Tobacco	45
Wood	40–45
Zinc oxide	36

sure–velocity relationship in the form

$$\Delta P = aU + bU^2 t \tag{2}$$

where ΔP = pressure drop across the bed
U = approaching stream velocity
t = time
a, b = constants to be determined

We shall now derive this relationship and attempt to obtain more information on a and b.

The clean cloth has velocities that produce essentially laminar flow around the cylinderlike cloth fibers. As an example, a flow velocity of 10 mm s^{-1} over a fiber of 100 μm diameter gives for standard air a Reynolds number of 6 $\times$ 10^{-4}. Under these circumstances the flow in the filter is assumed everywhere laminar and losses are linearly proportional to velocity. The resulting relation is given by D'Arcy's law:

$$\Delta P = k L \mu U \tag{3}$$

where L = bed thickness
μ = viscosity
U = approaching stream velocity
k = an empirical permeability coefficient (m^{-2})

The same result may be achieved by assuming that flow in a filter is similar to flow in a series of parallel channels in the filter medium. Several analytical models[29–32] predict the actual pressure loss in terms of filter variables. An empirical value of k has been suggested by Davies[31] as:

$$k_1 = \frac{64}{D^2} (1 - \epsilon)^{3/2}[1 + 56(1 - \epsilon)^3] \tag{4}$$

where D = effective fiber diameter
ϵ = porosity of the filter

and the relation is assumed to hold up to $\epsilon = 0.98$. The validity of the relationship is illustrated in Figure 5.12. D, the effective fiber diameter, is not necessarily the mean fiber diameter but must be determined by test. It is, however, the same diameter as that used in fiber efficiency calculations. Discrepancies between D and mean fiber diameter are larger at lower values of ϵ.

It has been determined that eq. 3 also describes pressure drop across a cake of dust on the filter surface. The values of k will, of course, be different (and typically much larger) from those for the filter cloth. In addition, the thickness of the cake, L_c, is a function of time. If we assume

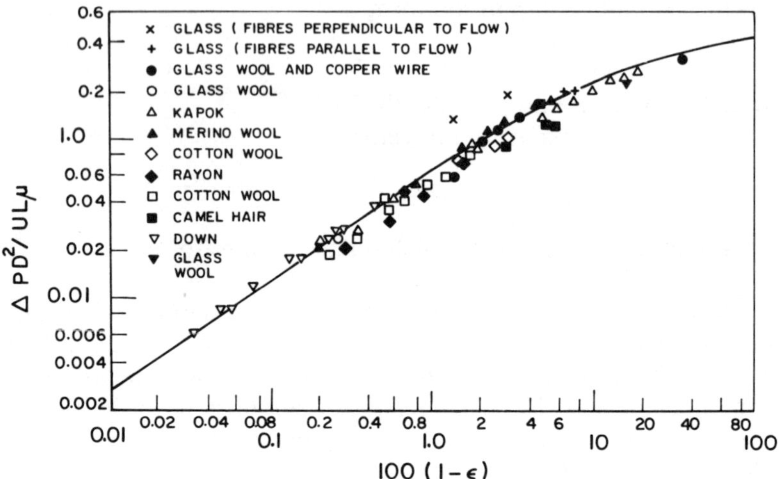

Figure 5.12. Correlation of mat density $(1 - \epsilon)$ with pressure drop and flow. $\bigcirc$, glass wool; $\bullet$, glass wool and copper wire; $+$, glass (fibers perpendicular to flow); $\times$, glass (fibers parallel to flow); $\triangle$, kapok; $\blacktriangle$, Merino wool; $\diamond$, cotton wool; $\blacklozenge$, rayon; $\square$, cotton wool; $\blacksquare$, camel hair; $\triangledown$, down; $\blacktriangledown$, glass wool.

that the dust has constant concentration, the change of thickness of the cake with time is linear and given by

$$L_c = \frac{C_a}{C_c} Ut \tag{5}$$

where C_a = concentration of dust in the gas (g m^{-3})

$\quad\;\; C_c$ = concentration of dust in the resulting cake (g m^{-3})

$\quad\;\; t$ = time from the start of cake buildup

Then, if the pressure drop is assumed to be the sum of the drop through the filter cloth ($k = k_1$) and the cake (where $k = k_2$), we may write from eqs. 3 and 5 that the total drop is

$$\Delta P = k_1 L\mu U + k_2\mu \frac{C_a}{C_c} U^2 t$$
$$= \Delta P_f + \Delta P_c \tag{6}$$

For fixed geometry and gas conditions, this is of the form of eq. 1 and ΔP_f is the pressure drop due to the filter and ΔP_c is the pressure drop due to the cake.

While k_1 may be effectively determined for a wide range of filter cloths, the effect of this term in actual filters is dominated by cake effects. k_2 is poorly documented and depends in practice on a wide range of variables. Since k_2 is a function of the cake, it will depend on filtration velocity, humidity, process variables that generated the dust, and so on. The major relevance of eq. 6 is therefore to illustrate the dependence of pressure drop upon time and face velocity.

Equation 6 indicates that the pressure drop is linear with time in the usual situation. Although this is true of most of the operating cycle, the relationship must be qualified in two ways. First, as previously mentioned, the pressure drop across the cloth itself is usually negligible and may be frequently ignored. Second, as pointed out in Section 5.1, a preliminary dust buildup occurs inside the filter medium and this preceeds normal caking.

The net result is summarized in Figures 5.13 and 5.14. Figure 5.13 illustrates schematically the variation of pressure drop with collected dust mass (w). Since even extreme amounts of cleaning do not remove all dust, the pressure will not return to ΔP_f but only to a residual value ΔP_r due to a permamently impregnated mass of dust, w_r. Ordinarily, $\Delta P_r >> \Delta P_f$. In addition, it may be seen that during the period when the cake is establishing itself, the pressure–mass relationship is not linear as eq .6 would imply. This is the period when the dust removed from inside the filter cloth is being replaced. In Figure 5.14 a typical pressure–time variation for an actual filter is plotted.[33] Again it may be seen that the actual residual pressure drop $\Delta P_r'$ is ordinarily greater than ΔP_f. After a period of cake establishment, the variation of pressure with time is linear. Proper choice

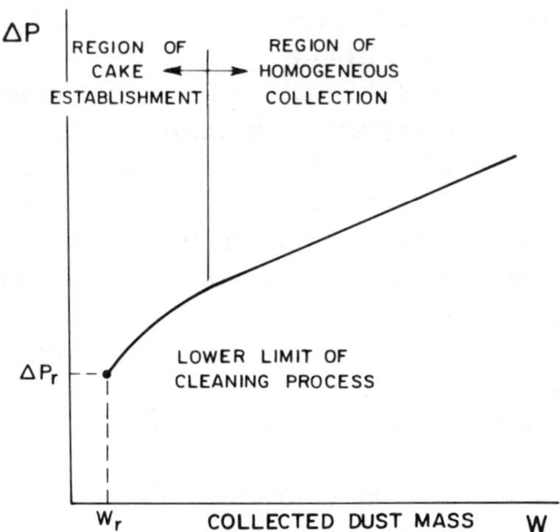

Figure 5.13. Typical pressure variation with collected dust mass, w, at constant face velocity (schematic).

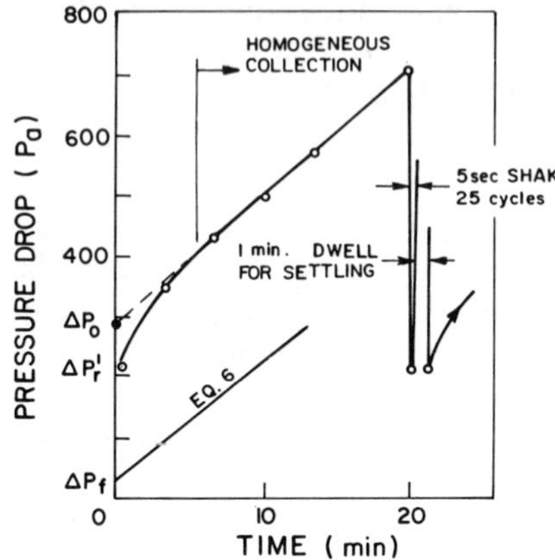

Figure 5.14. Typical pressure variation with time at constant face velocity (after Walsh and Spaite[33]).

of k_2 would give eq.6 the proper slope but not the proper intercept to predict the actual behavior. The pressure history over the majority of the constant velocity cycle is predicted by an equation of the form

$$\Delta P = \Delta P_0 + k_2\mu \frac{C_a}{C_c} U^2 t \tag{7}$$

where ΔP_o is the intercept of the extrapolated behavior during homogeneous dust deposit to the zero time point. Again, ΔP_o is not well documented, although attempts to evaluate both ΔP_r and ΔP_o are documented by Billings and Wilder[22] and Walsh and Spaite.[33]

The relative value of the actual residual pressure, $\Delta P_r'$, and the minimum achievable residual pressure from a large number of cleanings, ΔP_r, is a measure of the efficiency of the particular cleaning process. The amount of mechanical cleaning necessary to optimize $\Delta P_r'$ has been discussed by Walsh and Spaite.[33] While the efficiency of cleaning depends upon amplitude frequency and type, the processes tested by them required up to 600 cleaning strokes to obtain the ultimate possible cleaning. Ordinary mechanical cleaning processes consist of 10 to 100 cycles and achieve values of $\Delta P_r'$ close to ΔP_r.

In normal bag house operation, a single section or area of cloth or a single bag will not ordinarily run at constant velocity. Since cycling of bags or sections of bag house is done on a rotating basis in the normal case, the conditions imposed on the single bag will vary in time. Typical assumptions made are that the entire bag house passes a constant flow with time or that the entire bag house is exposed to a constant pressure at all times except for the short period of reverse cleaning cycle. Neither of these assumptions is strictly correct in most bag house operations.

However, both may be used with acceptable accuracy. The case of constant flow rate through a series of periodically cycled bag house sections, has been dealt with extensively by Crawford.[14] Under this condition the total flow will redistribute itself between sections, depending upon the relative resistance of each section. At a single instant in time the relative flow through each section may be straightforwardly calculated by the following procedure. At this particular instant the pressure drop across a given section i of n sections of filter cloth or elements of bag house is given by

$$\Delta P = k_{2_i} L_i \mu U_i \tag{8}$$

$$= k_{2_i} L_i \mu \frac{Q_i}{A_i}$$

where $k_{2i}(t)$ is the permeability coefficient of section i of the bag house, L_i the thickness of the cake or cloth in this section of the bag house, and U_i the face velocity at this instant in time in this section of the bag house. Q_i is the flow through this section and A_i the cross-sectional area of the section. If we define the quantity

$$R_i(t) = \frac{k_{2_i} L_i \mu}{A_i} \tag{9}$$

where R_i is the resistivity of the section, then placing eq. 9 in 8 gives

$$\Delta P_i = R_i Q_i \tag{10}$$

Equation 10 is an exact analogy of Ohm's law if ΔP is considered as analogous to the voltage, R_i is taken as the resistance, and Q_i is taken as analogous to the current. The problem then becomes one of calculating the flow and pressure drop across a series of resistances in parallel. As a result, the total pressure drop and total flow at an instant are given by

$$\Delta P = R \sum Q_i \tag{11}$$

where $\tag{12}$

$$\frac{1}{R} = \sum_{0}^{n} \frac{1}{R_i}$$

and the flow through each section is given by

$$Q_i = \frac{\Delta P}{R_i} \tag{13}$$

The consequences of assuming a constant total flow through the system may be easily seen by considering eq. 11. Since $\sum Q_i$ is assumed constant and R varies as a function of time due to the cleaning cycle, the pressure drop across the system must alter in time. Variations in R may be decreased by increasing n, the number of separate chambers or elements in the system, or by increasing the frequency of cleaning so that the net effect of variation over the whole system is decreased.

It is also possible to analyze the consequences of assuming a bag house system to be operating against constant pressure. As a result, the total flow will vary with time. Under these conditions a single bag will operate against a constant pressure drop. If we consider the homogeneous collection period of the collection cycle, eq. 7 indicates that for this period Q is proportional to $1\sqrt{t}$ for a constant pressure drop. This behavior is illustrated experimentally in Figure 5.15 based upon data tabulated by Strauss.[34] Again, for a given instant in time, eq. 11 will apply. Calculation of the instantaneous value of resistance will enable calculation of the individual and total flow rates. In many applications neither the assumption of constant pressure nor constant flow rate is appropriate. Determination of the actual operating condition is discussed in Section 5.6.

5.6. FAN AND DUCTING DESIGN

A complete air flow system that contains an industrial filter will consist of dust collection devices associated with the dust-producing region, ducting, fans or blowers, control devices, and a filter. Proper design requires that the entire system be determined with the objectives of minimum cost, minimum operating expense, low maintenance, and high collection efficiency.

Dust collection devices for air pollution control are normally of two

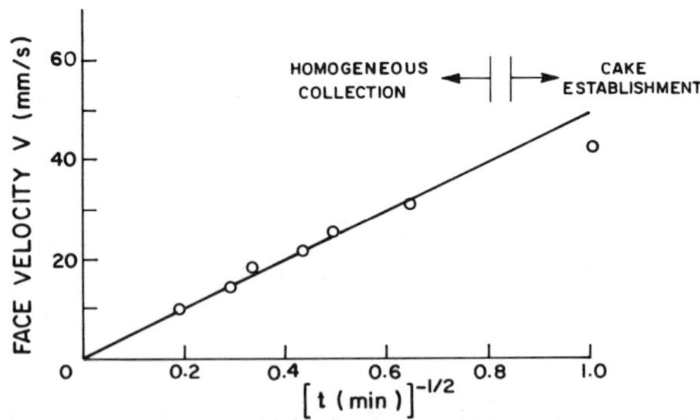

Figure 5.15. Face velocity as a function of time for constant pressure drop.

types. If dust is collected at the localized source of dust generation, the dust concentration in the associated air may be maximized. Since it is desirable to treat the minimum amount of air for dust removal, this type of device is ordinarily recommended.

If large volumes of air become dusty due to widely diffused dust generating processes or to collection of dusty air some time after dust generation, then, ordinarily, much larger volumes of air and lower dust concentrations must be handled. This is usually more expensive than the low-volume high-concentration situation. Good design normally requires that the dust be collected as soon after its generation as possible. The design of collection devices for air pollution control in working environments has been extensively treated. Full reviews of the subject are given by McDermott[35] and Rajhans and Bragg,[36] and extensive data on many designs is tabulated in Ref. 2.

Dust collection from processes is not ordinarily treated as a separate subject. However, collection from some process equipment, such as conveyor belts, furnaces, and spray booths, has been treated in the reviews cited above. More detailed analysis will depend upon the particular process.

In dust collection systems, fans may be installed either before or after the filter. Fans installed before the filter must, of course, be capable of running in a dusty atmosphere. Characteristics for the three most commonly used types of fans are shown in Figure 5.16. Forward curved blade fans or volume fans are normally used for large volumes of air at low pressure. As a result, they can be used at lower speeds than the other types and therefore usually have relatively quiet operation. These fans are relatively cheap. Since the horsepower curve rises steeply with increases in volume, this fan is appropriate for use only in locations where the resistance against the fan is fixed at the design value. This type of fan is not suitable for dust or fume handling because dust will collect on the curved blades. These fans are used extensively in heating and ventilating and air conditioning work but are not normally used for air pollution control devices. In summary, then, this type of fan should not be used to power industrial filters. They are, however, commonly used to drive flows in air conditioning apparatus that contains air conditioning-type filters.

Backward curved blade fans have a relatively smooth static pressure variation, with a peak near the zero volume point. An important aspect of these fans is that they are self-limiting. The horsepower required at constant rpm reaches a peak before the maximum volume is reached. As a result, running under low-resistance conditions does not endanger the motor, and the constantly varying demands of a complex duct system will

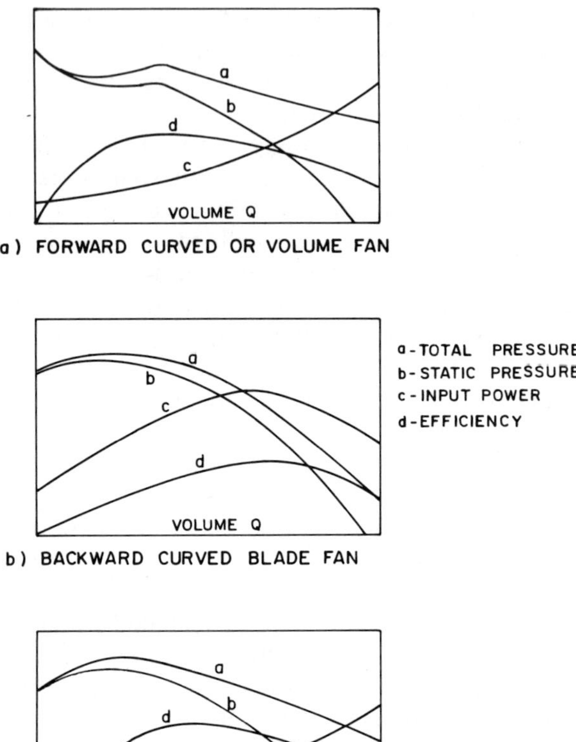

Figure 5.16. Fan characteristics for dust collection systems (constant rpm): (*a*) forward curved; (*b*) backward curved; (*c*) radial blade.

not result in overloading the motor if a fan of this nature is used. Ordinarily, fans of this type are large for a given application. This type of fan is very popular because of its self-limiting power requirement and is very commonly used to drive air pollution control systems. However, gases containing dust may not be sent through this fan as dust buildup will occur. As a result, this fan must be used in forced draft applications of downstream of the filter.

The radial blade fan is the fan type used for dusty gas streams with dusts. Many special designs have been developed for specific dust handling and pneumatic conveying problems. Efficiencies and costs are relatively less than for the other types. The design is not self-limiting on power.

Figure 5.17 illustrates the interaction of a fan and the system. ΔP_d is the resistance in the system of the ducting, dampers, and other passive elements. In the vast majority of applications the flow is turbulent and ΔP_d is proportional to Q^2. ΔP_d is independent of time and may be taken to include those losses in the filter other than that through the actual filter medium. Hence exit, entry, turning losses, and so on, can be included in ΔP_d. The loss due to the cloth itself is time-dependent and is shown as $\Delta P(t)$ on the diagram. The behavior of this quantity was described in Section 5.5. This pressure drop is linear with Q at an instant in time. The point where the fan characteristic intersects the sum of ΔP_d and $\Delta P(t)$ (total system resistance) is the operating point of the system. As a result, point $a(t)$ represents the operating point of the system. The resulting operating range of ΔP and Q are indicated schematically on the diagram. We can see that neither a constant-pressure nor a constant-volume-flow-rate model is exactly appropriate for real systems. A more accurate model may be achieved by analyzing the data in the form of Figure 5.17. For many systems, however, the variation in pressure and/or flow is relatively small. This is particularly true if system resistances other than the filter medium comprise the largest part of the system. In large or expensive systems, of course, a detailed analysis of all possible operating ranges is necessary. The dust to be conveyed must not fall to the walls of the duct. As a result, a minimum conveying velocity must be determined. These

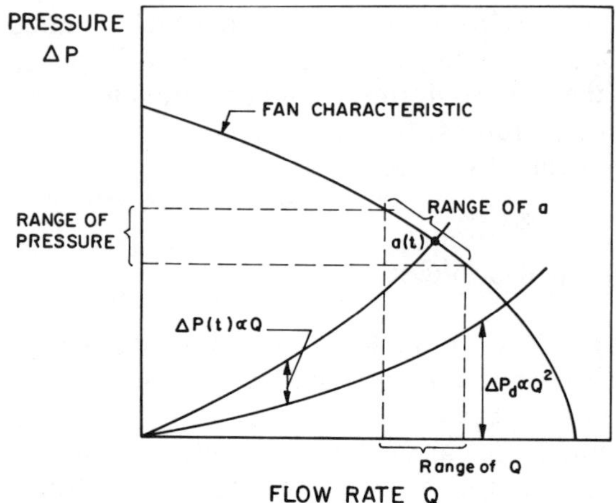

Figure 5.17. Interaction of dust system and fan (schematic).

values are available in the literature (Refs. 35 and 36) or may be calculated from basic principles.[41]

Ducting systems associated with air pollution control equipment have traditionally been designed to provide minimum capital cost and ease of maintenance. Increasingly, these systems are being designed to minimize total costs, including energy costs. As a result, there is considerable interest in developing minimum-pressure-drop ducting. The ducting associated with the equipment may be designed for minimum pressure drop by minimizing the number of loss-creating elements in the system. The following general rules summarize a considerable amount of more detailed knowledge.

1. Since losses are proportional to the square of velocities, lower velocities are to be preferred for lower losses.
2. Since the diffusion process is inherently inefficient, flow should be slowed down as infrequently as possible.
3. Smooth turns are to be preferred to sharp turns at any point where the flow must be changed in direction.
4. Efforts must be made to use all parts of the filter assembly with equal efficiency. As a result, attempts should be made in inefficient installations to distribute flow evenly over the entire filtering surface.

5.7. SPECIAL APPLICATIONS

Previous sections have concentrated upon the industrial bag filter, which is the most commonly used and most extensively studied air filtration device. There are, however, many filter types for special applications. These include the various forms of air conditioning filter, air cleaning equipment for clean rooms, filters for radioactive and toxic dusts, and demisting equipment.

5.7.l. Air Conditioning Filters

Air conditioning filters are intended to remove dusts at concentrations far below those required in industrial filters. Typical dust concentrations in air conditioning applications are in the range 1 to 0.01 mg m^{-3}. As a result, the cleaning cycle can be done manually or replacement filters can be installed. Low pressure drops of the order of 10 to 100 Pa are typical. These low pressures are necessary because the large amounts of air handled over long periods of time would otherwise require large amounts of energy. In addition, efficiencies are considerably less in normal air conditioning installations than in industrial filters. The 50% efficiency point on an air conditioning filter can be between 50 and 100 μm. Face velocities between 1 and 3 m s^{-1} are used compared to typical industrial filter velocities of 10 to 20 mm s^{-1}.

The most common type of filter for air conditioning is the viscous filter, where oils or other surface coatings of low volatility and high flash point are placed on the filter to increase the retention properties of the filter. Considerable increases in efficiency are possible by this procedure. Viscous filters are commonly made up in panels of approximately 0.5 m^2 and assembled into larger arrays for larger flow. Replacement or cleaning of these filters ordinarily occurs when the pressure drop is around 100 to 200 Pa. Metallic viscous filters can be steam-cleaned and recoated. A typical viscous filter efficiency curve is shown in Figure 5.18.

Continuously cleanable filters of this type are also available. Heavier filter media may be used in air conditioning applications if the cloth is arranged in such a way that multiple layers of cloth face the oncoming flow. Under these conditions the overall velocity may be as high as 2 to 3 m s^{-1} but the filter cloth velocity locally may be as low as 100 mm s^{-1}.

5.7.2. High-Efficiency Applications

There are many applications where extremely high efficiency cleaning at very low particle sizes is required. Under these circumstances the term

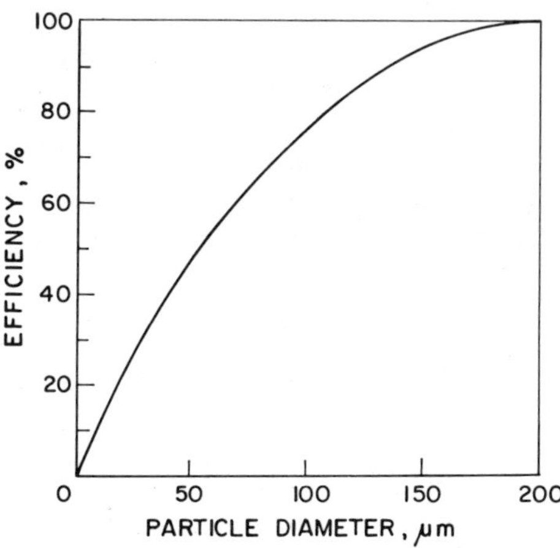

Figure 5.18. Collection efficiency of a typical viscous filter.

decontamination factor (D.F.) is used, where

$$\text{D.F.} = \frac{1}{1 - \eta} \qquad (14)$$

Decontamination values of 10^6 can be required in special radioactive applications. Efficiencies of this magnitude can be achieved with deep beds of filter media. The subject of extremely high efficiency air cleaning has been dealt with by White and Smith[1] and Keilholtz and Battle.[37] Such filters are frequently called "absolute" filters, although, of course, no filter can be perfectly efficient. Pressure drops of the order of 100 to 300

Table 5.9.   Properties of High-Efficiency Air Conditioning Filters (After Strauss[10])

	Face velocity (mm s^{-1})	Pressure (kPa)	Efficiency (%) D.O.P. or methylene blue
Asbestos bearing paper			
At start	27	0.20	99.980
After 205 min	27	0.24	99.993
Commercial "absolute" filters			
Esparto grass–asbestos	20	0.28	99.95
Glass fiber–asbestos	20	0.28	99.99
Glass fiber (suitable to 550°C)	20	0.28	99.99
Cellulose paper–asbestos	20	0.09	90
Esparto grass only (for use in photographic industry, where asbestos is undesirable)	20	0.06	65
Glass fiber pads			
3-µm fibers, loose 12-mm	170	0.01	63
1.3-µm fibers, loose 12-mm	150	0.45	91
1.3-µm fibers, loose 25-mm	140	0.38	99.4
12-mm, 3-µm followed by 12-mm, 1.3-µm	150	0.23	94
Vacuum cleaner bag paper	70	0.30	30
Woven glass fibers			
Fine-weave fabric	26	0.02	48
Coarse-weave fabric	26	0.005	22
Wool felt (reverse jet system)			
At start	100	0.20	30
After 840 hr (constant)	100	0.70	92

Pa are typical for face velocities of 20 to 150 mm s^{-1}. The efficiencies of this type of filter is commonly measured using either an aerosol of dioctyl phthalate (D.O.P.) with mean size around 0.3 μm or methylene blue with a mean size around 0.5 μm. American practice uses D.O.P. and British practice uses methylene blue.

The high efficiencies necessary for this type of filtration may be achieved by deep beds of collection material such as asbestos, fiberglass, cotton, or mixtures. Typical bed depths for this type of filter run from 100 to 500 mm. High efficiencies may also be obtained with papers made of asbestos, glass, and so on. These papers are commonly arranged in corrugated or honeycomb fashion to maximize space utilization. These papers may also be used in respirators. Typical behavior of these materials is summarized in Table 5.9. In applications such as the filtration of toxic dusts and radioactive particulate, high life expectancy measured in years may be necessary. Under these circumstances prefiltration and very light dust loading must be provided for.

5.7.3. Demisting

Cooling towers, spray columns, and similar equipment frequently requires the filtration of liquid droplets from a gas. Since the liquid drops can continuously agglomerate and run off, the device can produce a continuous process and the problems of normal particulate removal are not present. Demisters are normally of metal wire or metal wire/asbestos combinations. Demisting devices are also commonly used in air compressors where oil mists can cause the risk of explosion. For further information on this topic, see Chapter 4.

5.7.4. High-Temperature Applications

Many high-temperature applications for industrial filters exist. Fiber glass may be operated up to approximately 290°C, Teflon to 260°C, and Nomex to approximately 230°C. Stainless steel and other materials have been attempted experimentally[38] and mineral fibers on a metal core have been used.[34] These systems are intended for use in the range 350 to 400°C. Although glass and asbestos as well as other fibers are chemically stable at these higher temperatures, they suffer reduced strength and considerably reduced resistance to wear and flexure. As a result, many of the attempts to increase operating temperature revolve around ways of supporting and cleaning the medium.

SYMBOLS

a	constant in pressure relationship, eq. 2
A_i	cross-sectional area of section i of bag house
b	constant in pressure relationship, eq. 2
C_a	mass concentration of dust in the gas
C_c	mass concentration of dust in the cake
d	particle diameter
D	fiber diameter
D.F.	decontamination factor $= 1/(1 - \eta)$
i	typical bag house section
k	permeability coefficient
k_1	empirical value of permeability coefficient for cloth
k_2	empirical value of permeability coefficient for cake
k_{2_i}	permeability coefficient of section i of bag house
L	bed thickness
L_c	cake thickness
L_i	cake or cloth thickness of section i of bag house
m	particle mass
ΔP	total pressure drop across a bag house system
ΔP_c	pressure drop due to cake alone
ΔP_d	resistance in passive elements of filter system
ΔP_i	pressure drop across section i of bag house
ΔP_f	pressure drop due to filter cloth alone
ΔP_0	intercept of homogeneous cake pressure drop
ΔP_r	pressure drop after extreme amounts of cleaning
$\Delta P_r'$	pressure drop at start of cleaning cycle
Q_i	volume flow through section i of bag house
R	total resistivity of bag house system
Re	Reynolds number $= VD/v$
R_i	resitivity of section i of bag house
t	time
U	face velocity
U_i	face velocity in section i of bag house
V_e	entrance velocity to tube
V_f	face velocity in tube
w	collected dust mass
w_r	permanently impregnated dust mass
β	angle of internal friction of dust
η	filter efficiency
η_f	filtration efficiency as calculated by summing mechanisms
η_{fd}	filtration efficiency for collection by diffusion

η_{fi}	filtration efficiency for collection by impaction and interception
ϵ	porosity or void fraction
ν	kinematic viscosity
μ	dynamic viscosity

REFERENCES

1. White, P. A. F., and Smith, S. E., *High Efficiency Air Filtration*, Butterworths, London, 1964.

2. American Conference of Governmental Industrial Hygienists, *Industrial Ventilation*, 14th Ed., East Lansing, Mich., 1974.

3. Guilloud, R. L., *Proceedings of the User and Fabric Filtration Equipment Specialty Conference*, Air Pollution Control Association, Pittsburg, 1973, p.146.

4. Bergmann, L., *Proceedings of the User and Fabric Filtration Equipment Specialty Conference*, Air Pollution Control Association, Pittsburgh, 1973, p.85.

5. Ebaugh, W. C., *J. Ind. Eng. Chem.*, 1, 686 (Oct. 1909).

6. Ihlefeldt, H., *Staub*, 21(9), 448 (1961).

7. Löffler, F., "Collection of Particles by Fiber Filters," in W. Strauss (Ed.), *Air Pollution Control*, Pt. 1, Wiley, New York, 1971.

8. Natanson, G. L., *Kolloid Zh.*, 24, 52 (1962).

9. Sell, W., *Ver. Dtsch. Ing. Forschungsh.*, 347(1931).

10. Strauss, W., *Industrial Gas Cleaning*, Pergamon Press, Oxford, 1975, p. 278.

11. Kuwabara, S., *J. Phys. Soc. (Jap.)*, 14, 527 (1959).

12. Happel, J., *AIChE J.*, 5, 174 (1959).

13. Pich, J., "Theory of Aerosol Filtration," in *Aerosol Science*, C. N. Davies (Ed.), Academic Press, London 1966, Chap. 9.

14. Crawford, M., *Air Pollution Control Theory*, McGraw-Hill, New York, 1976, p. 439.

15. Bragg, G. M., and Hanna, B., Internal report, University of Waterloo, 1975.

16. Kirsh, A. A., Stechkia, I. B., and Fuchs, N. A., *Trans. Kolloidn. Zh.*, 31, 121 and 227 (1969).

17. Chen, C. Y., *Chem. Rev.*, 595 (1955).

18. First, M. W., and Drinker, P., *Arch. Ind. Hyg. Occup. Med.*, 5, 387 (1952).

19. Whytlaw-Gray, R., and Patterson, H. S., *Smoke*, Edward Arnold, London, 1932.

20. Wallis, G. B., *One-Dimensional Two-Phase Flow*, McGraw-Hill, New York, 1969.

21. Cadle, R. D., *Particle Size*, Reinhold, New York, 1965.

22. Billings, C. E., and Wilder, J., *Handbook of Fabric Filter Technology*, GCA Corporation, Bedford, Mass., 1970, pp. 3–23.

23. Herrick, R. A., Olsen, J. W., and Ray, F. A., *J. Air Pollut. Control Assoc.*, 16, 1 (1966).

24. Orr, C., Jr.,*Particulate Techology*, Macmillan, New York, 1966.

25. Clement, R. L., *Plant Eng.*, 15, 92 (1961).

26. *Buffalo Forge Company Bull. AHD-29*, Buffalo, New York.

27. *American Air Filter Co., Inc., Bull. 279c*, Louisville, Ky.

28. Frey, R. F., and Reinauer, T. V., *Air Eng.*, 30 (1964).

29. Carman, P. C., *Trans. Inst. Chem. Eng. (Lond.)*, 15, 150 (1937).

30. Spielman, L., and Goren, S., *Environ. Sci. Technol.*, 2, 279 (1968).

31. Davies, C. N., *Proc. Inst. Mech. Eng.*, 1B, 185 (1952).

32. Chen, C. Y., *Chem. Rev.*, 55, 595 (1955).

33. Walsh, G. W., and Spaite, P. W., *J. Air Pollut. Control Assoc.*, 12,57 (1962).

34. Strauss, W., *Industrial Gas Cleaning*, Pergamon Press, Oxford, 1975, p. 338.

35. McDermott, H. J., *Handbook of Ventilation for Contaminant Control*, Ann Arbor Science Publishers, Ann Arbor, Mich., 1976.

36. Rajhans, G. S., and Bragg, G. M., *Engineering Aspects of Asbestos Dust Control*, Ann Arbor Science Publishers, Ann Arbor, Mich. (1978).

37. Keilholtz, G. W., and Battle, G. C., Treatment of Radioactive Airborne Wastes from Reactors," in W. Strauss (Ed.), *Air Pollution Control*, Pt. 2, Wiley, New York, 1972.

38. Billings, C. E., Silverman, L., Levenbaum, L. H., Kurker, C., and Hickey, E. C., *J. Air Pollut. Control Assoc.*, 8(1), 58 (1958).

39. Clarenburg, L. A., *Aerosol Science*, 3, 461 (1972).

40. Bragg, G. M., and Pearson, B. H., *Ind. Eng. Chem.* Process Des. Dev., 18, 171 (1979).

41. Bragg, G. M., Proc. 4th Int. Con. Pneumatic Transp. Solids in Pipes, 1978, p. (2-13).

6

WASTE AIR CLEANING BY THERMAL DECOMPOSITION

R. Germerdonk

Lehrstuhl für Verfahrenstechnik, Universität Kaiserslautern, Germany

Translated from the German by P. Huttelmaier and B. L. Hümmel.

6.1. INTRODUCTION

The thermal decomposition of harmful substances, in particular decomposition in a flame, has gained importance due to increasing demands for pure exhaust air or gas emitted from industrial plants. This procedure, sometimes incorrectly described as burning of waste air, is an effective method for completely exterminating harmful organic substances in waste air. Moreover, it does not matter if the harmful organic substances occur in the form of vapor or dust. With other waste air or off-gas purification procedures, such as absorption and adsorption procedures, which have a long history, the harmful substances originally in the form of vapor are merely accumulated in the liquid or solid phase. With dust separation procedures, the separated dust occurs as a dry powder (e.g., with electrostatic gas purification procedures) or as a watery suspension (e.g., when separated in a venturi washer).

In case the harmful substance gained in the cleaning procedure cannot be used in the succeeding production procedure, an additional step is necessary. It has to be either exterminated (e.g., through burning) or it is delivered to the sewage system or appears as a waste deposit. Thus in the latter two cases, the elimination of waste air or gas problems creates additional sewage and waste deposit problems. Nevertheless, methods of thermal decomposition of harmful waste gas should be applied only in cases where more classical procedures of dust separation or the separation of harmful vapors fail. This is due to the fact that thermal decomposition (i.e., burning of harmful substances in a flame) is very expensive both in initial investment and more particularly in running costs. A large amount of high-quality energy is used, only to be transformed into waste heat. Moreover, harmful inorganic compounds may be created, such as carbon monoxide, nitrogen oxides, and possibly hydrochloric acid. Also, noise is produced.

In spite of this, thermal decomposition is still applied extensively. This is due to the following facts:

1. All harmful organic material is destroyed completely, independent of chemical properties.
2. In general, no wastewater—or waste deposit—problems are created, as opposed to most other waste air purification procedures.
3. A simple temperature control next to the combustion chamber is sufficient control assuming there is enough excess air in the flame. Thus the legally allowable emmission values are safely maintained.
4. Also, the final purification procedure for waste air is continuously and effectively maintained even while quantity and composition of waste air are fluctuating (e.g., during rearrangements of production procedures).

Points 1 and 4 are of interest to the production engineer, whereas 2 and 3 are of interest to the ecologist as well as the supervising authority.

However, the layout of processes where waste air is merely purified in a flame should be avoided, where possible, by the industrial engineer involved in the design and development of industrial plants. This is true in spite of the advantages of waste air combustion, considering its simple applicability and controllability. The reason for this is not only the high cost. Rather, the waste of high-quality energy in this process seems to be intolerable for future developments. Several possibilities for solving the problem of air purification by thermal decomposition. which are economically as well as ecologically justifiable, are considered in the following discussion.

Besides the problem of diminishing the excessive use of primary energy, questions considering control procedures and safety must be dealt with.

From that point of view it will be necessary to visualize thermal air purification as a complete process rather than a single step taking place during the reaction in the flame.

For convenience, the term "waste air or waste gas purification by thermal decomposition of the contained harmful substances" will be abbreviated to "thermal waste purification" or simply to "TWP." A TWP process, therefore, is a procedure in which harmful substances, contained in waste air or waste gas, are thermally decomposed. Temperatures are above 600°C and no catalysts are applied. Also, the use of waste air in existing combustion arrangements as combustion air is not discussed, since no additional knowledge of technical procedures is necessary in this use. Utilization of such possibilities should be self-evident for the engineer.

6.2. REACTION MECHANISMS

Conditions in the reaction zone (i.e., in the flame) have to be known if previous analysis of the process indicates that a waste gas or air problem is solvable only by means of a TWP procedure. This is a necessity for optional arrangements of a TWP procedure, where harmful substances are to be exterminated to allowable limits. For harmful substances that exist in the form of vapor, known relationships for gas-gas reactions are valid. No useful theoretical basis is yet available for dusts. In most cases, the material to be extinguished consists of harmful organic substances contained in waste air.

6.2.1. Definitions

To avoid misconceptions, some terms will be defined first.

WASTE AIR. Consists of a gas mixture containing sufficient oxygen to maintain combustion. This corresponds to $O_2 \rangle$ 14 to 16 vol. %, depending on burner and other operating conditions. The remainder consists mainly of nitrogen. In some cases small quantities of inert gases are present, such as carbon dioxide and water vapor (maximum 10 to 15 vol. %). Waste air may also contain harmful substances which oxidize in the flame, assuming that oxygen is present. These are vapors of solutions, organic microdusts, and odorous substances. Waste air may also contain harmful inorganic substances that can be thermally decomposed (e.g., H_2S and CO).

WASTE GAS. This mixture contains little oxygen; combustion without added air is not possible. The ratio between inert gases and ingredients that oxidize is arbitrary. Depending on the composition and temperature of the waste gas stream, the oxygen content should be ⟨8 to 10%.

WASTE HEATING GAS OR WASTE FUEL GAS. This is a waste gas containing small amounts of oxygen and large quantities of harmful substances that oxidize. At stoichiometric combustion in air, additional useful heat is set free in addition to the heat necessary for reaction.

INFLAMABLE WASTE AIR OR IGNITABLE WASTE AIR. This air contains oxygen as well as combustible harmful substances. Both their concentrations and their ratios correspond to the one necessary for combustibility or ignitability, respectively.

COMBUSTIBLE WASTE MATERIAL. Combustible or fluid waste fuel includes greater amounts of contaminants. Their viscosity is so low that they may be at least pressure-driven through a nozzle for atomization. This combustible mixture contains, besides hydrocarbons, harmful substances such as chlorine, nitrogen, or sulfur.

FLUE GASES. These are produced by the burning process in a TWP combustion arrangement. They consist of a gas mixture containing only such low concentrations of harmful, thermally decomposable substances that they are within emission standards. Other, nondecomposable contaminants may be contained in this flue gas at higher concentrations. These substances consist, for instance, of oxidized dusts, HCl or SO_2. Before emission, therefore, additional cleansing may be necessary (i.e., a water or alkaline washing process). This disagrees with German[1] and other guidelines.

6.2.2. Decomposition of Harmful Hydrocarbons

Oxidation of hydrocarbons of general composition H_mC_n is incomplete according to the conditions for reaction of a process for thermal waste air purification (i.e., low temperatures at combustion and short duration). The corresponding equation of gross reaction is

$$H_mC_n + \left(\frac{m}{4} + n\right)O_2 \rightarrow \frac{m}{2} H_2O + nCO_2 \tag{1}$$

In the light of the present discussion, the transformation of hydrogen is of lesser interest. For carbon, however, the following reaction steps are also of interest.[2]

$$C + \tfrac{1}{2}O_2 \rightarrow CO \tag{2}$$

and

$$CO + \tfrac{1}{2}O_2 \rightleftharpoons CO_2 \qquad\qquad (3)$$

The technical requirement is to decompose all harmful hydrocarbons according to eq. 1 only to a degree where the allowable remaining load in the flue gas, expressed in mg of organically bound carbon per m^3, is reached safely. At the same time, the carbon monoxide allowable in flue gas has to be met according to reactions 2 and 3.[12]

Equation 1 represents a rough equation, used for stoichiometric computations. It is a known fact that oxidation of even the simplest hydrocarbon (i.e., methane) takes place over many steps. Fristrom and Westenberg[3] considered 16 and Cremer[4] considered 25 steps.

For larger hydrocarbons the first decomposition steps lead to lower-order hydrocarbons or carbon–hydrogen–oxygen compounds. The relatively short time period, when passing through the combustion chamber of a TWP arrangement, does not allow complete decomposition to water or carbon dioxide.

In a TWP procedure it is desirable that all harmful organic substances should be decomposed completely. Substances produced in between are mostly of lesser interest. Only the gross result (i.e., the degree to which harmful organic substances are oxidized) and the time needed are of importance.

Figure 6.1 demonstrates the theoretical decrease of oxidation velocity in a waste air purification arrangement as compared to reaction velocities in a standard burner combustion (e.g., at 1500°C). As a matter or fact,

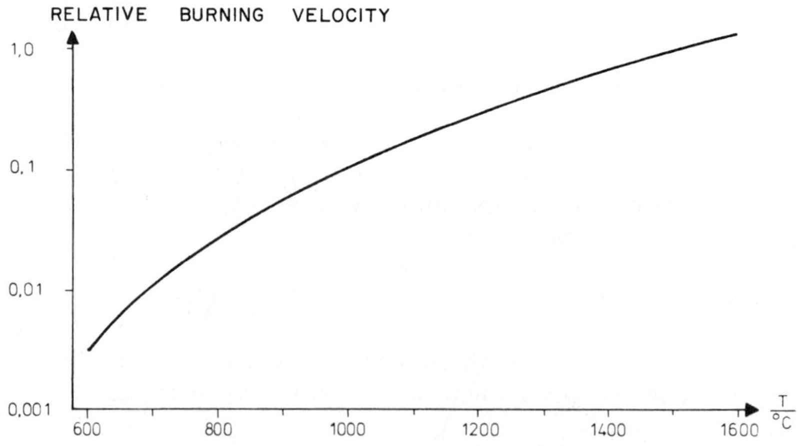

Figure 6.1. Decrease of burning velocity in TWP installations.

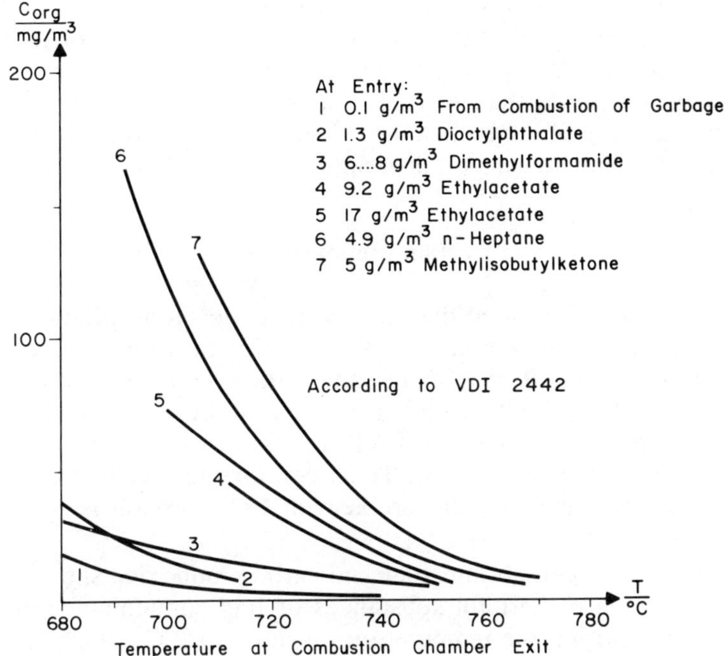

Figure 6.2. Remaining concentration of organic substances in purified waste gas in relation to the temperature at combustion chamber exit.

we know that the time required does not increase according to this theoretical curve. However, it can be recognized that the empirical rule by Rummel,[5] "mixed parts equal to burned parts," no longer applies. Considering the usual temperatures in TWP arrangements, between 650 and 850°C, the relatively low reaction velocity may be the decisive factor for low overall velocities. This is also depicted in Figure 6.2. Values of different, remaining contaminants in flue gases, measured in TWP arrangements with ceramic-lined combustion chambers, are compared to combustion chamber exit temperatures.

At this stage a fundamental difference between a normal combustion procedure and a TWP arrangement may be pointed out. At a normal combustion procedure, the fuel should be transformed with as little air as possible, such that the reaction enthalpy that is set free can be used for maximum heat production (e.g., for vapor pressure vessels). It is of no economical importance if some of the fuel (e.g., ⟨0.5%⟩ remains unburned. In normal burning procedures the amount of excess air can be kept low under such conditions, which increases the total combustion efficiency. The emission of a little unburned fuel is generally not a disturbing factor. In such arrangements an efficiency factor of maximum 99.5% is required. Comparing this to a TWP arrangement, generally a much higher efficiency than 99.9% is required. This corresponds to waste air containing 10 g of contaminant m^{-3} to be purified to less than 10 mg remaining contaminant m^{-3}. This requires at least 99.9% efficiency.

6.2.3.   The Formation of Carbon Monoxide

Equations 2 and 3 indicated that the complete decomposition of harmful substances containing carbon to the final product, carbon dioxide, causes particular difficulties. Basically, the same difficulties occur during burning of fossil fuels (e.g., in fuel-driven cars). But the allowable remaining amount of harmful substances in TWP arrangements is legally established and sometimes even supervised. Thus, contrary to locally fixed normal combustion arrangements, the problem of CO emission is particularly emphasized.

Considering the equilibrium state, no carbon monoxide should remain from the burning of harmful substances or fuel containing carbon in a TWP process, within the temperature range of 1000°C, which is usually kept there. This is demonstrated in Figure 6.3.

Decomposition is determined by reaction velocities rather than by the equilibrium state, as is known for oxidation of hydrocarbons. In the flame vicinity the velocity toward formation of carbon monoxide is much larger than, for example, the velocity of the continued reaction, according to eq. 3. The continued reaction, therefore, has to take place in the zone of the later reaction or after-burning zone in a TWP arrangement.

In the flame as well as in the after-burning zone there are radicals, which normally react much faster than molecules. Therefore, the reaction

$$CO + 0 \rightleftharpoons CO_2 \tag{4}$$

contributes more to the CO turnover than does reaction 3, in spite of the low concentration of radicals in the burning zone of TWP installations.

In flue gases from TWP burners, water vapor is always present. This means that the extremely fast reaction

$$CO + OH \rightleftharpoons CO_2 + H \tag{5}$$

still takes place in the burning zone behind the flame. In standard burners (i.e., at temperatures $\rangle$ 1250°C), reaction 5 causes the essential amount of turnover of carbon monoxide. The concentration of OH radicals which exist at the low temperature of the burning zone in a TWP installation is too small to achieve as fast a turnover as is measured in high-temperature flames. This is because only the reaction

$$H_2O \rightleftharpoons OH + \tfrac{1}{2}H_2 \tag{6}$$

can be considered in providing OH radicals. As Figure 6.3 shows, however, the equilibrium concentration of OH radicals of reaction 6 is very small. Nevertheless, the final combustion of carbon monoxide to dioxide is essentially controlled through the kinetics of reaction 5. The curve shown in Figure 6.4, obtained from experiments, determines constants for gross velocities of reaction 5.[6] It is important to notice that the influence of temperature should be very small in the zone below 1200°K, which is of particular interest for a TWP. According to this theory, it is apparent that the decomposition of carbon monoxide formed in the flame should depend essentially on the remaining concentrations of OH radicals as well as on the residence time.

The kinetic hampering of a continuous oxidation of carbon monoxide is additionally obvious from the fact that carbon monoxide may be completely decomposed on catalysts, which takes place efficiently even at low temperatures.

The minimum temperature of flue gases in the burning chamber of a TWP installation is generally determined by the remaining concentration

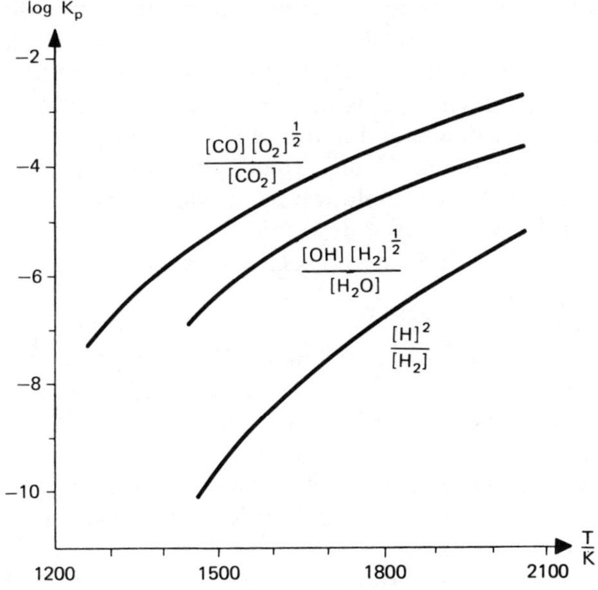

Figure 6.3. Equilibrium constants for combustion of carbon to CO_2.

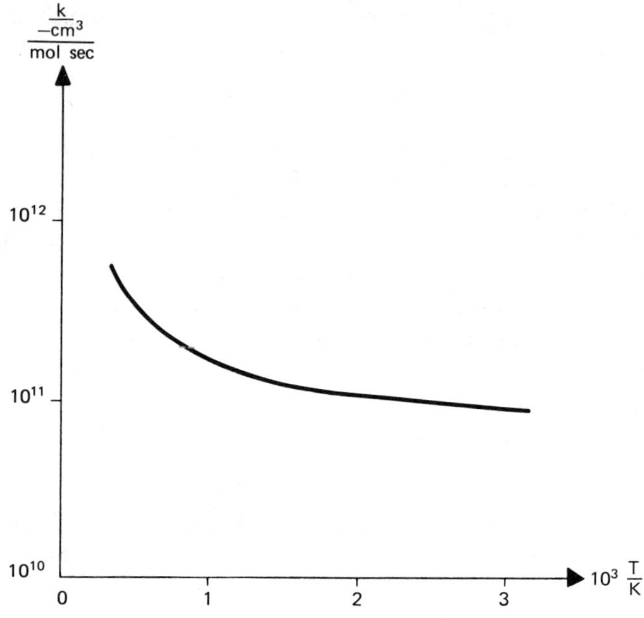

Figure 6.4. Velocity constants for CO + OH $\rightleftharpoons$ CO$_2$ + H.

of carbon monoxide. This temperature follows from the allowable concentration of emission normally applied (i.e., 100 mg C_{org} m^{-3} and 500 ppm CO), as shown in Figure 6.5. Some years ago a certain percentage of decrease in C_{org} emission was demanded in the "California formula," which, for example, demanded a 90% decrease independent of the concentration at the beginning. Today, final values for C_{org} and CO are prescribed. Figure 6.5, for example, refers to values for 1977 demanded in West Germany.[1]

Williams and Fine[7] derived the relationship for oxidation velocity of carbon monoxide:

$$\frac{d[\text{CO}]}{dt} = 1.3 \times 10^{14}[\text{CO}][\text{H}_2\text{O}]^{1/2} \exp\left(\frac{-15,100}{T}\right) \qquad (7)$$

where |CO| is expressed in moles cm^{-3} and t in seconds.

Equation 7 should be valid for low temperatures in TWP installations, up to 600°C. For the operation of TWP installations it follows that harmful hydrocarbons are in general much better decomposed than demanded. According to German "TA air" regulations[12] (the German code for clean air) the concentration of carbon monoxide is limited to 500 ppm. Thus there exists a fairly simple possibility of control compared to the usual control requirements for organically bound carbon in flue gases. This assumes, however, that all harmful substances in waste air are led through the zone where reactions take place. It implies that the TWP installation has to be designed properly as well as mounted and run properly. The reverse may also be considered; that is, if there is a measured higher concentration of hydrocarbons than allowable at proper temperatures (e.g., according to "TA air") in flue gases, the TWP installation is not properly designed or is not operated in the proper way.

6.2.4.   Formation of Nitrogen Oxides (NO_x)

The problem of nitrogen oxide emission has been discussed extensively. We shall not discuss the question of which emission group furnishes the largest proportion of NO_x or allowable emissions. We will deal only with the way nitrogen oxides are formed in a TWP, how emitted quantities can be reduced, and how nitrogen oxides that occur already as harmful substances in waste air can be exterminated by thermal processes.

   Nitrogen oxides that originate from TWP installations may derive from

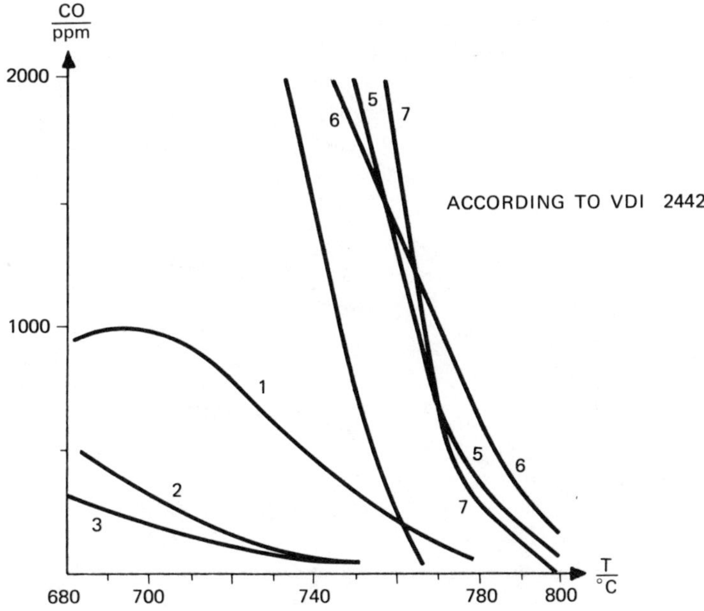

Figure 6.5. Remaining concentration of carbon monoxide in purified waste gas in relation to the temperature at the combustion chamber exit. (For numbers, see Figure 6.2.)

combustion air or fuel or from nitrogen oxides already present in the exhaust air. These three sources will be considered separately.

6.2.4.1.   Formation of NO_x from Nitrogen Contained in the Air

This mechanism has been thoroughly investigated. Quantities of nitrogen oxide, which are formed this way, can be estimated for TWP conditions.

Figure 6.6 demonstrates equilibrium curves[7] depending on nitrogen oxide and nitrogen dioxide contents, which are created according to the gross reactions

$$\tfrac{1}{2}O_2 + \tfrac{1}{2}N_2 \rightleftharpoons NO \tag{8}$$

and

$$[a1\tfrac{1}{2}O_2 + NO \rightleftharpoons NO_2 \tag{9}$$

NO_x formation from the air provides the largest possible values. However, it is realized that due to the low temperature range existing in TWP in-

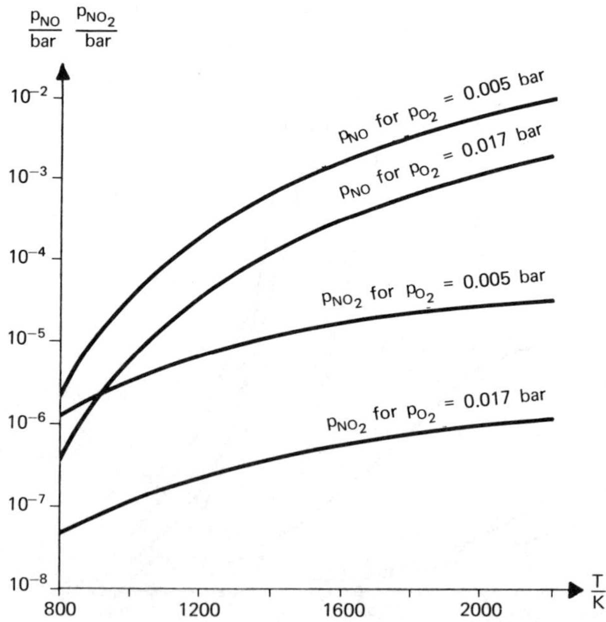

Figure 6.6. Equilibrium states during thermal formation of NO_x.

stallations, formation of nitrogen oxides is still negligible compared to standard burners, assuming that thermal equilibrium is reached.

The general assumption of negligible NO_x formations, however, requires that anywhere in the TWP process temperatures do not exceed values in the range of 1200°C. This condition cannot be completely satisfied, because in some restricted zones higher temperatures are required to stabilize the flame. From measurements on existing TWP installations it has been demonstrated, however, that formation of NO_x from nitrogen contained in air can be kept quite low (i.e., ≤ 100 mg m^{-3} flue gas). For this reason a thorough mixing in the combustion zone is necessary to achieve a fast reduction of local temperature extremes. As we shall see, there are other important reasons for thorough mixing in combustion zones of TWP installations.

6.2.4.2. Formation of NO_x from Nitrogen Bound to Fuel and Contaminants

The question of to what extent fuel-bound nitrogen contributes to NO_x emissions is not completely answered, as demonstrated in recent investigations.[8] These investigations refer mostly to temperature ranges that exist in standard burners (≥ 1200°C).

According to this, mainly nitrogen dioxide is formed from nitrogen fuel compounds directly within the reaction zone. However, this is decomposed in the hot zone behind the flame until an approximate equilibrium of $NO–NO_2$ is reached.

Considering the conditions in a TWP combustion chamber, it can be assumed that the second step is performed much more slowly. Also, one has to assume a larger NO_x content in flue gas, as compared to standard burners if nitrogen is introduced with fuel. On the other hand, NO emission should not be of any importance, because the velocity of decomposition is larger for NO than its velocity of formation. Figure 6.7[9] shows this relationship clearly, using the half-lives in both directions of reaction 8.

6.2.4.3. Decomposition of Nitrogen Oxides Supplied in Exhaust Air

According to reaction equation 8, it is practically impossible to decompose nitrogen oxide of concentrations well above equilibrium (see Figure 6.6) in normal TWP temperatures and at realistic residence times ($\langle 10$ s), as shown in Figure 6.7.

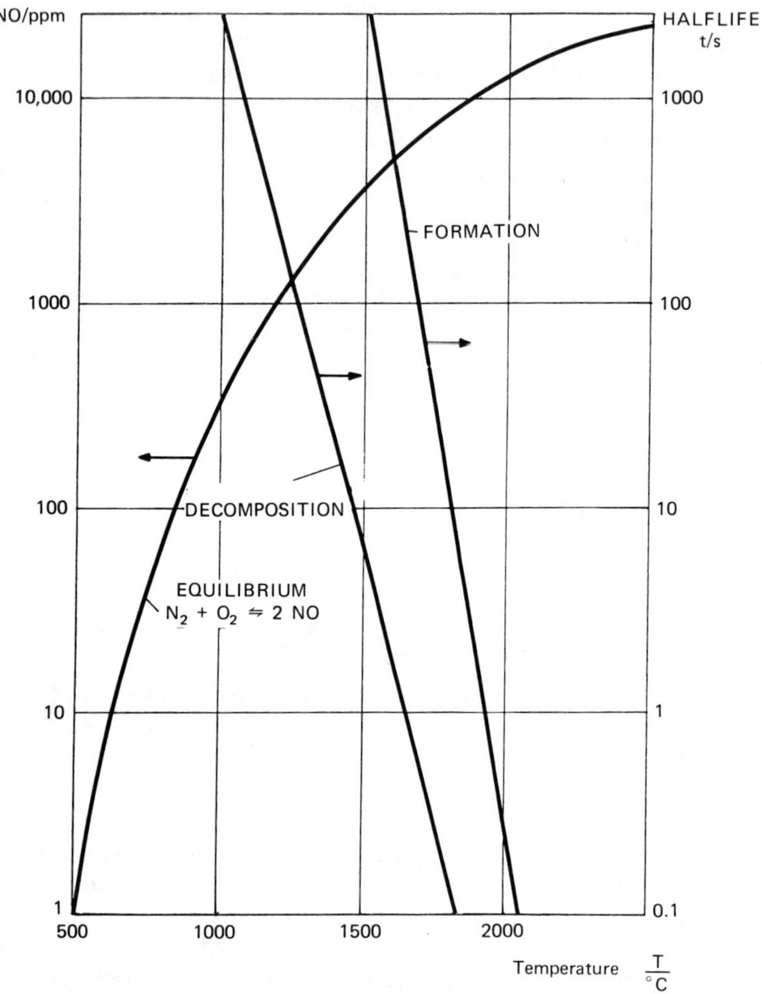

Figure 6.7. Half-lives for formation and decomposition of NO according to Biberacher et al.[9]

Values of the NO decomposition in Figure 6.7 were computed for an oxygen content of 21%, although in TWP installations, concentrations of only ≥3% in flue gases is required for complete oxidation of harmful organic substances so that there will be no significant increase in NO_x decomposition.

For significant reduction of NO, the reaction process has to be changed fundamentally. This is not possible at conditions in a standard TWP (i.e., at temperatures ⟨1000°C and oxygen contents ≳5% in the flue gas). Conditions for reduction have to be established; that is, practically all oxygen has to be bound in the form of hydrogen and carbon compounds. Only then do the gross reactions

$$2NO + 2H_2 \rightleftharpoons N_2 + 2H_2O \tag{10}$$

and

$$[a12NO + 2CO \rightleftharpoons N_2 + 2CO_2 \tag{11}$$

which are sufficiently fast for the decomposition of nitrogen oxide, have a noticeable influence on NO_x reduction.[10]

This means that waste air containing NO_x has to be used as combustion air in a flame run on a large fuel surplus (e.g., $\lambda \leq 0.85$). In an adiabatic process, reaction temperatures in these flames are still above 1300°C, even for reduced oxygen concentrations in waste air containing NO_x (16% $\lesssim O_2 \leq 21\%$). Thus gross reactions 10 and 11 occur very quickly.

If for such purposes standard fuels are used (natural gas, liquid fuels, coal), the NO_x reduction takes place primarily according to eq. 11. In that case a reducing flue gas occurs from the NO reduction burning chamber which still contains large quantities of CO. Carbon monoxide has to be oxidized in a following second stage. For this purpose it can be advantageous to use a second, NO_x-free waste air flow. It is always necessary to use two-stage combustion. The oxygen that occurs in waste air containing NO_x has first to be bound completely to the fuel, where parts can be decomposed to CO only. Thus the specific fuel demand is at least four times larger, as compared to a TWP procedure without NO_x extermination. Fuel consumption is even more unfavorable when applying waste air preheating. If the nitrogen oxide occurs in oxygen-deficient waste gas, however, the specific fuel demand is decreased.

In a thermal decomposition of NO_x by a two-stage process, with a CO surplus at the first stage, a temporary formation of cyanide has to be considered.[6] The formation of HCN does not usually cause any disturbance in the reaction zone of standard combustion chambers, since CN compounds are nearly completely decomposed at the high temperatures in burning zones. In TWP procedures for NO_x reduction, however, considerably larger HCN concentrations may occur. Therefore, temperature control, mixing zones for additional air, and residence times for the second burning stage have to be chosen such that the temporarily formed HCN is completely burned before the CO oxidation is completed. Otherwise, a remaining HCN emission and a new formation of NO may be encountered. Therefore, it is preferred to introduce the additional air into the second stage stepwise, to allow gradual mixing.

6.2.5. Formation of HCl and Chlorine

If contaminants in waste air or waste fuel also contain chlorinated hydrocarbons, they can only be decomposed to HCl and traces of free chlorine. It is assumed that the reaction

$$4HCl + O_2 \rightleftharpoons 2H_2O + 2Cl_2 \tag{12}$$

which is called the Deacon process, reaches an approximate equilibrium in TWP combustion chambers and determines the $HCl/\frac{1}{2}Cl_2$ ratio in flue gas.

Figure 6.8 shows the $HCl/\frac{1}{2}Cl_2$ ratio in relation to combustion chamber temperature and water vapor content in flue gas. An oxygen content of 5% by volume, which is common in TWP installations for complete decomposition of harmful substances, is assumed, as well as a total chlorine content of 20 g m_N^{-3} flue gas. The large influence of the water vapor content can be recognized.

When using waste fuels of large chlorine content (i.e., if large amounts of hydrogen atoms are replaced by chlorine atoms), flue gases will contain little water vapor and hence a relatively large amount of free chlorine. Since emission of free chlorine from industrial plants is limited in most

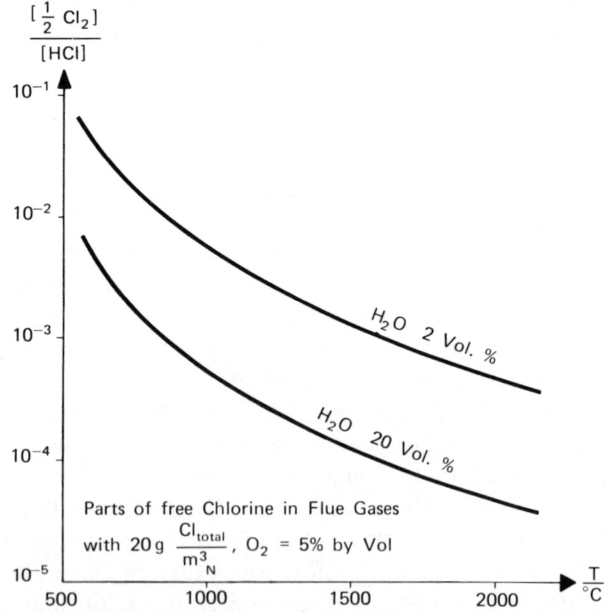

Figure 6.8. Equilibrium states during chlorine decomposition.

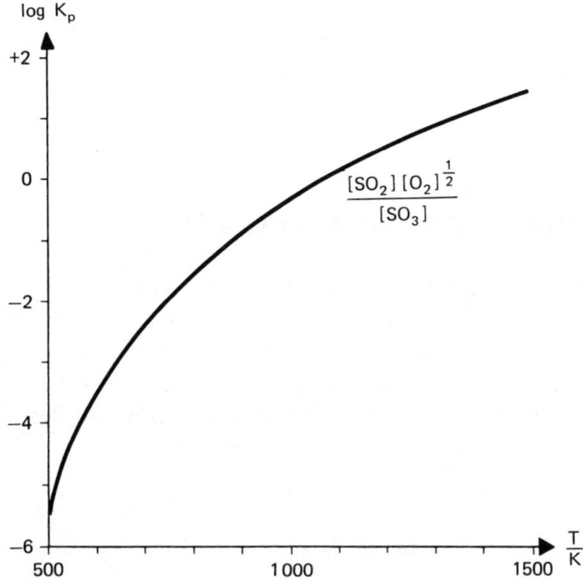

Figure 6.9. Equilibrium during the formation of SO_3.

countries, it follows that not only should HCl be removed by simple water washing, but also chlorine by alkaline in a second washing step. An alkaline washing with sufficiently high pH values ()9) for chlorine absorption would also absorb the carbon dioxide (i.e., the consumption of alkali is intolerably high). Therefore, the content of free chlorine in the combustion chamber should be kept sufficiently low by additional burning of Cl-free hydrocarbons or even by the addition of vapor (i.e., recycling from the flue gas quenching).

6.2.6. Formation of Sulfur Oxides

As shown in Figure 6.9, the equilibrium of the reaction

$$SO_2 + \tfrac{1}{2}O_2 \rightleftharpoons SO_3 \tag{13}$$

lies nearly completely on the side of sulfur trioxides for temperatures necessary to decompose harmful organic substances in TWP combustion chambers. Reaction 13, however, is so slow even in the temperature range of standard burners that catalytic effects by metal oxide dusts determine the quantity of SO_3 formation. Since hardly any metal oxides are formed from contaminants in waste air, sulfur trioxide could only form from waste fuels that contain stronger metal compounds (in particular, vanadium). Generally, we may assume that the total amount of sulfur contained in flue gas originating from waste air and fuel occurs in the form of sulfur dioxide.

6.3. STOICHIOMETRY AND HEAT CALCULATION

In the combustion chamber of a TWP plant, oxygen, contaminants, and additional fuel must react together, so that the heat released in the process is sufficient to heat up the entire mixture of waste air, waste gas, and fresh air from the respective inflowing temperatures to the disintegration temperature of the contaminants, and to compensate for heat losses. This results in the following heat balance equation:

$$\dot{M}_{WA}(C_A T_{WA\alpha} + X_{WA}Hu_{WA}) + \dot{M}_{WG}(C_G T_{WG\alpha} + X_{WG}Hu_{WG})$$

$$+ \dot{M}_{FA}C_A T_{FA\alpha} + \dot{M}_F(C_F T_{F\alpha} + Hu_F) = \dot{M}_{FG}(C_{FG}T_{FG\omega} \quad (14)$$

$$+ X_{FG}Hu_{FG}) + \dot{Q}_L$$

where $\dot{M}$ = mass flux
    C = specific heat
    T = temperature
    X = concentration of oxidizable contaminants in gases
    Hu = lower heat values of contaminants and fuel
    $\dot{Q}_L$ = heat-loss rate

Indices:

    WA = waste air
    WG = waste gas
    FA = fresh air
    F = fuel
    FG = flue gas
    α = inflow
    ω = outflow

For the calculation of TWP plants a specific heat of ~ 1.3 KJm$_N^{-3}$ for all air flows may be assumed. The value of X_{FG} is consistently negligible, as is the term $C_F T_F$.

At most TWP plants no waste gas, or compared with $\dot{M}_{WA}$, only a very small amount is introduced, and accordingly in standard operation (i.e., for the initial design) no fresh air supply is required. With that, eq. 14 is simplified to

$$\dot{M}_{WA}(C_A T_{WA\alpha} + X_{WA} Hu_{WA}) + \dot{M}_F Hu_F \approx \dot{M}_{FG} C_{FG} T_{FG\omega} + \dot{Q}_L \quad (15)$$

From this the specific fuel requirement b is obtained:

$$b = \frac{\dot{M}_F}{\dot{M}_{WA}} \approx \frac{C_{FG} T_{FG\omega} - C_A T_{WA\alpha} + (Q_L/\dot{M}_{WA}) - X_{WA} Hu_{WA}}{Hu_F - C_{FG} T_{FG\omega}} \quad (16)$$

whereby the heat loss with regard to the waste air quantity $\dot{Q}_F/\dot{M}_{WA}$ is set empirically at a value of 5 to 10% of Hu_F, with waste fuels up to 20%.

Apart from the heat balance, the necessary condition for a sufficient oxygen content in the flue gas for the complete oxidation of harmful substances must also be met. An empirical value for oxygen content in the flue gas, $(O_2)_{FG}$, of 3% is sufficient to obtain, in correctly constructed TWP combustion chambers, a complete decomposition of harmful substances provided that the temperature in the combustion chamber is sufficiently high that the carbon monoxide concentration in the flue gas becomes $(CO)_{FG} < 500 \text{ mgm}_N^{-3}$.

With high-grade additional fuels and $(O_2)_{WA} > 18\%$, $(O_2)_{FG}$ is always $> 3\%$. If the waste air is preheated before entering the TWP plant, $(O_2)_{FG}$ becomes larger yet.

If a waste fuel is at one's disposal and/or waste gas is to be decomposed also, an additional oxygen balance must be drawn up.

If the waste air and harmful substances have a composition of decomposable portions in a lump sum of $C_{m'}$, $H_{n'}$, $N_{j'}$, $Cl_{k'}$, $S_{l'}$, then an oxygen consumption $(\dot{O}_2)_{WA}$ of

$$(\dot{O}_2)_{WA} = \dot{M}_{WA} X_{WA}[m'C_{m'} + \tfrac{1}{4}n'H_{n'} + j'N_{j'} - k'\tfrac{1}{4}Cl_{k'} + l'S_{l'}] \quad (17)$$

is calculated, and for a waste gas, loaded with decomposable harmful substances with a lump-sum composition, $C_{m''}$, $H_{n''}$, $N_{j''}$, $Cl_{k''}$, $S_{l''}$, the analogous oxygen consumption $(\dot{O}_2)_{WG}$ of eq. 17a is obtained.

$$(\dot{O}_2)_{WG} = \dot{M}_{WG} X_{WG}[m''C_{m''} + \tfrac{1}{4}n''H_{n''} + j''N_{j''} - \tfrac{1}{4}k''Cl_{k''} + l''S_{l''}] \quad (17a)$$

Similarly, the oxygen consumption for the fuel $(\dot{O}_2)_F$ of composition C_m, H_n, N_j, Cl_k, S_l is

$$(\dot{O}_2)_F = \dot{M}_F[mC_m + n\tfrac{1}{4}H_n + jN_j - k\tfrac{1}{4}Cl_k + lS_l] \quad (18)$$

Now it is expedient to specify the gas quantities in standard m_N^3 (i.e., indirectly in kmoles). The concentrations, X, in volume percent are accordingly specified. The specification of the gas flows, expressed in m_N^3

hr^{-1} is given by $\dot{V}$. The complete inflowing oxygen stream then becomes $\dot{V}_{(O_2)\Sigma\alpha}$:

$$\dot{V}_{(O_2)\Sigma\alpha} = \dot{V}_{WA}O_{2WA} + \dot{V}_{WG}O_{2WG} + 0.21\dot{V}_{FA} \tag{19}$$

with the respective oxygen concentrations O_{2i} of the partial inflows.

Of this total flow the oxygen quantities 17, 17a, and 18, required for decomposition of contaminants and for complete burnout of fuels are consumed. The latter specifications were given in kg hr^{-1}. Hence the total oxygen consumption becomes

$$\dot{V}_{(O_2)\Sigma\text{react}} = [(\dot{O}_2)_{WA} + (\dot{O}_2)_{WG} + (\dot{O}_2)_F] \times \frac{22.41}{32} \tag{20}$$

The oxygen concentration in the flue gas O_{2FG} is then

$$O_{2FG} = \frac{\dot{V}_{(O_2)\Sigma\alpha} - \dot{V}_{(O_2)\Sigma\text{react}}}{\dot{V}_{FG}} \geq 0.03 \tag{21}$$

This should total a minimum of 3% (see above), so that complete decomposition of the harmful substances can be attained in the TWP plant.

The flue gas quantity $\dot{V}_{FG}$ results in

$$\dot{V}_{FG} = \dot{V}_{WA}[1 + X_{WA}(n'H_{n'} - k'Cl_{k'} - j'N_{j'})]$$
$$+ \dot{V}_{WG}[1 + X_{WG}(n''H_{n''} - k''Cl_{k''} - j''N_{j''})]$$
$$+ \dot{V}_{FA} - \dot{V}_{(O_2)\Sigma\text{react}}$$
$$+ \dot{H}_F[2nH_n + mC_m + jN_j + \tfrac{1}{2}kCl_k + lS_l] \tag{22}$$

This relation applies to gas as the supplementary fuel. Liquid fuels have to be converted to the corresponding standard volume of the combustion products CO_2, H_2O, NO_2, HCl, and SO_2.

Generally, the flue gas volume is not calculated by using eq. 22 since the waste air quantity $\dot{V}_{WA}$ is not accurately known. For the basic design of TWP plants, including retroactive connection of heat exchangers or purification plants (flue gas washers), it is usually sufficient that the fixed fuel quantity, shown in eqs. 14 and 16 ($\dot{M}_F$ in kg hr^{-1} and m_N^3 hr^{-1}), multiplied by the empirical factors 1.1 m_N^3 kg^{-1} or 1.1 m_N^3 m_N^{-3}, be added to the inflowing waste air/waste gas volumes. With this the rule-of-thumb formula is obtained:

$$\dot{V}_{FG} \approx \dot{V}_{WA} + \dot{V}_{WG} + \dot{V}_{FA} + 1.1\dot{M}_F m_N^3 \, kg^{-1} \tag{23}$$

6.4.　PREHEATING

As eq. 16 shows, the specific supplementary fuel requirement, *b*, depends proportionally on the difference

$$T_{FG\omega} - T_{WA\alpha}$$

This means, on the one hand, that to economize on fuel usage, the flue gas temperature in TWP plants should be as low as possible. However, as described in Section 6.2.3, there are limits due to the resulting inadmissibly high CO emission or incomplete decomposition of harmful substances. On the other hand, the temperature of the waste air flowing to the TWP plant should be as high as possible. Therefore, it is desirable to preheat the waste air before it reaches the burner, utilizing waste heat as much as possible.

In the operation of TWP plants, the waste heat of the flue gas is usually the only source available. Therefore, to economize on fuel, the preheating of waste air by means of heat exchange with the flue gases is increasingly significant. Such arrangements for the economizing of high-quality energy through heat exchange are specified as heat economizing controls.

If not only waste air with ~21% O_2 but also waste gas occurs and if, therefore, a fresh air supply is required, various heat-economizing controls can be chosen (see Figure 6.10*b* to *d*).

FIGURE 6.10a.　This is a TWP procedure without waste heat utilization. Even though this calls for the greatest fuel requirements, the installation costs are the smallest. In exceptional cases (e.g., in short-term operations or with occasional ignitable waste air) this can be, at present, the most economical solution. Whether such procedures should be employed in the future is questionable, since a large quantity of high-quality energy is consumed.

FIGURE 6.10b.　Waste air is preheated by flue gases in a heat exchanger. Fuel requirements with preheating b_{ph} for operation, according to Figure 6.10*a*, results in the economizing quota

$$\frac{b_{ph}}{b} \approx \frac{T_{FG\omega} - T_{WA\alpha_{ph}}}{T_{FG\omega} - T_{WA\alpha_0}} \tag{24}$$

with the subscripts ph = preheating and 0 = without preheating.

Since the waste air to be purified usually occurs at temperatures $< 150°C$, noticeable fuel economy can be achieved by preheating to only a few hundred degrees. However, opposed to this saving are the increased investment and maintenance costs, together with the increased pressure loss of the heat exchanger. The optimization strategy for the evaluation

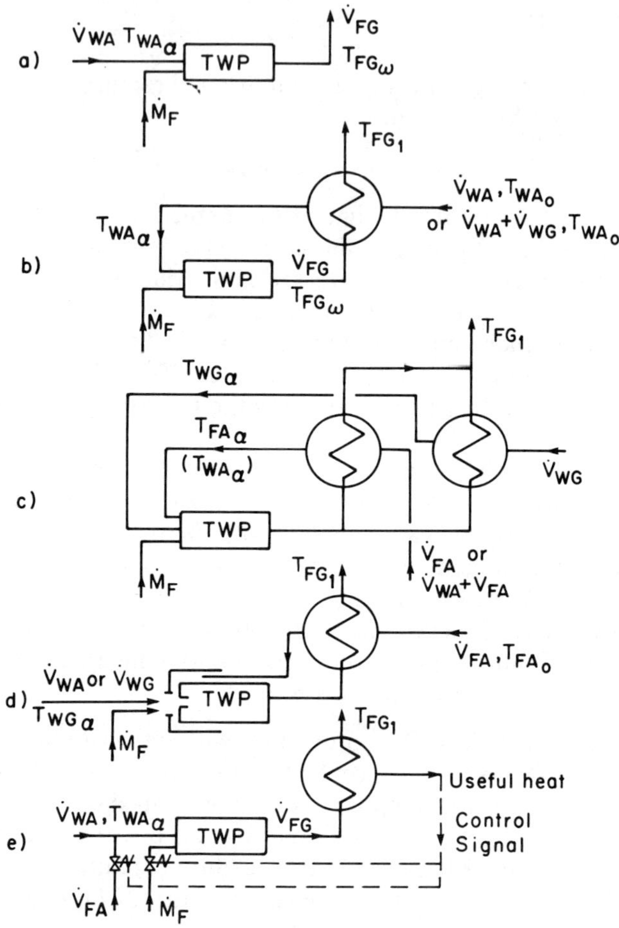

Figure 6.10. Heat economizing controls for TWP installations.

of the economical preheating temperature is the same as in every heat exchanger of a heat-economizing control. It must also be considered that with higher preheating, better heat-resistant steels would be required for the heat exchangers.

As indicated, waste gas may be mixed with waste air. In this case, however, eq. 21 must be considered.

FIGURE 6.10c.   If larger amounts of waste gas occur, so that a corresponding quantity of fresh air has to be supplied, it is usually more beneficial for the flame stability to separately preheat the fresh air and lead it into the burner. If waste air occurs at the same time, a portion of the fresh air can be replaced by waste air with $\sim$21% O_2. However, for control purposes the TWP process should be planned so that a small portion of fresh air, as determined by the waste gas flow. is constantly required.

Arrangement 6.10c has the advantage that the supplementary fuel can be completely burned up with fresh air or waste air. As a result, the load of unburned fuel in the flue gas, which would otherwise contribute to an increase of the C_{org} concentration in a TWP plant, could be eliminated in a simple way. If the fuel is burned in the waste gas/air mixture, the oxygen concentration will be much lower and the burnout would slow down accordingly.

FIGURE 6.10d.   In this arrangement preheating of waste gas is relinquished. At the same time the waste air is flowing toward the burner through the cooling mantle of the combustion chamber, which is not insulated and thus is further preheated.

In Figure 6.11 the additional fuel requirements for each 1000 m^3 of waste air is given for variable heat values of the contaminants in the waste air and for variable preheating temperatures. The illustration applies to

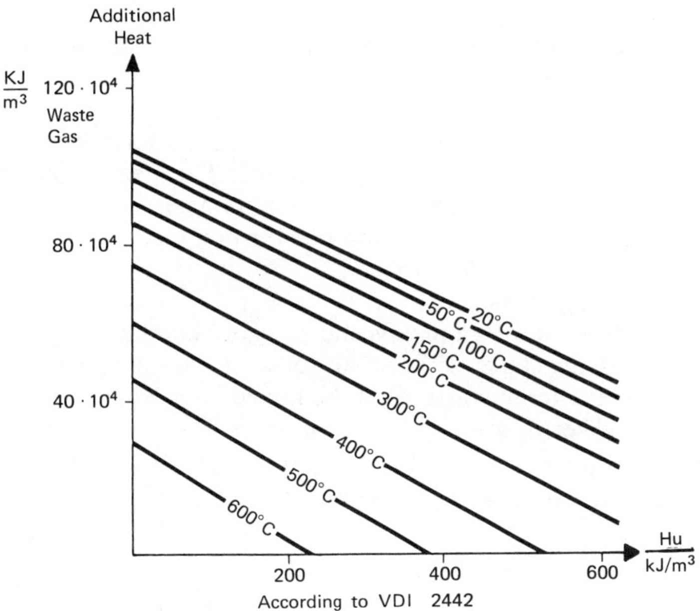

Figure 6.11. Additional fuel requirements. Lines are for values of constant waste air temperature or prewarming temperature.

a combustion chamber temperature of 750°C in the TWP plant and represents a good working base for estimates.

6.5.   DESIGN OF BURNERS AND COMBUSTION CHAMBERS

In contrast to standard burners and combustion chambers for industrial heat production, in a TWP plant all oxidizable harmful substances introduced as waste air, waste gas, and occasionally also in waste fuel should, if possible, become completely converted. In power plants some quantity of unconverted fuel is admitted to reduce λ and, consequently, heat losses in the flue gas. In TWP plants, as mentioned earlier, fuel and organically harmful substances should be converted up to)99.9%. As shown in Section 6.2, this conversion has to be achieved under even more unfavorable reaction conditions than in standard combustions. A TWP plant should, therefore, be viewed rather as a reactor than as a combustor.

Therefore, the following conditions have to be fulfilled in a TWP plant:

1.   Blending of the entire waste air and entire waste gas into the reaction zone.
2.   Blending of the entire waste and supplementary fuel into the reaction zone.
3.   Sufficient duration time of waste air and waste gas in the reaction zone at decomposition temperatures.

The design of the burner (i.e., the supplying and mixing device) is essentially determined by requirements 1 and 2. The measurements for the combustion chamber than follow from requirement 3.

For the installation of burner and combustion chamber, the following must be established:

(i)   The fuels should be quickly and possibly completely converted in a first-burning stage, with less air (or waste air) at high temperatures, and the remaining waste air to be mixed later with the hot, burned-out flue gases in a second reaction stage.
(ii)   Or the entire waste air, along with the fuel, can be converted in one stage (i.e., with high air excess).

6.5.1.   Choice of Burner Type

With condition (i) various commercial burners can be installed. In the second stage, the hot viscous flue gas must be well mixed with the remaining cold waste air stream. A disadvantage of this is the expensive two-stage construction of the combustion chamber. With (ii) the difficulty lies in finding a burner that mixes well and, at the same time, supplies a stable flame at a low temperature in the combustion chamber. Several types of this burner are well known, so that this method is chosen most of the time. However, in some cases method (i) can be more beneficial, for example if hard-to-burn waste fuel is to be fed in or if the oxygen content of the waste air fluctuates over a wide range.

Only method (i) is applicable if the waste air contains nitrogen oxide along with organic contaminants or if the harmful substances contain much nitrogen. In this case the first stage has to be operated with an air shortage (see above). An approved construction for natural gas is the so-called short flame burner of the line type. Here, similar to the gas heater of a hot water tank, the natural gas streams through a horizontal arrangement of jets and then blends with the waste air that flows uncontrolled from below through conically arranged ports. A hot flame results in which the natural gas is completely burned out. The remaining waste air is mixed in through the higher ports. The maximum flow rate for such a burner element is around 500 m_N^3/hr^{-1}. For a large waste air system, several such units of jet arrangements and conical air supply ports are assembled on a horizontal level.

If, on the other hand, waste fuel gases or waste fuels are to be burned or if waste gas is to be supplied right away to the first combustion stage, special burners or jetting equipment have to be installed.

For dilute fuel gases or if gases with constantly varying heats of combustion and ignition velocities are used as supplementary fuel, spiral burners are suitable. In these devices the fuel flows through a spiral wound transverse to the main axis, which opens downstream. The fuel is mixed with the approximately stoichiometric waste air flowing through the slots of the spiral duct. As a result, the mixing passages are short and at least at one spot of the spiral the instream velocity will always be equal to the ignition velocity. Thus quick blending and a good flame stability is obtained even under highly fluctuating operating conditions.

Instead of the spiral, tangential, widely spaced gas injection channels of various diameters and therefore various injection speed are also suitable. Burners with greater twist are usually not as good here since their backflow area changes with the fluctuation in operating conditions. As a result, the flame stability for the entire operating area is no longer guaranteed and burnout of the contaminants could be incomplete.

The usefulness of liquid waste fuels as supplementary heating fuels is dependent on good atomization over a broad range of mass flows since the fuel quantity depends on the waste air quantity, its content of harmful substances (i.e., heating fuel content), and the fluctuating heating quality of the supplementary fuel. For this application only dual flow nozzles with regulated backflow are suitable. Air can be used as an atomization medium provided that no additional problem arises due to soot, Cl_2, or NO_x formations. Otherwise, steam has to be used for atomization.

With combustion method (ii), the one-stage decomposition of harmful substances, the demands on good blending of fuel, stabilization, and flame stability are considerably higher. Here, too, spiral burners for natural gas and burners with dual flow nozzles for liquid fuels are suitable.

In any case, standard gas burners or dual fuel oil burners have to be able to blend well, for instance, with large twist. In this way the entire waste air is to be mixed completely into the reaction zone. This mixing is especially important here since, for fuel economy, the temperatures have to be kept as low as possible. Therefore, the risk is taken that the harmful substances of any portion of waste air not directly mixed into the reaction zone will not become sufficiently decomposed. However, the disadvantage of burners that blend well is that their range of waste air flow, where stable flames are achieved, may be low. Therefore, with high fluctuations of waste air flow, method (i) might be the favored process.

6.5.2.   Combustion Chamber Layout

In TWP plants the combustion chamber is the reaction area in which harmful substances are to be destroyed. The following conditions must be met:

1.   The chamber must be adapted to the chosen burner.
2.   It must be able to preserve the reaction temperature (i.e., to provide sufficient insulation.)
3.   It must correctly direct the flue gas flow.
4.   It must form a thermal buffer in some cases.

The combustion chambers of TWP plants are usually cylindrical, with a coaxially arranged burner. Occasionally, a tangential burner is used. An exception are the line burners with a rectangular burner shaft.

For installations of the one-step TWP system (ii), a twist is usually required to achieve complete mixing. Such TWP plants are also referred to as combustors if the twist is made very large so that a central low-pressure area develops with a strong backflow. With this type of flow an optimal blending is obtained whereby the residence time can be as short as possible. This way, with residence times, in the case of cold gas, of $\langle 0.1$ s at temperatures of $\sim 830°C$, complete decomposition of harmful substances is still achieved. This corresponds to a residence time of $\langle 0.03$ s in regard to the hot flue gas. The requirements, found among some environmental regulations (e.g., that of the German T. A. Luft), for considerably greater residence times seem to apply to the more unfavorable combustion chamber layouts. It also seems unjustified to assign recalculation of residual contents of the harmful substances to the CO_2 contents, obtained from standard burnings if no false air (i.e., fresh air) is supplied and if the oxygen content of the waste air is $\rangle 16\%$.

To achieve adequate blending and sufficiently high burnout rates, the ratio of length L to diameter D should be $1.5\langle L/D\langle$ max. 4.5 in TWP combustion chambers. The effective residence time of $\rangle 0.3$ s (for cold waste air ≥ 1 s) is then sufficient even if the burners or jets are no longer new.

Two-step combustion chambers for method (i) can, at the first stage, be planned for effective residence times of 0.2 s, since the temperatures are much higher here than in standard TWP plants. The blending and subsequent reaction of the high-temperature flue gas from the first stage with the remaining waste air in the second stage requires a further effective residence time of at least 0.3 s, and preferably 0.5 to 0.7 s, depending on the construction of the waste air entrance. To improve mixing, the secondary waste air can be jetted into the enlarged combustion chamber through several tangential jets. Some types of burning chambers have additional built-in or retrofitted flow obstructions to create swirl.

The wall of the combustion chamber is made of ceramic materials that can stand large temperature changes or of durable steel. With the steel type, the waste air is led to the outside through a double mantle in order to cool the wall of the combustion chamber directly and to preheat the waste air at the same time. Steel combustion chambers are lighter, less expensive, and less endangered by explosions. Ceramic-coated combustion chambers act as heat buffers in strongly fluctuating operating conditions, and they stabilize the flame along the walls.

6.6. MEASUREMENT AND CONTROL TECHNIQUES

As mentioned before, for correctly installed TWP plants, control of the flue gas temperature should be sufficient to guarantee the required residual concentration of the contaminants. Because of the danger inherent in burners as well as the possible uncontrolled emission of unburned contaminants, TWP plants require more expensive measurement and control techniques. The customary monitoring and control systems for safe operation of burners (including security against gas and air shortage, ignition and flame control, prewashing control, preignition control, quick closure with a blockage for ignition recurrence, and pressure controls for the oil and atomization medium) are also required at TWP plants. Added to this standard control system is one of the following:

1. Control of the entry variables $\dot{V}_{WA}$, $T_{WA\alpha}$, $\dot{V}_{WG}$, T_{WG}, (T_{WA0}), X_{WA}, X_{WG} for feed-forward regulation of the fuel quantity.
2. Control of the fuel and supplementary air flow by the flue gas temperature and the O_2 content in the flue gas, as sketched in Figure 6.12.

CONTROL METHOD 1. This method, as a feed-forward control, makes possible faster adaption of the supplementary fuel amount to intermittent load changes. This operating system becomes especially simple and safe to operate if it is restricted to the regulation of the waste air/supplementary

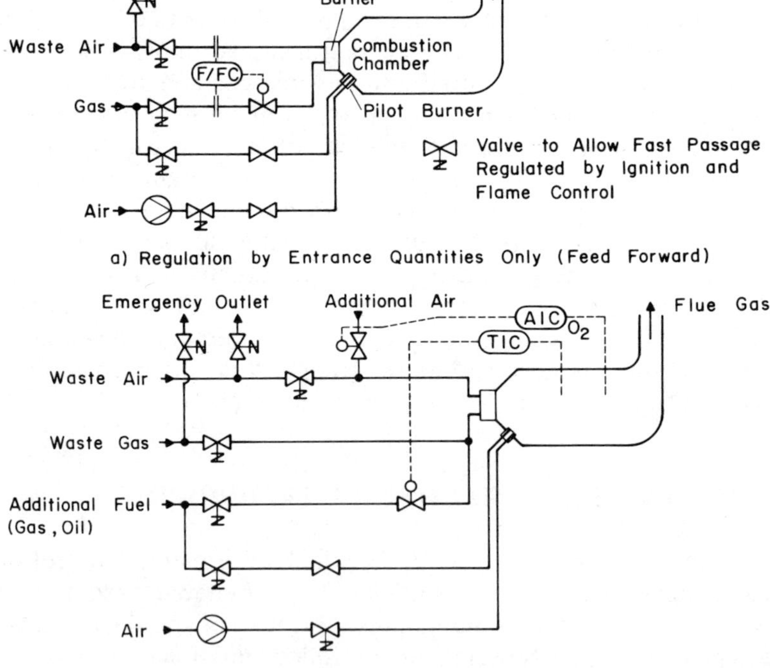

Figure 6.12. Control principles of TWP installations.

fuel proportion only. A disadvantage, however, is that to guarantee adequate decomposition of harmful substances, even in extreme cases (e.g., highest waste air quantity with lowest content of harmful substances, highest waste air quantity with lowest oxygen content and highest content of organic harmful substances, lowest waste air quantity with highest content of harmful substances and lowest oxygen content), the highest fuel requirement (eq. 16) as well as the highest supplementary air requirement (eq. 19) has to be kept. This involves an increased fuel requirement.

CONTROL METHOD 2.   This method is optimal as far as fuel requirements are concerned. It stipulates, however, for safe operation and fast response in the measuring and regulating of temperatures. If additional air is required, the problem of oxygen regulation with sufficiently short response periods is added.

Figure 6.12 shows both control systems 1 and 2 for the simplest case of single-stage combustion without preheating. A more complicated case will be dealt with in the concluding example calculation.

6.7. TWP PROCEDURE FOR OCCASIONALLY EXPLOSIVE WASTE AIR

6.7.1. Problem Statement

In this section a problem will be dealt with that, theoretically, should not exist (i.e., a TWP procedure for waste air which, at times, is loaded with contaminants above the lower explosion limit).

As shown in Section 6.4, we try, by means of heat exchange, to feed in as little additional fuel as possible. In many cases, however, fuel has to be added to the waste air in order to obtain the decomposition temperature. Therefore, for further economy, one may attempt to direct minimal quantities of waste air into the combustion chamber, which means, with a given quantity of contaminants, a higher concentration in the waste air. There are limitations to this because concentrations of combustible materials in the waste air have to be kept below the lower explosion limits X_{ex} so as not to endanger people or the operation. $X_{max} < 0.5X_{ex}$ applies as a guideline.

According to this arrangement, the problem of purifying explosive waste air by thermal decomposition should not exist at all. However, as experience has shown, keeping the concentration of harmful substances in waste air below the lower explosion limit cannot always be realized.

This applies especially to discontinuous operations. Because of possible unavoidable production and regulation errors, one has to consider the possibility of explosive mixtures. TWP procedures, therefore, must be safely operable even if waste air is explosive at times.

Figure 6.13 shows schematically the time variation of waste air flow and of concentrations of harmful substances in a discontinuous operation, for example the production of chemicals in an intermittently filled reactor. First, we deal with the case where the oxygen concentration of the waste air lies in the range 16 to 21%.

The time variation of flow and concentration illustrates the three extreme cases:

Largest flow with greatest burden	greatest combustion chamber burden and danger of flashback
Largest flow with light burden	greatest fuel requirement
Smallest flow with highest burden	greatest danger of flashback

In the last case the flow velocity may be less than the flashback velocity in the burner. In the worst case the flow goes to zero and the concentration of contaminants lies within the explosion range.

Danger of flasback means here that, without additional protective measures, the flame from the combustion chamber can flash back to the plant through the pipeline, which is filled with ignitable waste air.

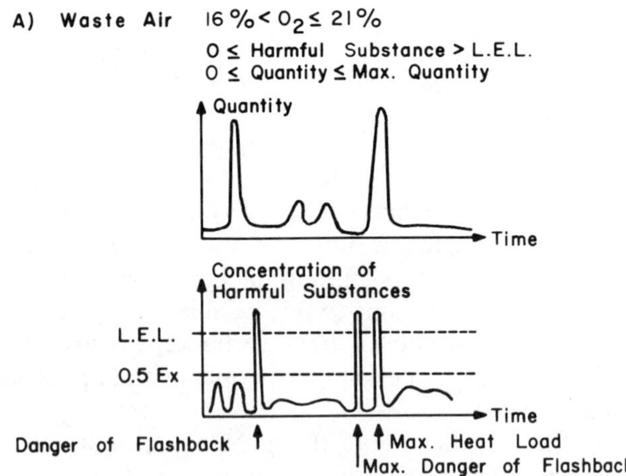

Figure 6.13. Thermal purification of waste air capable of flashback. L.E.L., lower explosion limit.

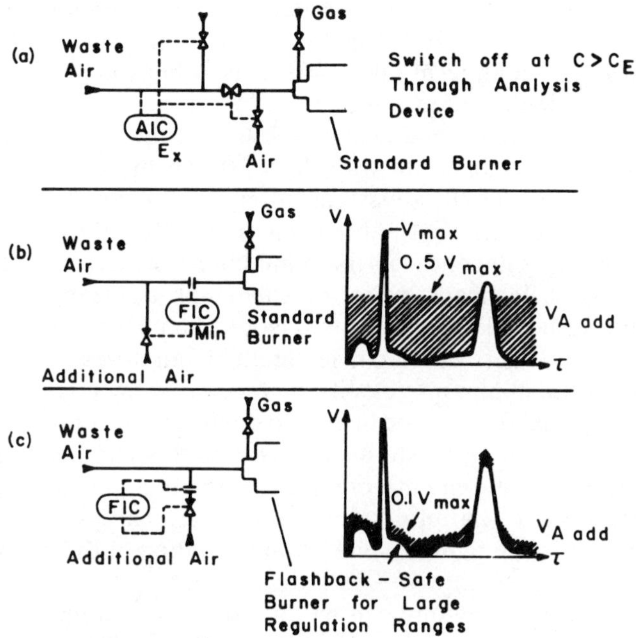

Figure 6.14. Possible flashback prevention procedures.

6.7.2. Safety Philosophy and Fundamental Considerations

Figure 6.14 shows various possibilities for preventing flashback into the plant. Unfortunately, experience has shown that, with this high risk, sufficient safety can be achieved only by means of at least two independent security systems. The one generally applied safety precaution against the flashback from an explosion consists of the installation of a flame screen or of a liquid-type flame barrier. These are considered further in Section 6.7.6.

To this flame barrier a second security system, working on a fundamentally different physical principle, should be added. Only in this way can sufficient safety be obtained.

For this second security system several possibilities are available of which three are indicated in Figure 6.14. In case (a) we determine, by means of an analysis within the waste air duct, whether the waste air is at explosive concentrations or near the danger zone. In this case it has to be switched over and the ignitable waste air discharged through an emergency outlet, provided that this is permissible. Alternatively, large amounts of fresh air can be fed in so that the concentration within the waste air/fresh air mixture is safely reduced below the point of explosion. The existing available analysis equipment, however, has not yet proven sufficiently safe in long-term operation. It is either too slow in response rate or not sufficiently safe to operate.

Case (b) shows another practical possibility for the second safety measure. Here fresh air is mixed in with the waste air through a flow control so that the system is either always at full flow or, depending on the construction of the burner, up to about one-half to two-thirds of full flow. The additional flow depends on the minumum velocity in the burner required to make it flashback-proof. Flashback-proof means here that the flame cannot flash back from the combustion chamber through the burner into the waste air duct, which carries ignitable mixtures.

To maintain the minimum velocity by feeding in fresh air requires that the entire additional fresh air be heated to the decomposition temperature. With heavily fluctuating waste air flow rates, this can mean a multiplication of the fuel requirement compared to the fuel amount required for the decomposition of harmful substances alone. This is represented by the shaded area in the figure, which is an indication of the additional fuel requirement.

Case (c) indicates how one can try to reduce the intolerably high fuel consumption of case (b) and still achieve a flashback-proof procedure. Here a portion of fresh air, only ~10% of the maximum waste air flow, needs to be constantly supplied to the waste air. One can see that the area which indicates the additional fuel requirement can then become considerably smaller than in case (b). In one specific case this system provides a fuel saving of $60,000 or more per year at one plant.

6.7.3.   Large Flashback-Proof Burners

The principle of a large flashback-proof burner is shown in Figure 6.15. Curve (a) is the laminar ignition speed as a function of the concentration.[2] Curve (b) is the flashback limit which applies to a standard burner, where the turbulence leads to an increase of the ignition speed. Curve (c) is the flashback limit for an improved flashback-proof burner. If C_{max} is the highest occurring load of harmful substances in the waste air, then for burner of type (c) it is sufficient to constantly supply an amount of unloaded fresh air, corresponding to the velocity V_0. The mixture of a constant fresh air amount with the harmful substances load O and an increasing amount of waste air with C_{max} load provides, then, for the burner mouth a work line c–c with the indicated safety distance (shaded) from the flashback curve (c).

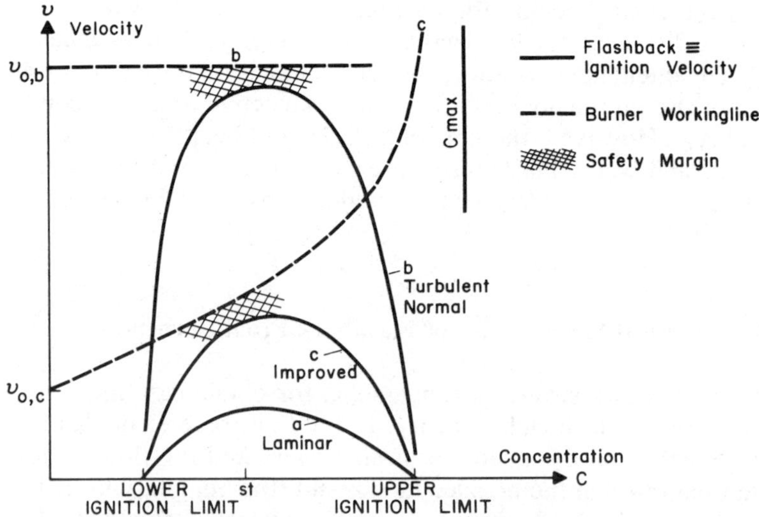

Figure 6.15. Working line for a flashback-safe process.

If the velocity V_0 has to be at about 10% of the maximum velocity to reach economically acceptable fuel requirements, the flashback curve (c) has to lie correspondingly low. For this, as schematically shown in Figure 6.16, the velocity profile near the wall is shown as a solid line for turbulent flow. In standard burners this has to be changed to the dashed line by means of transition to a laminarized flow. This has the effect that

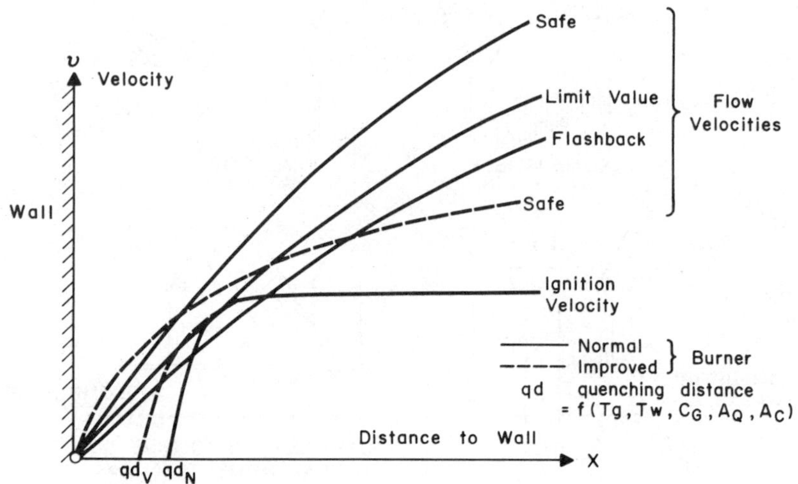

Figure 6.16. Safe and unsafe velocity profiles against flashback.

even in the critical region for flashback (i.e., near the wall outside the quenching distance), the local gas velocity is greater than the local ignition speed. The quenching distance, though, depends on the gas and wall temperatures and on the concentration and convective transport coefficients A_C and A_Q. However, the concentrations and heat flux, for given contaminants and wall temperatures, are hard to change. Therefore, the desired flashback security can be achieved only by a suitable velocity profile near the wall.

6.7.4.   Practical Performance of Flashback-Proof Burners

Figure 6.17 shows several arrangements for obtaining a more favorable flow for greater flashback security. In the first three examples flow curvature is achieved by means of straighteners and the flow is uniformly accelerated through the nozzle. As a result, the boundary layer becomes thin and the desired velocity gradient near the wall is obtained. However, as is well known, with such laminar flows stable flame fronts can seldom be achieved and suitable flame sustainers have to be provided. Also, relatively large combustion chambers have to be provided for the complete

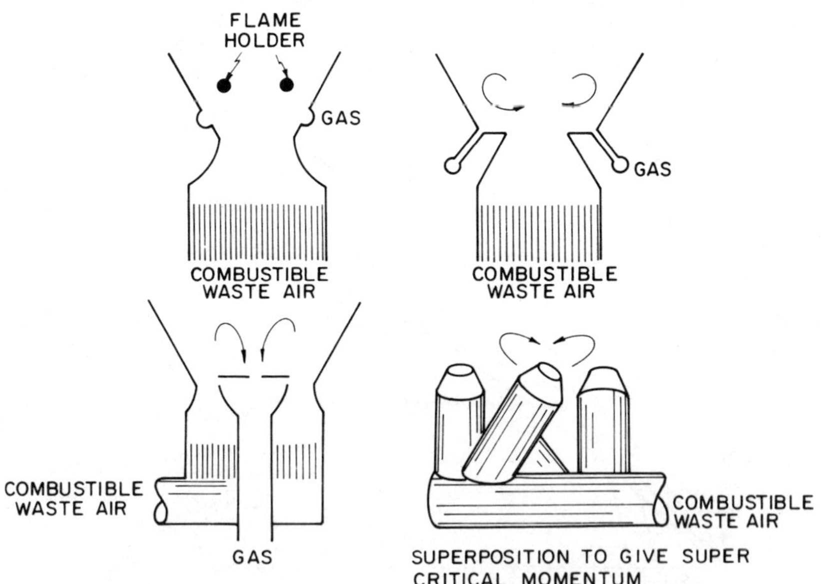

Figure 6.17. Construction types of burners with backfire-safe jets.

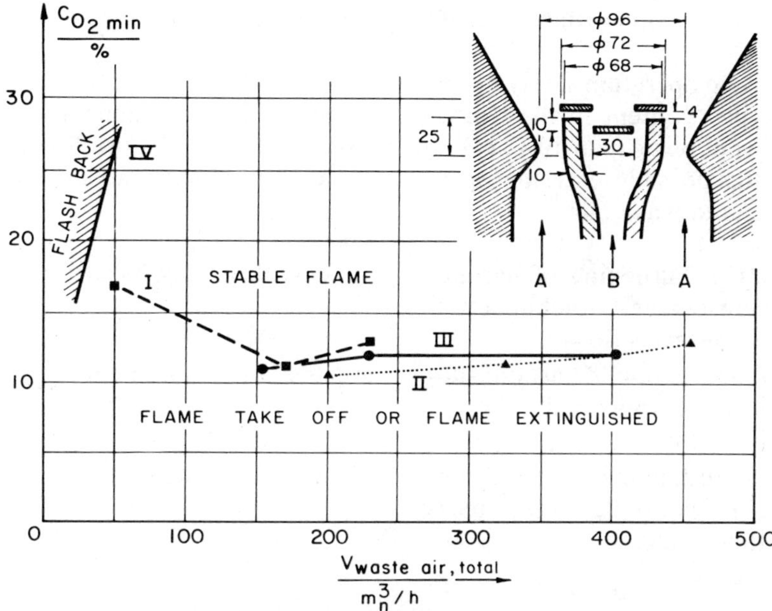

Figure 6.18. Flashback-safe burner for a high flux ratio of waste air and varying O_2 content. A, waste air; B, natural gas, including premixed air. I, 5 m_N^3 hr^{-1} natural gas with 10 m_N^3 hr^{-1} premixed air; II, 12.5 m_N^3 hr^{-1} natural gas with 25 m_N^3 hr^{-1} premixed air; III, 10 m_N^3 hr^{-1} natural gas with 20 m_N^3 hr^{-1} premixed air; IV, flashback when waste air contains natural gas stochiometrically.

burnout of all harmful substances. Especially interesting in this respect is the third arrangement (due to a Swedish suggestion), with additional fuel feeding into the stagnation area of the central body, for flame stabilization. Such burners could also be flashback-proof-operated with a stoichiometric air/natural gas mixture as simulated explosive waste air over a range 100 to ~10% of the maximal load of 1000 m^3 hr^{-1}. Even with ethylene/air mixtures, a flashback-proof range of almost 1:10 has still been achieved.

Another, very interesting construction of a flashback-proof burner is the capillary tube burner developed at the Gaswärmeinstitut in Essen. This burner type has the great advantage that from the individual vorticity-free jets a flow with supercritical vorticity can be generated which permits high combustion chamber burdening.

Figure 6.18 shows the measured performance graph of a flashback-proof burner with central body that, even with waste air containing only 10 to 11% by volume of oxygen, can still supply a stable flame with complete burnout of harmful substances.

6.7.5.   Waste Gas with Varying Oxygen Content

In regard to operation safety, especially difficult situations can arise where the oxygen content in the waste air, loaded with contaminants, can at times drop lower than 10%, even occasionally to zero. Figure 6.19 illustrates the possible extreme operating conditions at the transition from waste gas to waste air.

$O_2 \sim 21\%$, burdening of harmful substances stoichiometric, small mass flow — greatest flashback hazard

$O_2 \to 0$, burdening of harmful substances maximal, largest mass flow — greatest air requirement

$O_2 \to 0$, burdening of harmful substances $\to 0$, largest mass flow — greatest fuel requirement

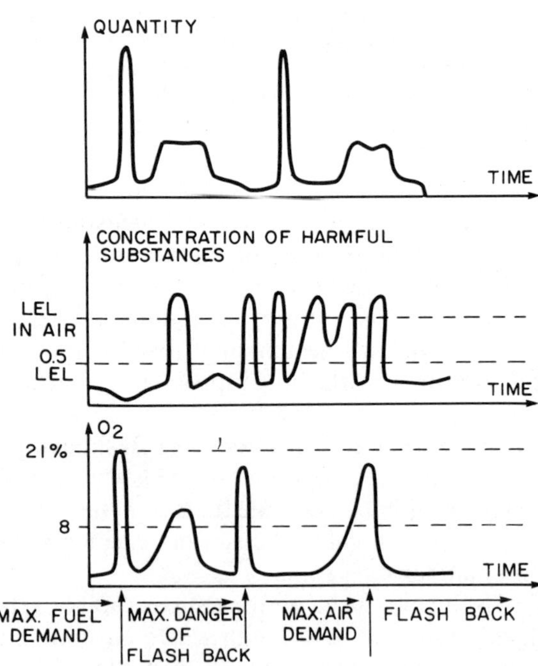

Figure 6.19. Extreme operational conditions of thermal waste gas purification installations.

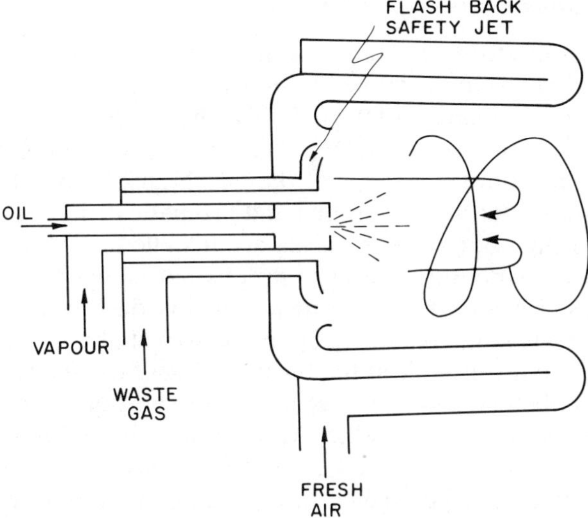

Figure 6.20. Injection of waste gas in a turbulent pulsating combustion chamber. The waste gas can be temporarily explosive.

Such gases can no longer be fed into a burner as combustion air, so additional air is required. This supplementary air can again be waste air burdened with harmful substances; however, it must contain ~21% oxygen. This occasionally inflammable and occasionally oxygen-deficient gas then has to be led to the burner through separate jets. This possibility is outlined in Figure 6.20. In regard to flashback-proof jets, the considerations given to occasionally inflammable waste air also apply here. In addition to this, it has be remembered that even in the case of maximum air requirements for the oxidation of the contaminants in the waste gas, a stable flame and complete burnout have to be achieved. For this the successful flashback-proof jetting of such waste gases into the highly turbulent flow of combustors has been proven.

6.7.6.   Flame Barriers

The flame barriers mentioned previously serve to suppress the flame front, which runs through the waste air duct from the burner back to the plant. They act as a security system independent of the flashback-proof burner. Types in operation are dry or liquid flame barriers, or equipment for the injection of fire extinguishers.

a.   Dry Flame Barriers or Flame Screens

These screens are formed as disks wound from one obliquely corrugated iron band and one smooth refined steel band, which form waste gas paths that are triangular channels of 0.2 to 0.7 mm hydraulic width. This channel width is smaller than the quenching distance of cold walls for the various inflammable harmful substances of the waste air (0.2 mm for hydrogen, 0.7 mm for saturated hydrocarbons and aromatics). For safety reasons, usually more than one of these disks are installed, in series. The effectiveness of this is doubtful because it does not compensate for any statistical errors. Because of the principle of this flashback safety device, the channel walls have to be cold upon the arrival of a flame. This is also the case with cold waste air at the backflashing of the flame. The hot flue gas that flows behind a flame after a larger explosion in the combustion chamber is no longer combustible. At this time it is not known if the band safety device, in the case of large flue gas quantities, which can stream back at first and later be pushed to the burner by possibly combustible waste air, may heat the filters to such a degree that they themselves may become a source of ignition directed backward toward the plant.

With suitable and approved constructions and with very short pipe lengths for freely burning flames, flame screens are safe from detonation. They are also safe from flashback if the flame burns directly behind the screen. The screen is then sufficiently cooled by the waste air that flows behind.

For this form of application (free burning) band safety devices have also been developed. However, different operating conditions exist if the band safety device is built into a closed conduit, for example in the waste air pipe in front of a burner. Therefore, in closed conduit systems with reduced heat radiation, an additional temperature control in the conduit, immediately behind the screen, is required. If this temperature increases, the system must be turned off or different measures have to be taken (e.g., the feeding in of N_2 to extinguish the flame behind the screen). If there is a possibility for a detonation between burner and flame screen, there is increased danger of backflashing for present designs. In this case a different safety device has to be found or the burner construction has to be changed.

If the waste air contains dust or harmful substances that tend to deposit, a danger of obstruction exists. In this case the pressure loss has to be watched and spare screens made available for cleansing.

Marble packings are sometimes provided instead of flame screens. Here, instead of definite steel channels, the passages between ceramic or glass marbles are used to extinguish a backward-propagating flame by means of heat withdrawal. With marble packing (marble diameter 6 to 10 mm and bed depth 200 to 500 mm) only the heat-retaining capacity is utilized. Cooling of the bed, which is directly heated by a standing flame, by heat conduction through up-stream equipment is not effective. Temperature control is therefore absolutely essential. However, an advantage of these marbles is that there is less danger of heating the entire device to ignition temperature with backflowing flue gases. Also, glass marbles are a corrosion-proof and there is less danger of obstruction.

b. Liquid Flame Barriers

In this device the gas is broken into bubbles which must rise through a layer of water before they get into the burner. Here the water serves as a heat conductor to extinguish the flame. Special designs can arrange that various waste air currents be led into one burner without influencing each other and without the possibility of backflow from the flue gas. The development of such liquid flame barriers is not yet complete. However, with correct planning they should offer great security against flame backflash even under unfavorable operating conditions. A disadvantage is that a wastewater problem can arise.

c. Injection of Extinguishers

This method, developed in coal mining against dust explosions, utilizes the pressure increase connected with a passing flame. By means of fast-response sensors, extinguishing materials (e.g., halones) are shot by propulsive charges into the pipe that carries ignitable waste air. Whether this method can also be applied in other situations and whether it is capable of suppressing flashback is not known. A requirement for safe operation, in addition to the dependability of sensors and extinguisher cartridges, is that the sensors not be covered with dirt or crusts. A disadvantage is that the plant must be shut down immediately after a flashback until the extinguisher cartridges are replaced.

At present only a few plants have been built for thermal decomposition of harmful substances in occasionally explosive waste air. However, according to the considerations given here, with the twofold flashback safety device, the results show the following:

1. Up to now, flashback has been prevented.
2. Compared with the customary procedure, avoidance of critical operation conditions by mixing in large amounts of fresh air saves fuel.

6.8.   EXAMPLE

The following example is intended to illustrate the procedures and measurements required for the installation of a TWP plant. It will become evident that only in exceptional cases is an optimal solution available that will satisfy all demands of the environment, especially minimum waste of high-quality energy, at an economical cost. Furthermore, the solution chosen will depend greatly on marginal conditions, such as legal regulations for particular harmful substances, possible combining with other procedures, the state of present technology, and familiarity of personnel with new procedures, as well as cost and availability of supplementary fuels.

In this example several solution methods are given and an explanation provided as to why a certain procedure is chosen. In this way, an attempt is made to demonstrate how the variable marginal conditions determine the procedure. Furthermore, several viewpoints are discussed for which it is difficult to find a systematic presentation.

6.8.1.   Problem Satement

Gases present are:

1. *WG*	0 to 200 m_N^3 hr^{-1} waste gas Average value: 20 m_N^3 hr^{-1}	with 5 vol. % O_2 15 vol. % HCl 2 vol. % $C_3H_4Cl_2$ Remainder N_2
2. *WA*	500 to 5000 m_N^3 hr^{-1} waste air Average value: 2000 m_N^3	with 0.05 vol. % $C_3H_4Cl_2$ 15 g H_2O m^{-3}
3. *WF*	Waste fuel (a mixture of chloroacetic acid) can be blended with oil, 0.5 ton hr^{-1}	*Composition*: 50% $C_2H_3O_2Cl$ by weight 30% $C_2H_2O_2Cl_2$ by weight 20% $C_2H\ O_2Cl_2$ by weight

Since a TWP is to be considered, we shall investigate if and at what expense waste fuel is applicable here. The requirements are:

$$C_{org} \leq 100 \text{ mg m}^{-3}$$

$$CO \leq 200 \text{ ppm}$$

$$Cl_2 \leq 10 \text{ ppm in the flue gas}$$

6.8.2. Calculation of Air Requirement and of Flue Gas Composition

6.8.2.1. Waste Gas

The harmful substance is dichlorpropylene (DCPy).

Maximum O_2 requirement of $O_{2WG} = 0$

Using eq. 17 with $m = 3$, $n = 4$, $k = 2$ gives

(O_2) $(m^3\ m_N^{-3}{}_{WG}) = 3 + 1 + \frac{1}{2} = 3.5$

$(O_2)_{WGmax} = 200 \times 0.02 \times 3.5 = 14\ m^3\ hr^{-1}$

$(O_2)_{WGaverage} \qquad\qquad\qquad = 1.4\ m^3\ hr^{-1}$

Cl_2 formation neglected

HCl deposit

$k = 2 \rightarrow 2\ m^3\ HCl\ m^{-3}\ DCPy$

$HCl_{WGmax} = 200\ (0.02{\cdot}2 + 0.15) = 38\ m^3\ hr^{-1}$

$HCl_{WGaverage} \qquad\qquad\qquad = 3.8\ m^3\ hr^{-1}$

6.8.2.2. Waste Air

Oxygen requirement:

$(O_2)_{WA\ max} = 5000 \times 0.0005 \times 3.5 = 8.75\ m^3\ hr^{-1}$

$(O_2)_{WA\ average} \qquad\qquad\qquad = 3.75\ m^3\ hr^{-1}$

$(O_2)_{WA\ min} \qquad\qquad\qquad\ \ = 0.88\ m^3\ hr^{-1}$

$HCl\ deposit_{max} = 5000 \times 0.0005 \times 2 = 5\ m^3\ hr^{-1}$

$HCl\ deposit_{average} \qquad\qquad\qquad = 2\ m^3\ hr^{-1}$

$HCl\ deposit_{min} \qquad\qquad\qquad\ \ = 0.5\ m^3\ hr^{-1}$

Because of the HCl/Cl_2 ratio, the H_2O quantity is important.

Waste gas:

The reaction equation for DCPy is

$C_3H_4Cl_2 + 3.5\ O_2 \rightarrow 3\ CO_2 + H_2O + 2\ HCl$

$H_2O_{WG\ max} = 200 \times 0.02 \times 1 \quad = 4\ m^3\ hr^{-1}$

$H_2O_{WG\ average} \qquad\qquad\qquad\ \ = 0.4\ m^3\ hr^{-1}$

Waste air: 15 g water vapor $m^{-3} \triangleq 1.87$ vol. %

$H_2O_{WA\ max} = 5000 \times (0.0005 \times 1 + 0.0187) = 96\ m^3\ hr^{-1}$

$H_2O_{WA\ average} \qquad\qquad\qquad\qquad = 38.4\ m^3\ hr^{-1}$

$H_2O_{WA\ min} \qquad\qquad\qquad\qquad\quad = 9.6\ m^3\ hr^{-1}$

6.8.2.3. *Waste Fuel*

The waste fuel is a mixture of chloroacetic acid (CAA).

Substance	Hu	O_2 Requirement
Mono-$C_2H_3O_2Cl$	716 kJ mole^{-1}	1.5 moles mole^{-1}
Di-$C_2H_2O_2Cl_2$	555 kJ mole^{-1}	1
Tri-C HO_2Cl_3	389 kJ mole^{-1}	$1 + 1H_2$

Waste fuel per kg:
  5.29 moles mono-CAA
  2.33 moles di- CAA
  1.22 moles tri-CAA $\rightarrow Hu_m = 5555$ kJ kg_{WF}^{-1}

This heating value is just adequate to pass for waste fuel. Difficulties have to be expected (ignition, control area).

Checking:

$$(O_2)_{WF\ min} = (1.5 \times 5.29 + 1 \times 2.33 + 1.22) \times \frac{22.4}{1000}$$

$$= 0.257 \text{ m}^3 \text{ kg}_{WF}^{-1}$$

Air requirement $\rightarrow \dfrac{(O_2)_{F\ min}}{0.21} = 1.22$ m^3 kg$_{WF}^{-1}$

For the atomization of waste fuels and for Cl_2 reduction in the flue gas 0.5 kg kg_{WF}^{-1} steam is used.
  $\rightarrow$ Minimum flue gas quantity: $O_{2FG} = 0$

$$CO_2 = (5.29 + 2.33 + 1.22) \times 2 \times \frac{22.4}{1000} = 0.396 \text{ m}_N^3 \text{ kg}_{WF}^{-1}$$

$$H_2O = 5.29 \times \frac{22.4}{1000} + 0.5 \times \frac{22.4}{18} = 0.741 \text{ m}_N^3 \text{ kg}_{WF}^{-1}$$

$$HCl = (5.29 + 2 \times 2.33 + 3 \times 1.22) \frac{22.4}{1000} = 0.305 \text{ m}_N^3 \text{ kg}_{WF}^{-1}$$

$$N_2 = 0.257 \times \frac{0.79}{0.21} \qquad\qquad V_{FG\ min\ WF} \quad \begin{array}{l} = 2.409 \text{ m}_N^3 \text{ kg}_{WF}^{-1} \\ \hline = 0.967 \text{ m}_N^3 \text{ kg}_{WF}^{-1} \end{array}$$

  for $O_{2FG} \geq 0.05$

Supplementary air quantity $\dot{V}_{A\ add}$:

$$V_{A\ add} = V_{FG\ min} \times \frac{0.05}{0.21 - 0.05} = 0.753 \text{ m}_N^3 \text{ kg}_{WF}^{-1}$$

With this the flue gas quantity, which occurs during combustion of the waste fuel, results in $V_{FG\,WG} = 3.16\ m_N^3\ kg_{WF}^{-1}$. With C_{pM} estimated at $1.60\ kJ\ m_N^{-3}\ K^{-1}$, the theoretical temperature increase, $\Delta\Theta_{th\,WF}$, obtained with combustion becomes

$$\Delta\Theta_{th\,WF} = \frac{Hu_{WF}}{V_{FG'} \times C_{pM}} = \frac{5555}{3.16 \times 1.6} = 1098\ K$$

Based on experience, this value is too small. Additional fuel or air preheating is required.

6.8.2.4. *Waste Fuel Plus Supplementary Fuel*

Since additional fuel is required, the waste air amount that can be used for the oxidation of the waste fuel must be calculated. If a waste air surplus occurs, the supplementary fuel can be oxidized with it. Otherwise, additional fresh air is needed.

The air requirement for waste fuel is

$$(V)_{A\,WF} = \frac{O_{2\,WF\,min}}{0.21} + V_{A\,add}$$

$$= 1.22 + 0.753 = 1.973\ m_N^3\ kg_{WF}^{-1}$$

For 500 kg hr^{-1} waste fuel, we require 986.5 m^3 air hr^{-1}. The supply is

$$\dot{V}_{WA\,average} = 2000\ m^3\ hr^{-1}$$

For combustion of supplementary fuel an average of 1013.5 m_N^3 hr^{-1} of waste air is available.

As supplementary fuel, oil is required for 1013 m^3 average waste air + 20 m^3 hr^{-1} waste gas. The oil requirement is

$$\text{Oil: } H_u = 42 \times 10^3\ kJ\ kg^{-1}$$

We start with a first calculation of the $\frac{1}{2}Cl_2/HC_1$ ratio.

Assume: equilibrium of Deacon process not obtained

$Cl_2 = 3Cl_2^*$ from Figure 6.8

$H_2O \geqq 20$ vol. %

For the calculation of the HCl content in the flue gas, the additional fuel balance over H_u is required first.

Waste fuel for 48% of the waste air:

$$\text{Supplementary fuel for } 52\% \rightarrow 500 \times \frac{0.52}{0.48} \times \frac{5555}{42 \times 10^3} = 71.6\ kg\ hr^{-1}$$

as a minimum value
   According to eq. 23:
   $71.6 \times 1.1 = 78$ m^3 hr^{-1} volume increase in $\dot{V}_{FG}$.
   $\dot{V}_{FGgas} \approx 500 \times 3.16 + 1013 + 20 + 78 = 2691$ m^3_N hr^{-1}
   $HCl_{gas} = 0.305 \times 500 + 2 + 3.8 = 158.3$ m^3_N hr^{-1}
     $HCl \triangleq 5.88$ vol. %
      $Cl_2 \leq 3$ ppm

$$\frac{\frac{1}{2}Cl_2}{HCl} = \frac{1.5 \times 10^{-6}}{58.8 \times 10^{-3}} = 2.5 \times 10^{-5}$$

which will not be achieved.
   Evaluation of the data for the process $4HCl + O_2 \rightleftharpoons 2H_2O + 2Cl_2$ results in the following values:

$$P_{Cl_2} \text{ bar}$$

$$900°C = 0.0004 \triangleq 400 \text{ ppm} \triangleq 0.7\% \, Cl_{gas}$$
$$950°C = 0.0003 \triangleq 300 \text{ ppm} \triangleq 0.5\% \, Cl_{gas}$$
$$1500°C = 0.00007 \triangleq 70 \text{ ppm} \triangleq 0.1\% \, Cl_{gas}$$
$$2000°C = 0.00002 \triangleq 20 \text{ ppm} \triangleq 0.03\% \, Cl_{gas}$$

   According to this, even at a temperature of 2000°C in the TWP plant and under the assumption that complete balance occurs, the required permissible Cl_2 concentration in the flue gas cannot be obtained if the waste fuel also has to be destroyed.
   The chloroacetic acid mixture is an especially harmful substance, which is difficult and expensive to destroy. In this case the higher Cl_2 emission has to be taken into account. Hence parallel to the possibility of destroying the waste fuel in a combined process, the use of pure heating oil for fuel can be considered.
   For the simultaneous destruction of the waste fuel (case A), because of the high chlorine content, a flue gas temperature of $T_{FG} = 950°C$ is chosen. Air preheating is problematic here, because of danger of HCl corrosion in the heat exchanger on part of the flue gas. This results in a temperature increase $\Delta\Theta$ in the combustion of $\Delta\Theta = 930$ K.
   Assume: heat losses $\triangleq 10\% \, \dot{Q}$

$$\dot{V}_{FG \, average} = 2690 \text{ m}^3_N \text{ hr}^{-1}$$

$C_p = 1.6$ kJ m^{-3} K^{-1}
The total fuel heat Q_F to be supplied is

$$\dot{Q}_F = 1.1 \times 2690 \times 1.6 \times 930 = 4.4 \times 10^6 \text{ kJ hr}^{-1}$$

Of this, from the waste fuel,

$$\dot{Q}_{WF} = 500 \times 5555 = 2.778 \times 10^6 \text{ kJ hr}^{-1}$$

Therefore, from the supplementary oil,

$$\dot{Q}_{F\,add} = 1.622 \times 10^6 \text{ kJ hr}^{-1}$$

This requires

$$\frac{1.622 \times 10^6}{4.2 \times 10^4} = 38.6 \text{ kg hr}^{-1}$$

of heating oil at average values.

6.8.2.5. *Heating Oil Only without Waste Fuel*

The $HCl_{gas} = 5.8 \text{ m}^3 \text{ hr}^{-1}$ on average then comes $\triangleq 0.276$ vol. %. Section 6.23 gives 850°C, $Cl_2 = 2$ ppm.

With this process all demands of environmental protection or control can be complied with. Furthermore, preheating is also possible now. A preheating temperature of 400°C is chosen. This results in $\Delta\Theta_B = 450$ K, and the heat supplied with the heating oil is

$$\dot{Q}_{F\,add} = 1.1 \times 2100 \times 1.6 \times 450 = 1.66 \times 10^6 \text{ kJ hr}^{-1}$$

corresponding to
39.6 kg hr^{-1} on average
giving CO $\langle$ 200 ppm (see Figures 6.3 and 6.4)
Average flue gas composition:
39.6 kg hr^{-1} oil, corresponding to 34.8 kg hr^{-1} C $\triangleq$ 65 m^3 hr^{-1} CO$_2$
3.96 kg hr^{-1} H $\triangleq$ 44 m^3 hr^{-1} H$_2$O

This gives $\triangleq 87$ m^3 O$_2$ consumption for 420 m^3 supplied. This results in an oxygen content in the flue gas of

$$(2000 - 87 + 65 + 44 + 20) = 0.163 O_2 \triangleq 16.3\% \; O_2$$

6.8.2.6. *Extreme Cases*

The most unfavorable operating conditions still have to be checked.

a. *With Waste Fuel*

Because of the difficulty of regulating the waste fuel burner, it should run with a constant load; a constant 500 kg hr^{-1} of waste fuel + 250 kg hr^{-1}

of vapor constantly required for optimal combustion with 5% O_2 in the flue gas, 987 m³ hr⁻¹ waste air (see above).

$\dot{V}_{WA\ min}$ = 500 m³ hr⁻¹ → regulating fresh air to
987 m³ hr⁻¹, constant
Waste gas max. 200 m³ hr⁻¹
This amount can no longer be added without dangerously lowering the temperature in the reaction zone.
With an additional oil burner for the constant basic load:
Assume: range of satisfactory atomisation for oil and flame stability 1 : 10 provided. $\dot{V}_{WA\ max}$ = 5000 m³ hr⁻¹
Hence for oil burners,
5000 − 987 = 4013 m³ hr⁻¹ max. + $V_{WG\ max}$ = 4213 m³_N hr⁻¹
The heat requirement calculation then supplies

$$\frac{4213 \times 1.1 \times 1.6 \times 930}{42 \times 10} = 164.2 \text{ kg oil hr}^{-1} \text{ maximum}$$

Lower limit → oil$_{min}$ = 16.4 kg hr⁻¹

plus $\dot{V}_{FA\ min} \sim 250 \text{ m}^3 \text{ hr}^{-1} + \dot{V}_{WG\ max}$

$$\approx 450 \text{ m}^3 \text{ hr}^{-1}$$

This is still possible.
Chosen values: 16.4 kg oil hr⁻¹ for a constant supporting flame.
This can be regulated up to 164 kg hr⁻¹.

b.   With Supplementary Fuel Operation Only

Maximum oil requirement

$$V_{WA}\ 5000 \text{ m}^3 \text{ hr}^{-1} + V_{WG}\ 200 \text{ m}^3 \text{ hr}^{-1} = 5200 \text{ m}^3 \text{ hr}^{-1}$$

If the heat exchanger is set to give 400°C preheating on average, the preheating temperature drops with larger waste air quantities.

To review: Maximum and minimum operation calculated with constant property values and heat transmission varying approximately as $W^{0.8}$ gives an operating temperature gradient in the heat exchanger of $\Delta\Theta_m \approx \Theta_{FG} - \Theta_{WA\omega}$ at the heat exchanger exit. With this the heat flux in the heat exchanger, transferred through area A, in normal operations equals

$$\dot{Q}_{MN} = \alpha_{MN}A(\Theta_{FG} - \Theta_{WA\omega N}) = \dot{V}_{WA}C_p(\Theta_{WG\omega N} - \Theta_{WG\alpha})$$

With maximum waste air quantity, we obtain correspondingly:

$$\dot{Q}_{max} = \alpha_{max} A (\Theta_{FG} - \Theta_{WA\omega\,max}) = \dot{V}_{WA} C_p (\Theta_{WA\omega\,max} - \Theta_{WA\alpha})$$

$$\alpha_M = \frac{\dot{V}_{WA_M} C_p}{A} \frac{\Theta_{WA\omega M} - \Theta_{WA\alpha}}{\Theta_{FG} - \Theta_{WA\omega M}}$$

$$\alpha_{max} = \alpha_M \left(\frac{\dot{V}_{WA\,max}}{\dot{V}_{WA_M}} \right)^{0.8}$$

$$\frac{\dot{V}_{WA_M} C_p}{A} \frac{\Theta_{WA\omega M} - \Theta_{WA\alpha}}{\Theta_{FG} - \Theta_{WA\omega M}} \left(\frac{\dot{V}_{WA\,max}}{\dot{V}_{WA_M}} \right)^{0.8} A$$

$$= \dot{V}_{WA\,max} C_p \frac{\Theta_{WA\omega\,max} - \Theta_{WA\alpha}}{\Theta_{FG} - \Theta_{WA\omega\,max}}$$

With a preheating factor Θ:

$$\frac{\Theta_{WA\omega M} - \Theta_{WA\alpha}}{\Theta_{FG} - \Theta_{WA\alpha M}} = \Theta$$

$$\left(\frac{\dot{V}_{WA_M}}{\dot{V}_{WA\,max}} \right)^{0.2} \Theta = \frac{\Theta_{WA\,max} - \Theta_{WA\alpha}}{\Theta_{FG} - \Theta_{WA\omega\,max}}$$

If another factor ψ is added, which takes into consideration the dependence of the heat-transfer coefficient on the power of the Reynolds number,

$$\left(\frac{\dot{V}_{WA_M}}{\dot{V}_{WA\,max}} \right)^{0.2} = \psi_{max}$$

This eventually becomes

$$\frac{\psi\Theta \times \Theta_{FG} + \Theta_{WA\alpha}}{1 + \psi\Theta} = \Theta_{WA\omega\,max}$$

For this example,

$$\psi_{max} = \left(\frac{2000}{5000} \right)^{0.2} = 0.833$$

$$\Theta_{max} = \frac{400 - 20}{850 - 400} = 0.844$$

With this the preheating temperature with maximum waste air quantity drops to

$$\Theta_{WA\omega\,max} = 362°C$$

and the additional heat requirement rises to

$$\dot{Q}_{\text{add max}} = (5000 + 200) \times 1.6 \times (850 - 362) \times 1.1$$

$$= 4.47 \times 10^6 \text{ kJ hr}^{-1}$$

$$\triangleq 106.3 \text{ kg oil hr}^{-1}$$

At a regulation ratio of 1 : 10 for the oil burner, this results in

$$\dot{Q}_{\text{add min}} \triangleq 10.6 \text{ kg oil hr}^{-1}$$

which corresponds to a minimum waste air quantity of $\sim$160 m^3 hr^{-1} or, for an approximately stoichiometric single flame, $\langle 500$ m^3 hr^{-1} = $\dot{V}_{WA\,\text{min}}$.

This gives $\dot{Q}_{\text{add}_{\text{min}}}$ on the part of the waste air flow.

The preheating temperature $\dot{V}_{WA\,\text{max}}$:

$$\psi_{\text{min}} = \left(\frac{2000}{500}\right)^{0.2} = 1.32$$

$$\Theta_{\text{min}} \equiv \Theta_{\text{max}}$$

$$\Theta_{WA\omega_{\text{min}}} = 457°C$$

$$\dot{Q}_{\text{add}_{\text{min}}} = 500 \times 1.6 \times 1.4(850 - 457) = 4.4 \times 10^5 \text{ kJ hr}^{-1}$$
$$\downarrow$$

larger losses

$$\text{oil}_{\text{min}} \triangleq 10.5 \text{ kg oil hr}^{-1} \text{ or, for regulation reasons, oil}_{\text{min}} = 10.6 \text{ kg hr}^{-1}$$

HCl emission in the most unfavorable case:

$$V_{WG\,\text{max}} \rightarrow \text{HCl } 38 \text{ m}^3 \text{ hr}^{-1}$$

$$V_{WA\,\text{max}} \rightarrow \frac{5 \text{ m}^3 \text{ hr}^{-1}}{43 \text{ m}^3} \text{ in } 5300 \text{ m}^3 \text{ hr}^{-1} \triangleq 0.8 \text{ vol. \% HCl}$$

Otherwise,

$$\dot{V}_{WG\,\text{max}} \rightarrow 38 \text{ m}^3 \text{ hr}^{-1} \text{ HCl}$$

$$\dot{V}_{WL\,\text{min}} \rightarrow \frac{0.9 \text{ m}^3 \text{ hr}^{-1}}{38.9 \text{ m}^3 \text{ hr}^{-1}} \text{ in } 710 \text{ m}^3 \text{ hr}^{-1} \triangleq 5.5 \text{ vol. \% HCl}$$

which is worse.

6.8.3.   Arrangements of TWP Installations

6.8.3.1.   Case A with Waste Fuel

A two-stage burner with constant waste fuel supply is chosen where a constant oil-waste air support flame burns in the first stage. Into the second stage, the remaining waste air, waste gas, and required additional oil is supplied in a regulated manner. This is the standard method of layout.

Alternatively, oil can be mixed into the CAA to improve the heating value and burning properties. This alternative is not discussed further since greater control problems and probably a larger overall oil consumption is to be expected. Addition of water vapor to lower the Cl_2 emission seems to be a requirement for all cases.

The equilibrium of $HCl/\frac{1}{2}Cl_2$ could be influenced in a positive way by adding water to the waste fuel. However, it is difficult to mix water with the CAA mixture to provide an emulsion that gives stable ignition; it also increases corrosion, and the addition of water increases heat consumption.

From this follows the flow scheme shown in Figure 6.21. The following amounts are supplied to the waste fuel burner:

500 kg hr^{-1} CAA mixture

250 kg hr^{-1} vapor at 5 bar

987 m_N^3 hr^{-1} waste air or 500 m_N^3 waste air

  + 487 m_N^3 hr^{-1} fresh air

This yields

$$\dot{V}_{FA_1} = 0 \div 487 \; m_N^3 \, hr^{-1}$$

The solidification point, Θ_i, of a CAA mixture is above 50°C, and heating of the piping is required. Because of a high boiling point, this could be achieved by using the 5-bar vapor. However, the use of low-pressure vapor is of more advantage, since corrosion is decreased. From this we obtain the arrangement in Figure 6.22, which demonstrates the control and atomization of waste fuel mixtures.

Since it is difficult to burn waste fuels completely and without soot, a flame with a supercritical vortex is installed. Burning and ignition properties of waste fuels differ widely; it is therefore an advantage to supply the combustion air through two concentrically located swirl chambers. By means of a throttle located in the weak vorticity exterior air supply ring, the magnitude of the central momentum can be adjusted to create an optimum backflow in the center of the flame.

The ignition and support burner, operated under constant load, is arranged such that the atomizing zone of the waste fuel burner is not in-

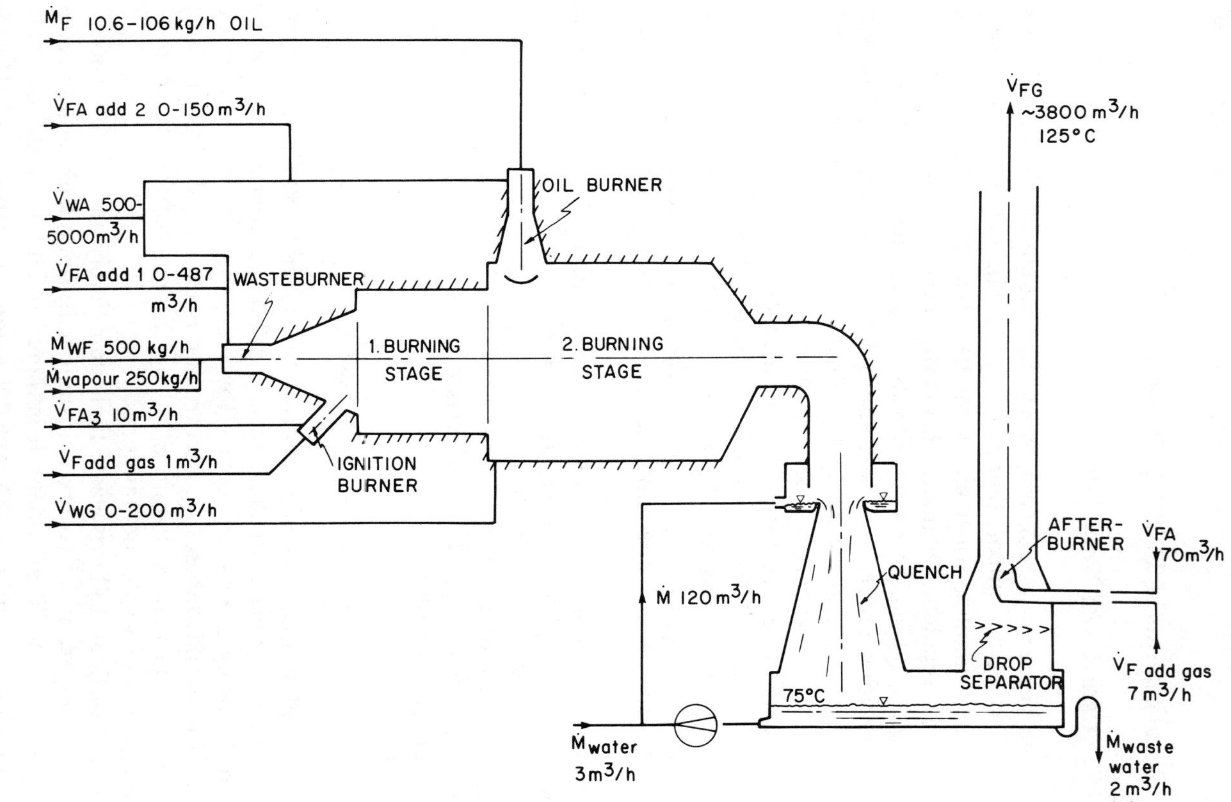

Figure 6.21. Flow scheme for two-stage combustion with waste fuel and quenching.

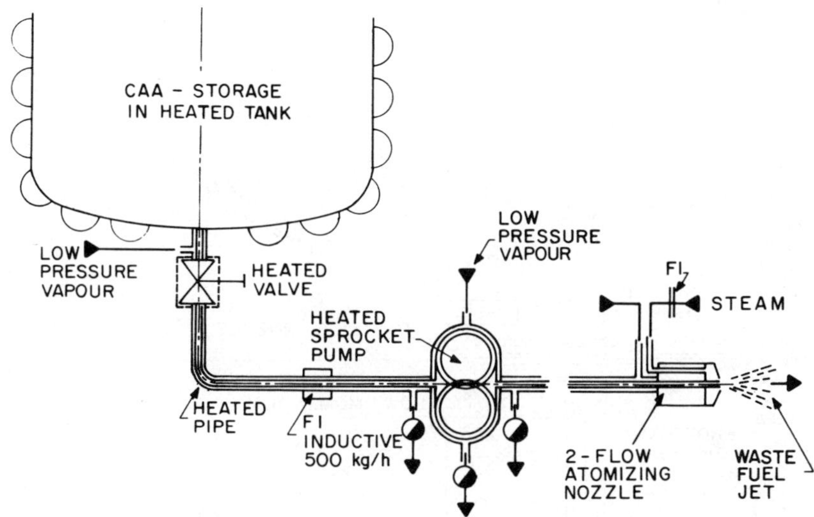

Figure 6.22. Proposed arrangement for the injection of waste fuel CAA into the TWP combustion chamber.

fluenced. However, the burnout zone has to receive sufficient additional energy to secure complete ignition. From this we obtain the arrangement shown in Figure 6.23.

The second burner, used for the remaining waste air and waste gas, is operated on additional heating oil and can be designed as a standard burner. Because of the control ratio for oil of 1 : 10 and the large amount of air surplus, only burners with a controlled backflow valve or a dual fuel nozzle can be considered.

The total burnout of harmful substances to be reached is very large. Therefore, it seems risky to arrange this second burner directly parallel to the waste fuel burner. This may severely disrupt the flow in the waste fuel flame during fluctuations in the amount of waste air. It is better to arrange the latter burner downstream relative to the waste deposit burner. To improve the mixing in the combustion chamber as well as the burnout, the second burner should fire tangentially into the combustion chamber (see Figure 6.21).

A simpler possibility of additional energy supply using a three-flow nozzle is demonstrated in Figure 6.24. Axisymmetric flames are created, which result in optimum use of additional ignition energy; also, control expenditures are smaller. Moreover, a smaller support flame can be used, created, for example, by a simple natural-gas ignition burner, which, however, has to be operated constantly.

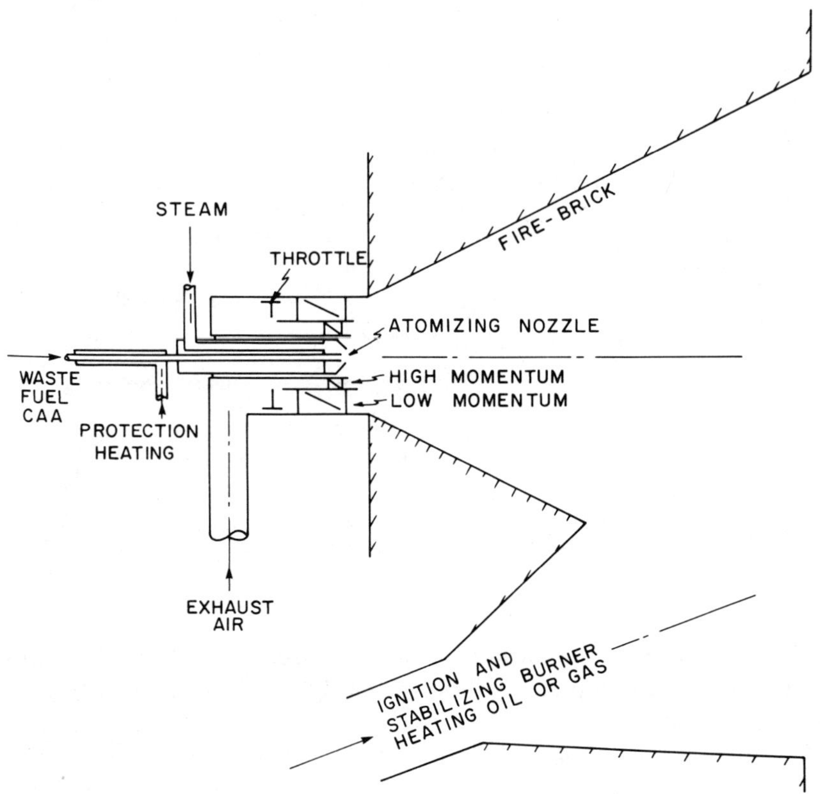

Figure 6.23. Waste fuel burner with additional stabilizing burner.

The proper selection of such three-flow nozzles, however, is possible only through experiments. In addition, the water vapor partial pressure is decreased in the decomposition zone. Thus a larger chlorine emission is to be expected, as in the arrangement of Figure 6.23.

Waste gas should not be injected centrally into the waste fuel flame. This results in blocking of the backflow zone, causing deterioration of flame stability and burnout.

6.8.3.2. Case B without Waste Fuel

Any oil or natural gas burner may be chosen assuming that the control range is sufficiently large and that a thorough blending of the total combustion air with the flame is possible. In this example a combustor is provided which is equipped with a dual-fuel nozzle for the additional heating oil. The large momentum of the waste air secures thorough burnout of all harmful substances.

Waste gas is supplied at the outer backflow zone of the combustor. Experience shows that a combustor flame is particularly sensitive to the supply of waste gas in the inner backflow zone (Figure 6.25).

6.8.3.3.   *Combustion Chamber Dimensions*

For case A, a residence time of $\tau_A = 0.3$ s (calculated for the hot gas) is assumed for the largest amount of waste air. For the chosen burner construction it is then expected that the allowable emission limits are maintained except for chlorine. With an effective combustion chamber temperature of 850°C,

$$\dot{V}_{FG_{\max}\Theta} = \dot{V}_{FG_{\max}} \frac{850°C + 273°C}{273°C} = 23,850 \ m^3 \ hr^{-1}$$

$$\triangleq 1.98 \ m^3 \ \text{combustion chamber volume}$$

Chosing a ratio of $L/D = 3.5$ (L = length, D = diameter of combustion chamber), we get $D = 0.9$ m and $L = 3.1$ m.

For case B, a combustor of sheet metal construction is assumed with

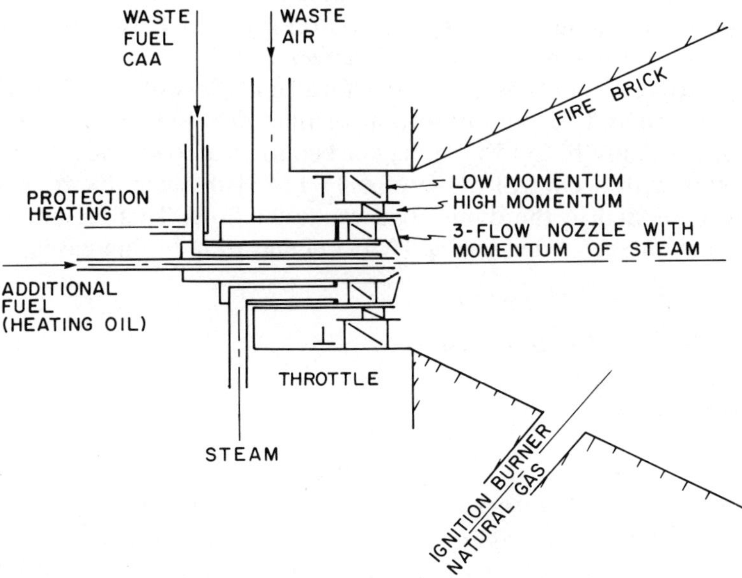

Figure 6.24. Waste fuel burner with central fuel supply.

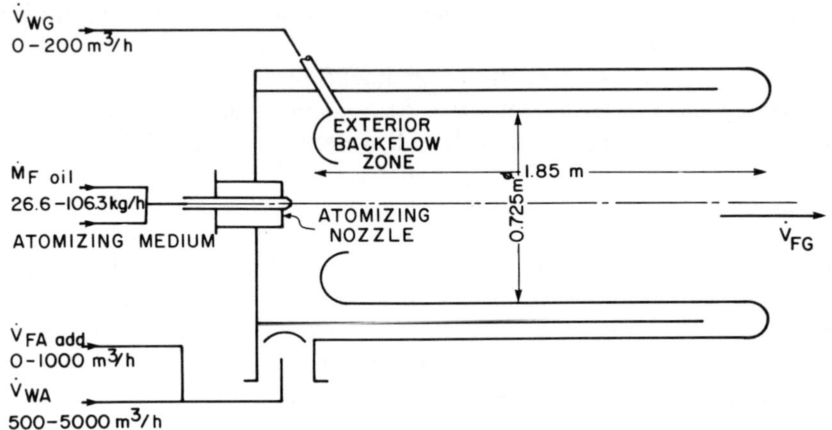

Figure 6.25. Process B without waste fuel.

residence time τ_B = 0.12 s (see Figure 6.25). The combustion chamber volume is then 0.75 m³.

For an L/D ratio of 2.5, which suits the flow conditions in the combustor, one obtains the dimensions D = 0.725 m and L = 1.85 m. This construction is particularly simple. However, experiments are still required to determine if a sufficient decomposition of harmful substances is possible throughout the complete control range of 1 : 10. From present experience, use is limited to control ranges 1 : 4.

If the proportion of time allotted to a waste air flow of $<\sim$1500 m³ hr⁻¹ is small compared to the proportion of time devoted to a larger waste air flow (e.g., only 10 to 15%), it may be economical to provide the simpler combustor with a control range of only 1 : 4. However, fresh air might have to be added to the diminished waste air flow. This technique also has the advantage of decreasing HCl concentration in flue gases.

6.8.4.   Flow and Control Charts

6.8.4.1.   Case A

For case A, waste fuel atomizing is performed using a three-fuel nozzle, as demonstrated in Figure 6.26. Safety switches and heating of the piping, which are required for every arrangement, are not shown.

As a means of control the effective combustion chamber exit temperature is used. An additional control of O_2 concentration to guarantee burn-

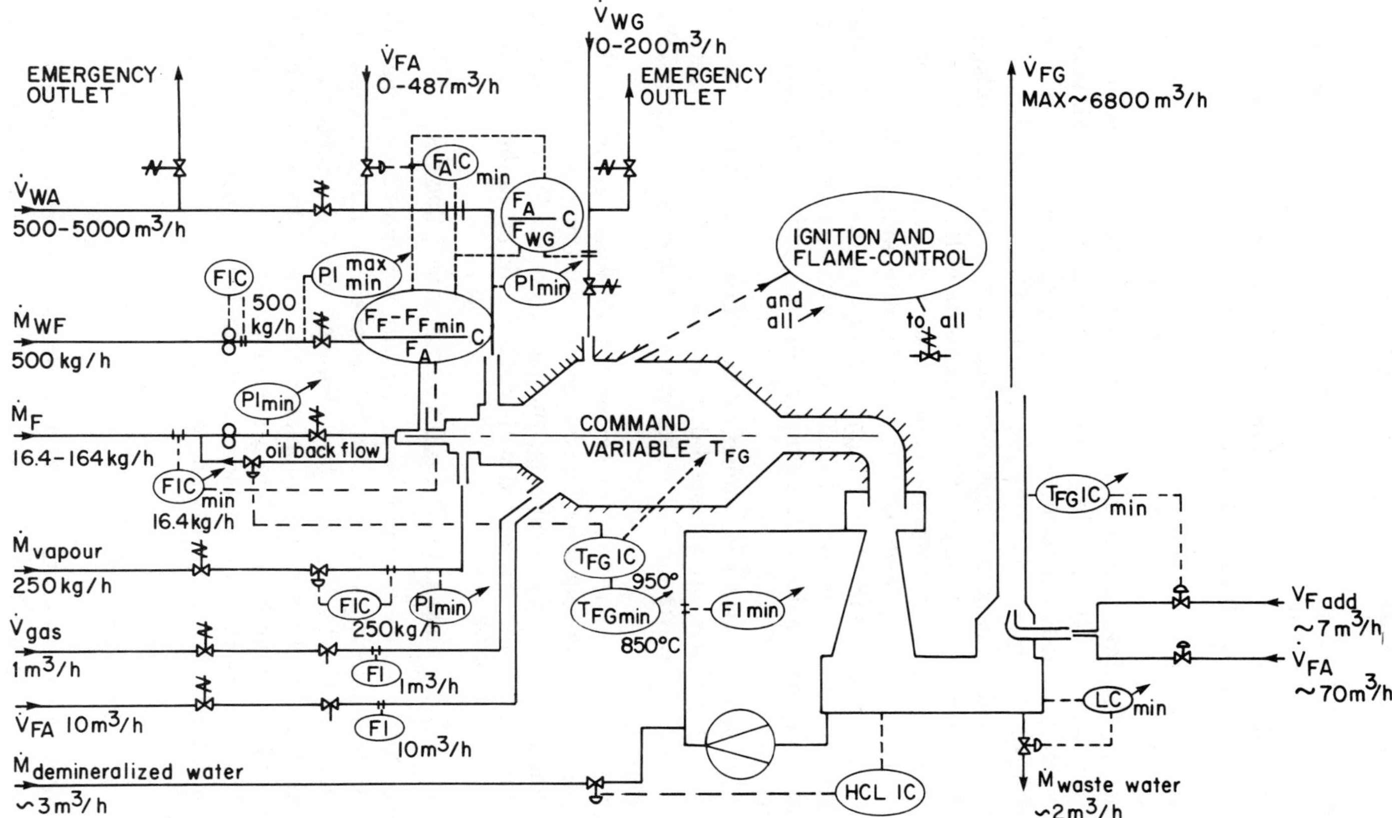

Figure 6.26. Flow and control chart for process A, including waste fuel and additional fuel in a three-fuel nozzle with quenching.

out of harmful substances is not required. Control of supplementary fuel where a minimum flow of 16.4 kg hr^{-1} is required is provided through a throttle valve located in the backflow line. A scrubber for the flue gases is necessary due to a high HCl concentration in the outlet gases.

6.8.4.2.   *Case B*

For case B, the flow scheme is much simpler, as shown in Figure 6.27. If a flow of waste air $\langle 1500$ m^3 hr^{-1} only occurs during a short time period, the simpler combustor construction can be chosen. In this case a controlled fresh air supply has to be provided until 1500 m^3 hr^{-1} of waste air are reached. This is also desirable because then the heat exchanger is not endangered by too high a HCl concentration. Because of the fresh air control, the HCl concentration in flue gases can always be kept below 1%. Thus flue gas washing might be avoided, depending on government regulations. As a means of controlling the oil quantity, the combustion chamber exit temperature is used again. In some cases a temperature limit control has to be provided, depending on the properties of the heat exchanger.

The way in which emergency relief valves or emergency outlets, shown in the two flowcharts, are arranged depends on the regulatory situation, as does the way waste air and waste gas are discharged when a TWP installation fails.

If scrubbing is required, as for example in flowchart A, hydrochloric acid results. The procedure and costs for its discharge again depend on environmental laws. By scrubbing, the gases are cooled down to approximately 75°C and are loaded with water vapor up to the saturation point.

$$H_{FG} \approx 1500 \text{ kJ m}_N^{-3}, \quad x \approx 0.5 \text{ kg water m}_N^{-3} \text{ flue gas}$$

If these moist cold gases cannot be mixed with hot gases before they are lead into a chimney, a heater has to be installed. A heat exchanger for gases between the TWP combustion chamber and the scrubber cannot be used, because of corrosion. Supplementary energy has to be used for heating, for example by installing a burner that fires into the scrubbed gases. The cleansing of hydrochloric acid from gases therefore leads to additional waste of high-quality energy.

6.8.5.   Cost Estimates

The following cost estimates should demonstrate the order of magnitude of capital investments as well as operating costs. The data given are ref-

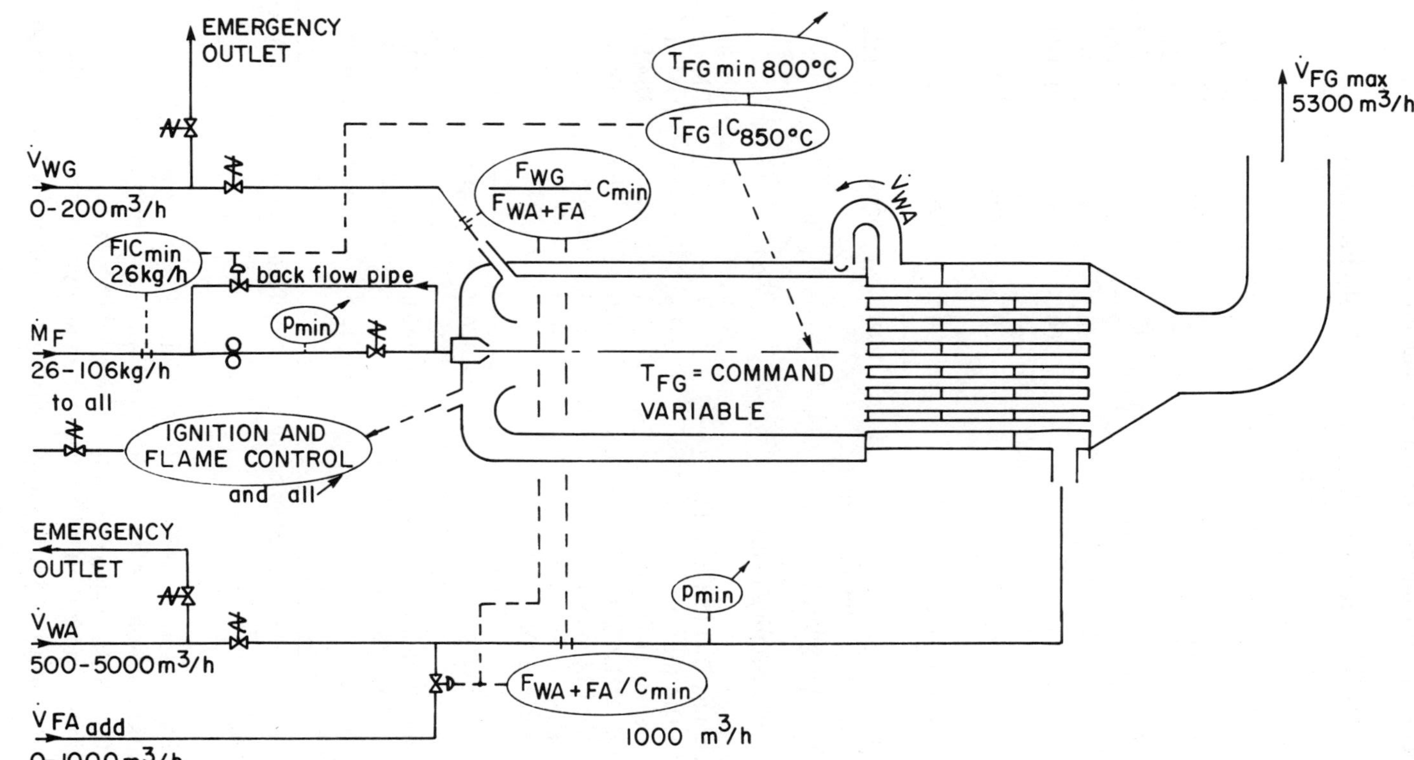

Figure 6.27. Flow and control chart for process B with sheet metal construction and downstream heat exchanger.

Table 6.1.   Comparative Cost Estimates in Deutsche Marks

Building Costs	Procedure		
	A	B	
Burner	30	8	$\times$ 10^3 DM
Combustion chamber	250	80	10^3 DM
Scrubber	200	0	10^3 DM
After burner	10	0	10^3 DM
Heat exchanger	0	80	10^3 DM
Control system	240	60	10^3 DM
Building	200	50	10^3 DM
	930	278	10^3 DM
Piping and erection, 100%	930	278	10^3 DM
Total investment	1860	556	$\underline{10^3\ \text{DM}}$
Total debt service for environmental plants, 20% annually			
Assuming 6000 operating hours			
Debt service	62.0	18.5	DM hr^{-1}
Supplementary fuel (250 DM ton^{-1})	9.7	15.0[a]	DM hr^{-1}
Vapor (25 DM ton^{-1})	7.8	1.5	DM hr^{-1}
Water softening (1 DN m^{-3})	3.0	—	DM hr^{-1}
Energy (0.1 DM kWhr^{-1})	8.0	1.5	DM hr^{-1}
Services (50 DM man-hr^{-1})	25.0	5.0	DM hr^{-1}
Neutralization and preparation of waste water (5 DM m^{-3})	10.0	0	DM hr^{-1}
	126.2	41.5	
Extermination of highly chlorine hydrocarbons (280 DM ton^{-1})	$\div$ 140.0	—	DM hr^{-1}
Total operation costs	$\div$ 13.8	$+41.5$	DM hr^{-1}

[a] For computation an average of 60 kg oil hr^{-1} is assumed. This is due to a reduced control range, as in Figure 6.27.

erence values based on 1978. Costs for high-quality energy and costs for extermination of solid or liquid industrial wastes by burning will be subjected to the largest relative changes (see Table 6.1).

Accordingly, there would be a gain when applying a combined thermal extermination system for waste air, waste gas, and waste fuel. This gain is derived by omitting the enormously high costs that arise in exterminating special waste deposits in commercially operated waste treatment plants. Highly chlorinated hydrocarbons are substances that are completely exterminated only with considerable effort. On the other hand, if these materials can be removed, as may be achieved in the future, the saving for example A would be smaller. If procedure A as compared to B is in reality less economical, the decision depends on environmental regulations for a particular plant.

From the example it becomes obvious that the invested capital as well as operation costs, even for "normal" TWP installations, are so high that by careful planning, significant savings can be achieved. The costs of a TWP procedure, however, are always so high that it should be used only as a last result.

6.9. SUMMARY

Thermal decomposition of a waste air or gas flow in a flame is very effective. However, it is an expensive and energy-consuming procedure. Conditions for reactions, which imply complete decomposition while maintaining a minimum effort, were discussed and criteria for the technical process arrangements are given (i.e., burner, combustion chamber, and control system). Safe treatment of temporarily explosive waste air has been discussed. The concluding part demonstrated by a detailed example how such installations could be arranged, the influence of government authorities, and expected operation costs. In this context the burning of waste deposits, which is interrelated to waste air purification, was also discussed.

SYMBOLS

A	exchanger size
b	specific fuel requirement
c	specific heat
C_{org}	organic carbon
H_u	heating value
$\dot{M}$	mass flux
$\dot{Q}$	heat rate
t	time
T, θ	temperature
V	gas flow rate
X	concentration

Subscripts

AIC	quality: analyzed, indicated, and controlled
F	fuel
FA	fresh air
FG	flue gas
FIC	flow rate indicated and controlled
TIC	temperature indicated and controlled
WA	waste air
WG	waste gas
WF	waste fuel
α	inflow
ω	outflow

Controls

min/max   minimum/maximum value controlled

REFERENCES

1. VDI Richlinien 2442, VDI-Verlag, Düsseldorf, in press.
2. Günther, R., *Verbrennung und Feuerungen*, Springer-Verlag, New York, 1974.
3. Fristrom, R. M., and Westenberg, A. A., *Flame Structure*, McGraw-Hill, New York, 1965.
4. Cremer, H., *Chem. -Ing. -Tech.*, 44, 8 (1972).
5. Rummel, K., *Arch. Eisenhüttenw.* 10. pp. 505–510, 541–548 (1936); 11. pp. 19–30, 113–123, 163–181, 215–224 (1937).
6. Chigier, N. A. Ch., *Progress in Energy and Combustion Science*, Vol. 1, Pergamon Press, New York, 1976, pp. 33–45.
7. Lewis, B., and v. Elbe, G., *Combustion—Flames and Explosions of Gases*, 2nd ed., Academic Press, New York, 1961.
8. 15th Symp. (Int.) Combust., The Combustion Institute, Pittsburgh, 1974, pp. 1039–1103.
9. Biberacher, G., Greulich, H., Hess, K., and Stichel, R., *Verfahrenstechnik*, 5, 108 (1971).
10. Germerdonk, R., *Chem. -Ing. -Tech.*, 44(5), 332 (1972).
11. Kiang, K. H., *Chem. Eng. Prog.*, 72, 37 (Dec. 1976).
12. *TAL Technische Anleitung zur Reinhaltung der Luft*, 1974 Gesetzblatt der Bundesrepublik Deutschland.

INDEX